Populations, Species, and Evolution

Populations, Species, and Evolution

An Abridgment of *Animal Species and Evolution*

Ernst Mayr

The Belknap Press of
Harvard University Press
Cambridge, Massachusetts
and London, England

Sixth Printing, 1977

ISBN 0-674-69013-3

Library of Congress Catalog Card Number 79-111486

Printed in the United States of America

To my friend Theodosius Dobzhansky,
foremost architect of the
evolutionary genetics of today

Preface

In the Preface of *Animal Species and Evolution* (1963), I wrote that it was "an attempt to summarize and review critically what we know about the biology and genetics of animal species and their role in evolution." The result was a volume of XIV + 797 pages. Ever since its publication, I have been urged to provide an abridged edition that would be more handy for class use and for the reader who does not want the detailed documentation of the large volume. To satisfy this demand is the object of *Populations, Species, and Evolution.* The essential discussions of the original volume have been retained, but the massive citations of the literature and discussions of many peripheral subjects have been eliminated. The specialist as well as the teacher will still need the fully documented treatment in *Animal Species,* but the general reader will find the new volume easier to use.

At first, there was no intention to combine a revision of *Animal Species* with the preparation of this abridgment. Yet, during the work on many of the chapters a rather extensive revision became inevitable, and certain chapters, particularly 2, 8, 9, 10, 15, and 17 were almost completely rewritten. No conscious effort was made, however, to incorporate the entire literature published since 1962. To do so would have nullified the endeavor to provide an abridgment. Furthermore, a number of excellent books have been published recently that review current research. The present volume should be used in conjunction with such volumes as Verne Grant, *The Origin of Adaptations* (1963), E. B. Ford, *Ecological Genetics* (1964), B. G. Campbell, *Human Evolution* (1966), and Bruce Wallace, *Topics in Population Genetics* (1968).

Animal Species had three dominant themes: (1) the species is the most important unit of evolution; (2) individuals (and not genes) are the target of natural selection, hence the fitness of "a" gene is a nebulous if not misleading concept; and (3) the most important genetic phenomena in species are species-specific epistatic systems that give species internal cohesion. These three theses are now (in 1970) far more widely, or at least more consciously, accepted than they were in 1963, and there was no need, therefore, to modify this basic platform in the abridgment.

A detailed typewritten criticism of *Animal Species*, prepared by Professor R. Alexander (Michigan) and a graduate class (D. J. Futuyma, D. L. Hoyt, R. J. Jehl, B. G. Murray, D. Otte, A. E. Pace, R. T. Vinopal, and M. J. West) was most helpful in eliminating errors and inducing me to tighten up loose arguments. In the Preface of *Animal Species* I thanked those who gave me most generous help in the preparation of the original volume. Mrs. Sara S. Loth undertook the preparation of the manuscript (typing, bibliography, illustrations, permissions, and so forth) in the most competent manner, and Mrs. Nancy Clemente devoted exceptional care to the editing of the typescript. The volume owes a great deal to their dedicated efforts.

E. M.

November 1969

Contents

Tables

Figures

Populations, Species, and Evolution

1 · Evolutionary Biology

The theory of evolution is quite rightly called the greatest unifying theory in biology. The diversity of organisms, similarities and differences between kinds of organisms, patterns of distribution and behavior, adaptation and interaction, all this was merely a bewildering chaos of facts until given meaning by the evolutionary theory. There is no area in biology in which that theory does not serve as an ordering principle. Yet this very universality of application has created difficulties. Evolution shows so many facets that it looks alike to no two persons. The more different the backgrounds of two biologists, the more different their attempts at causal explanation. At least, so it was, until the 1930's, when the many dissenting theories were fused into a broad unified theory, the "modern synthesis." But even it has grown and matured since then.

Many of the earlier evolutionary theories were characterized by heavy emphasis, if not exclusive reliance, on a single factor (Table 1.1). The modern synthetic theory selected the best aspects of the earlier hypotheses and combined them in a new and original manner. In essence a two-factor theory, it regards the diversity and harmonious adaptation of the organic world as the result of a steady production of variation and of the selective effects of the environment.

Attempting to explain evolution by a single-factor theory was the fatal weakness of the pre-Darwinian and most nineteenth-century evolutionary theories. Lamarckism with its internal self-improvement principle, Geoffroyism with its induction of genetic change by the environment, Cuvier's catastrophism, Wagner's evolution by isolation, De Vries' mutationism, all tried to explain evolution by a single principle, excluding all others. Even Charles Darwin occasionally fell into this error, yet on the whole he was the first to make a serious effort to present evolutionary events as due to a balance of conflicting forces. The current theory of evolution—the "modern synthesis," as Huxley (1942) has called it —owes more to Darwin than to any other evolutionist and is built around Darwin's essential con-

Table 1.1. Theories of evolutionary change.

A. Monistic (single-factor explanations)
  1. Ectogenetic: changes directly induced by the environment
    (a) Random response (for example, radiation effects)
    (b) Adaptive response (Geoffroyism)
  2. Endogenetic: changes controlled by intrinsic forces
    (a) Finalistic (orthogenesis)
    (b) Volitional (genuine Lamarckism)
    (c) Mutational limitations
    (d) Epigenetic limitations
  3. Random events ("accidents")
    (a) Spontaneous mutations
    (b) Recombination
  4. Natural selection
B. Synthetic (multiple-factor explanations)
  1b + 2a + 2b = most "Lamarckian-type" theories
  1b + 2b + 2c + 4 = some recent "Lamarckian" theories
  1b + 3 + 4 = late Darwin, Plate, most nonmutationists during first three decades of 20th century
  3 + 4 = early "modern synthesis"
  1a + 2c + 2d + 3 + 4 = recent "modern synthesis"

cepts. Yet it incorporates much that is distinctly post-Darwinian. The concepts of mutation, variation, population, inheritance, isolation, and species, still quite nebulous in Darwin's day, are now far better understood and more rigorously defined.

The development of the modern theory was a slow process. Evolutionary biology was at first in the same situation as sociology, psychology, and other vast fields still are today: the available data were too voluminous and diversified to be organized at once into a single comprehensive theory. Looking back over the history of the many false starts gives a valuable insight into the process of theory formation. One important lesson is that some sets of data may not have significance until certain concepts are clarified or principles established. For instance, the true role of the environment in evolution could not be understood until the nature of small mutations and of selection was fully comprehended. Polygenes could not be analyzed and understood until the laws of inheritance had been clarified with the help of conspicuous mutations. The process of speciation (multiplication of species) could not be understood until after the nature of species and of geographic variation had been clarified. Discussions of variation among early evolutionists were utterly confused because they failed to make a clear distinction between geographical "variety" (geographical race) and individual variety. The analysis of quantitative characters was futile until the principles of particulate inheritance were fully understood.

Genetics, morphology, biogeography, systematics, paleontology, embryology, physiology, ecology, and other branches of biology, all have illuminated some special aspect of evolution and have contributed to the total explanation where other special fields failed. In many branches of biology one can become a leader even though one's knowledge is essentially confined to an exceedingly limited area. This is unthinkable in evolutionary biology. A specialist can make valuable contributions to special aspects of the evolutionary theory, but only he who is well versed in most of the branches of biology listed above can present a balanced picture of evolution as a whole. Whenever a narrow specialist has tried to develop a new theory of evolution, he has failed.

The importance of eliminating erroneous concepts is rarely given sufficient weight in discussions of theory formation. Only in some cases is it true that the new, better theory vanquishes the old, "bad" one. In many other instances it is the refutation of an erroneous theory that vacates the field for new ideas. An excellent illustration of this is Louis Agassiz's neglect of what seem to us most convincing evolutionary facts because they were inconsistent with his well-organized, harmonious creationist world view. Darwin, who had started the voyage of the *Beagle* with views similar to those of Agassiz, began to think seriously about evolution only after he had found overwhelming evidence that was completely irreconcilable with the creationist explanation of the diversity of animals and plants. Or, to cite another example, as long as spontaneous generation and the instantaneous conversion of one species into another were universally accepted, even for higher animals and plants, there was no room for a theory of evolution. By insisting on the fixity of species, Linnaeus did more to bring about the eclipse of the concept of spontaneous generation than did Redi and Spallanzani, who disproved it experimentally. Indirectly, Linnaeus did as much to prepare the ground for a theory of evolution as if he had proposed such as theory himself.

More important for the development of the synthetic theory than the rejection of ill-founded special theories of evolution was the rejection of two basic philosophical concepts that were formerly widespread if not universally held: preformism and typological thinking. *Preformism* is the theory of development that postulates a preformed adult individual in miniature "boxed" into the egg or spermatozoon, ready to "unfold itself" during development. The term evolution is derived from this concept of unfolding, and this connotation continued well into the post-Darwinian period. It was perhaps the reason Darwin did not use the term "evolution" in his *Origin of Species.* Transferred from ontogeny to phylogeny, evolution meant the unfolding of a built-in plan. Evolution, according to this view, does not produce genuine change, but consists merely in the maturation of immanent potentialities. This, for instance, was Louis Agassiz's theory of evolution (Mayr 1959c). Some of the orthogenetic and finalistic theories of evolution are the last remnants of this type of thinking. Mutationism was the most extreme form of reaction to these orthogenetic concepts. The current theory compromises by admitting that genotype and phenotype of a given evolutionary line set severe limits to its evolutionary potential (Table 1.1, *A2c, d*), without, however, prescribing the pathway of future evolutionary change.

Typological thinking is the other major misconception that had to be eliminated before a sound theory of evolution could be proposed. Plato's concept of the *eidos* is the philosophical codification of this form of thinking. According to this concept the vast observed variability of the world has no more reality than the shadows of an object on a cave wall, as Plato puts it in his allegory. Fixed, unchangeable "ideas" underlying the observed variability are the only things that are permanent and real. Owing to its belief in essences this philosophy is also referred to as *essentialism* and its representatives as *essentialists* (typologists). Most of the great philosophers of the seventeeth, eighteenth, and nineteenth centuries were influenced by the idealistic philosophy of Plato and the modifications of it by Aristotle. The thinking of these schools dominated the natural sciences until well into the nineteenth century. The concepts of unchanging essences and of complete discontinuities between every *eidos* (type) and all others make genuine evolutionary thinking well-nigh impossible. I agree with those who claim that the essentialist philosophies of Plato and Aristotle are incompatible with evolutionary thinking:

> The assumptions of population thinking are diametrically opposed to those of the typologist. The populationist stresses the uniqueness of everything in the organic world. What is true for the human species, that no two individuals are alike, is equally true for all other species of animals and plants . . . All organisms and organic phenomena are composed of unique features and can be described collectively only in statistical terms. Individuals, or any kind of organic entities, form populations of which we can determine the arithmetic mean and the statistics of variation. Averages are merely statistical abstractions; only the individuals of which the populations are composed have reality. The ultimate conclusions of the population thinker and of the typologist are precisely the opposite. For the typologist, the type (*eidos*) is real

> and the variation an illusion, while for the populationist the type (average) is an abstraction and only the variation is real. No two ways of looking at nature could be more different (Mayr 1959b:2).

The replacement of typological thinking by population thinking is perhaps the greatest conceptual revolution that has taken place in biology. Many of the basic concepts of the synthetic theory, such as that of natural selection and that of the population, are meaningless for the typologist. Virtually every major controversy in the field of evolution has been between a typologist and a populationist. Even Darwin, who was more responsible than anyone else for the introduction of population thinking into biology, often slipped back into typological thinking, for instance in his discussions on varieties and species.

Clarification of Evolutionary Concepts

A comparison of current evolutionary publications with those of only twenty or twenty-five years ago shows what great conceptual progress has been made in this short period. Since much of this volume is devoted to reporting this progress, I will barely mention some of these advances in this introductory discussion. Our ideas on the relation between gene and character have been thoroughly revised and the phenotype is more and more regarded not as a mosaic of individual gene-controlled characters but as the joint product of a complex interacting system, the total epigenotype (Waddington 1957). Interactions and balances among opposing forces are stressed to an increasing extent (Chapter 10). Virtually every component of the phenotype is recognized as a compromise made in response to opposing selection pressures.

The realization that the DNA of the chromosomes carries a program of information has led to great clarification. The phenomena of ontogeny and physiology are now interpreted as manifestations of the decoding of the information embodied in the genotype. Phylogeny, on the other hand, and all the phenomena involving evolutionary change are considered the production of ever-new programs of information.

Let me cite some other advances in our understanding. Natural selection is no longer regarded as an all-or-none process but rather as a purely statistical concept. Isolation has been revealed as a dual phenomenon, either the separation of populations by environmental barriers or the maintenance of the genetic integrity of gene pools by isolating mechanisms. The environment is restored to its place as one of the most important evolutionary factors, but in a drastically different role than it held in the various

"Lamarckian" theories. The new role of the environment is to serve as principal agent of natural selection.

Open Problems

The development of the evolutionary theory is a graphic illustration of the importance of the *Zeitgeist.* A particular constellation of available facts and prevailing concepts dominates the thinking of a given period to such an extent that it is very difficult for a heterodox viewpoint to get a fair hearing. Recalling this history should make us cautious about the validity of our current beliefs. The fact that the synthetic theory is now nearly universally accepted is not in itself proof of its correctness. It will serve as a warning to read with what scorn the mutationists (saltationists) in the first decade of this century attacked the contemporary naturalists for their belief in gradual changes and in the immense importance of the environment. It never occurred to the saltationists that their own typological and antiselectionist interpretation of evolution could be much further from the truth than the late Darwinian viewpoint of their adversaries. Mutations do not guide evolution, nor are their effects on the phenotype always sufficiently drastic to be visible. Recombination makes far more new phenotypes available for selection than does mutation, and the kinds of mutations and recombinations that can occur in a given organism are severely restricted. These statements are entirely consistent with the synthetic theory, but they may be quite startling to those who are unaware of the modern developments and who are still fighting the battle of the last generation.

The essentials of the modern theory are to such an extent consistent with the facts of genetics, systematics, and paleontology that one can hardly question their correctness. The basic framework of the theory is that evolution is a two-stage phenomenon: the production of variation and the sorting of the variants by natural selection. Yet agreement on this basic thesis does not mean that the work of the evolutionist is completed. The basic theory is in many instances hardly more than a postulate and its application raises numerous questions in almost every concrete case. The discussions throughout this volume are telling testimony of the truth of this statement.

Modern research is directed primarily toward three areas: evolutionary phenomena that do not yet appear to be adequately explained by the synthetic theory, such as stagnant or explosive evolution; the search for various subsidiary factors that, although inconspicuous on casual inspection, exercise unexpected selection pressures; and, perhaps most important, the interplay among genes and between genotypes and environment resulting in the phenotype, the real target of natural selection.

Most contemporary arguments concern the relative importance of the various interacting factors. One will get highly diverse answers if one asks a number of contemporary evolutionists the following questions:

How important are random events in evolution?

How important is hybridization in evolution?

What is the effect of interpopulation gene flow?

What proportion of new mutations are beneficial?

What proportion of genetic variability is due to balanced polymorphism?

Other areas in which there is still wide divergence of opinion are the importance of phenotypic plasticity, the pathway to adaptation, evolutionary mechanisms in higher and lower organisms, the origin of sexuality, and the origin of life. It must be stressed, for the benefit of nonevolutionists, that none of the arguments going on in these areas touches upon the basic principles of the synthetic theory. It is the application of the theory that is sometimes controversial, not the theory itself. And with respect to application we still have a long way to go. There are vast areas of modern biology, for instance biochemistry and the study of behavior, in which the application of evolutionary principles is still in the most elementary stage.

The Major Areas of Evolutionary Research

Important contributions to our understanding of the evolutionary process have been made by virtually every branch of biology. During the past one hundred years most of the research has been concerned with a number of discrete areas, progress within which has been unequal:

The fact of evolution,

The establishment of phylogenies,

The origin of discontinuities (speciation),

The material of evolution,

Rates of evolution,

Causes of evolution, and

The evolution of adaptation.

The amount of attention given to each of these areas has changed with time. To establish unequivocally the fact of evolution was after 1859 the first concern of the young science of evolutionary biology. The study of phylogeny soon became predominant, at least in zoology. Indeed, even today there still are some zoologists to whom the term "evolution" signifies little more than the determination of homologies, common ancestors, and phylogenetic trees. The interest of most evolutionary biologists, however, has shifted to a study of the causes and mechanisms of evolutionary change and to an attempt at determining the role and relative importance of various factors. The different responses to these factors displayed by different types of organisms are also receiving increasing attention. Evolutionary biology is beginning to become truly comparative.

Each branch of evolutionary biology occupies a special niche and is uniquely qualified to illuminate some special problem. The geneticist is mainly concerned with the individual, the stability or mutability of loci, the modification of the phenotype, the interaction of parental genes in the production of the phenotype and the effect of this interaction on fitness, in short, all the problems concerning the gene and its interaction with other genes and with the environment. The development of population genetics led to an expansion of the geneticist's field of interest from the gene to the gene pool of the population.

The contribution of genetics to the understanding of the process of evolution has not yet been evaluated objectively. The assumption made by some geneticists, that it was quite impossible to have sensible ideas on evolution until the laws of inheritance had been worked out, is contradicted by the facts. Everyone admits that Darwin's evolutionary theories were essentially correct; yet his genetic theories were about as wrong as they could be. Conversely, the early Mendelians, the first biologists (except for Mendel himself) who truly understood genetics, misinterpreted just about every evolutionary phenomenon. Some of their contemporaries among the naturalists, on the other hand, though they did not understand genetics and even believed in some environmental induction (Geoffroyism), presented a remarkably correct picture of speciation, adaptation, and the role of natural selection. It would be going too far to claim that it is immaterial whether one believes the De Vriesian or the Lamarckian theory of the source of genetic variation, yet it is true that it is less important for the understanding of evolution to know how genetic variation is brought about than to know how natural selection deals with it. Replacing the erroneous belief in blending inheritance with the theory of particulate inheritance is the greatest single contribution of genetics. This advance has been the basis of all subsequent developments. The genetic material presented in Chapter 7–10 and 17 shows to what extent the modern genetic theory can explain many phenomena that the naturalist had long known and correctly described, but was unable to interpret.

The study of long-term evolutionary phenomena is the domain of the paleontologist. He investigates rates and trends of evolution in time and is interested in the origin of new classes, phyla, and other higher taxa. Evolution means change and yet it is only the paleontologist among all biologists who can properly study the time dimension. If the fossil record were not available, many evolutionary problems could not be solved; indeed, many of them would not even be apparent.

The taxonomist, who deals primarily with local populations, subspecies, species, and genera, is concerned with the region that lies between the areas of interest and competence of the geneticist and of the paleontologist, overlapping with both but approaching problems in the area of overlap from a somewhat different viewpoint. The species, the center of the taxonomist's interest, is one of the important levels of integration in the organic world. Neglect of this level in much of our biological curriculum is puzzling. We do not even have a special term for the study of the species, corresponding to cytology, the study of cells; histology, the study of tissues; and anatomy, the study of organs. Yet the species is not only the basic unit of classification, but also one of the most important units of interaction in ecology and ethology. The origin of new species, signifying the origin of essentially irreversible discontinuities with entirely new potentialities, is the most important single event in evolution. Darwin, who devoted so much of his life to the systematics of species, fully appreciated the significance of this level, as he made clear in the choice of title for his classic *On the Origin of Species.*

The emphasis in the present volume is deliberately placed on those aspects of evolution that involve the species. Other aspects, of greater interest to the geneticist or paleontologist, and adequately treated by Dobzhansky (1951), Simpson (1953a), Rensch (1960a), and other modern writers, are discussed only incidentally. Evolutionary biology has become far too vast a field to be covered adequately in a single volume.

The basic structure of this volume is as follows. The *characteristics of species* is dealt with in Chapters 2–6; the *structure and genetics of populations* in Chapters 7–10; the (population) *structure and variation of species* in Chapters 11–14; and the *multiplication of species* in Chapters 15–18. Chapter 19 is devoted to a discussion of the role of *species in transpecific evolution* and Chapter 20 to a review of the possible consequences of our findings for *man.*

2 · Species Concepts and Their Application

Darwin's choice of title for his great evolutionary classic, *On the Origin of Species,* was no accident. The origin of new "varieties" within species had been taken for granted since the time of the Greeks. Likewise the occurrence of gradations, of "scales of perfection" among "higher" and "lower" organisms, was a familiar concept, though usually interpreted in a strictly static manner. The species remained the great fortress of stability and this stability was the crux of the antievolutionist argument. "Descent with modification," true biological evolution, could be proved only by demonstrating that one species could originate from another. It is a familiar and often-told story how Darwin succeeded in convincing the world of the occurrence of evolution and how—in natural selection—he found the mechanism that is responsible for evolutionary change and adaptation. It is not nearly so widely recognized that Darwin failed to solve the problem indicated by the title of his work. Although he demonstrated the modification of species in the time dimension, he never seriously attempted a rigorous analysis of the problem of the multiplication of species, of the splitting of one species into two. I have examined the reasons for this failure (Mayr 1959a) and found that foremost among them was Darwin's uncertainty about the nature of species. The same can be said of those authors who attempted to solve the problem of speciation by saltation or other heterodox hypotheses. They all failed to find solutions that are workable in the light of the modern appreciation of the population structure of species. An understanding of the nature of species, then, is an indispensable prerequisite for the understanding of the evolutionary process.

Species Concepts

The term *species* is frequently used to designate a class of similar things to which a name has been attached. Most often this term is applied to living organisms, such as birds, fishes, flowers, or trees, but it has also been used for inanimate objects and even for human artifacts. Mineralogists speak of species of minerals, physicists of nuclear species; interior decorators consider tables and chairs species of furniture. The application of the same term both to organisms and to inanimate objects has led to much confusion and an almost endless number of species definitions (Mayr 1963, 1969); these, however, can be reduced to three basic species concepts. The first two, mainly applicable to inanimate objects, have considerable historical significance, because their advocacy was the cause of much past confusion. The third is the species concept now prevailing in biology.

1. *The Typological Species Concept*

The typological species concept, going back to the philosophies of Plato and Aristotle (and thus sometimes called the essentialist concept), was the species concept of Linnaeus and his followers (Cain 1958). According to this concept, the observed diversity of the universe reflects the existence of a limited number of underlying "universals" or types (*eidos* of Plato). Individuals do not stand in any special relation to one another, being merely expressions of the same type. Variation is the result of imperfect manifestations of the idea implicit in each species. The presence of the same underlying essence is inferred from similarity, and morphological similarity is therefore, the species criterion for the essentialist. This is the so-called morphological species concept. Morphological characteristics do provide valuable clues for the determination of species status. However, using degree of morphological difference as the primary criterion for species status is completely different from utilizing morphological evidence together with various other kinds of evidence in order to determine whether or not a population deserves species rank under the biological species concept. Degree of morphological difference is not the decisive criterion in the ranking of taxa as species. This is quite apparent from the difficulties into which a morphological-typological species concept leads in taxonomic practice (see Chapter 3). Indeed, its own adherents abandon the typological species concept whenever they discover that they have named as a separate species something that is merely an individual variant.

2. *The Nominalistic Species Concept*

The nominalists (Occam and his followers) deny the existence of "real" universals. For them only individuals exist; species are man-made abstractions. (When they have to deal with a species, they treat it as an individual on a higher plane.) The nominalistic species concept was popular in France in the eighteenth century and still has adherents today. Bessey (1908) expressed this viewpoint particularly well: "Nature produces individuals and nothing more . . . species have no actual existence in nature. They are mental concepts and nothing more . . . species have been invented in order that we may refer to great numbers of individuals collectively."

Any naturalist, whether a primitive native or a trained population geneticist, knows that this is simply not true. Species of animals are not human constructs, nor are they types in the sense of Plato and Aristotle; but they are something for which there is no equivalent in the realm of inanimate objects.

From the middle of the eighteenth century on the inapplicability of these two medieval species concepts (1 and 2 above) to biological species became increasingly apparent. An entirely new concept, applicable only to species of organisms, began to emerge in the later writings of Buffon and of many other naturalists and taxonomists of the nineteenth century (Mayr 1968).

3. *The Biological Species Concept*

This concept stresses the fact that species consist of populations and that species have reality and an internal genetic cohesion owing to the historically evolved genetic program that is shared by all members of the species. According to this concept, then, the members of a species constitute (1) *a reproductive community.* The individuals of a species of animals respond to one another as potential mates and seek one another for the purpose of reproduction. A multitude of devices ensures intraspecific reproduction in all organisms (Chapter 5). The species is also (2) *an ecological unit* that, regardless of the individuals composing it, interacts as a unit with other species with which it shares the environment. The species, finally, is (3) *a genetic unit* consisting of a large intercommunicating gene pool, whereas an individual is merely a temporary vessel holding a small portion of the contents of the gene pool for a short period of time. These three properties raise the species above the typological interpretation of a "class of objects" (Mayr 1963: 21). The species definition that results from this theoretical species concept is: *Species are groups of interbreeding natural populations that are reproductively isolated from other such groups.*

The development of the biological concept of the species is one of the earliest manifestations of the emancipation of biology from an inappropriate philosophy based on the phenomena of inanimate nature. The species concept is called biological not because it deals with biological taxa, but because the definition is biological. It utilizes criteria that are meaningless as far as the inanimate world is concerned.

When difficulties are encountered, it is important to focus on the basic biological meaning of the species: A species is a protected gene pool. It is a Mendelian population that has its own devices (called isolating mechanisms) to protect it from harmful gene flow from other gene pools. Genes of the same gene pool form harmonious combinations because they have become coadapted by natural selection. Mixing the genes of two different species leads to a high frequency of disharmonious gene combinations; mechanisms that prevent this are therefore favored by selection. Thus it is quite clear that the word "species" in biology is a relational term. *A* is a species in relation to *B* or *C* because it is reproductively isolated from them. The biological species concept has its primary significance with respect to sympatric and synchronic populations (existing at a single locality and at the same time), and these—the "nondimensional species"—are precisely the ones where the application of the concept faces the fewest difficulties. The more distant two populations are in space and time, the more difficult it becomes to test their species status in relation to each other, but also the more irrelevant biologically this becomes.

The biological species concept also solves the paradox caused by the conflict between the fixity of the species of the naturalist and the fluidity of the species of the evolutionist. It was this conflict that made Linnaeus deny evolution and Darwin the reality of species (Mayr 1957). The biological species combines the discreteness of the local species at a given time with an evolutionary potential for continuing change.

The Species Category and Species Taxa

The advocacy of three different species concepts has been one of the two major reasons for the "species problem." The second is that many authors have failed to make a distinction between the definition of the species category and the delimitation of species taxa (for fuller discussion see Mayr 1969).

A *category* designates a given rank or level in a hierarchic classification. Such terms as "species," "genus," "family," and "order" designate categories. A category, thus, is an abstract term, a class name, while the organisms placed in these categories are concrete zoological objects.

Organisms, in turn, are not classified as individuals, but as groups of organisms. Words like "bluebirds," "thrushes," "songbirds," or "vertebrates" refer to such groups. These are the concrete objects of classification. Any such group of populations is called a *taxon* if it is considered sufficiently distinct to be worthy of being formally assigned to a definite category in the hierarchic classification. *A taxon is a taxonomic group of any rank that is sufficiently distinct to be worthy of being assigned to a definite category.*

Two aspects of the taxon must be stressed. A taxon always refers to specified organisms. Thus *the* species is not a taxon, but any given species, such as the Robin (*Turdus migratorius*), is. Second the taxon must be formally recognized as such, by being described under a designated name.

Categories, which designate rank in a hierarchy, and taxa, which designate named groupings of organisms, are thus two very different kinds of phenomena. A somewhat analogous situation exists in our human affairs. Fred Smith is a concrete person but "captain" or "professor" is his rank in a hierarchy of levels.

The Assignment of Taxa to the Species Category

Much of the task of the taxonomist consists in assigning taxa to the appropriate categorical rank. In this procedure there is a drastic difference between the species taxon and the higher taxa. Higher taxa are defined by intrinsic characteristics. Birds is the class of feathered vertebrates. Any and all species that satisfy the definition of "feathered vertebrates" belong to the class of birds. An essentialist (typological) definition is satisfactory and sufficient at the level of the higher taxa. It is, however, irrelevant and misleading to define species in an essentialistic way because the species is not defined by intrinsic, but by *relational* properties.

Let me explain this. There are certain words that indicate a relational property, like the word "brother." Being a brother is not an inherent property of an individual, as hardness is a property of a stone. An individual is a brother only with respect to someone else. The word "species" likewise designates such a relational property. A population is a species with respect to all other populations with which it exhibits the relationship of reproductive isolation—noninterbreeding. If only a single population existed in the entire world, it would be meaningless to call it a species.

Noninterbreeding between populations is manifested by a gap. It is this gap between populations that coexist (are sympatric) at a single locality at a given time that delimits the species recognized by the local naturalist. Whether one studies birds, mammals, butterflies, or snails near one's home town, one finds each species clearly delimited and sharply separated from all other species. This demarcation is sometimes referred to as the species delimitation *in a nondimensional system* (a system without the dimensions of space and time).

Anyone can test the reality of these discontinuities for himself, even where the morphological differences are slight. In eastern North America, for instance, there are four similar species of the thrush genus *Catharus* (Table 2.1), the Veery (*C. fuscescens*), the Hermit Thrush (*C. guttatus*), the Olive-backed or Swainson's Thrush (*C. ustulatus*), and the Gray-cheeked Thrush (*C. minimus*). These four species are sufficiently similar visually to confuse not only the human observer, but also silent males of the other species. The species-specific songs and call notes, however, permit easy species discrimination, as observationally substantiated by Dilger (1956). Rarely do more than two species breed in the same area and the overlapping species, $f + g$, $g + u$, and $u + m$, usually differ considerably in their foraging habits and niche preference, so that competition is minimized with each other and with two other thrushes, the Robin (*Turdus migratorius*) and the Wood Thrush (*Hylocichla mustelina*), with which they share their geographic range and many ecological requirements. In connection with

Table 2.1. Characteristics of four eastern North American species of *Catharus* (from Dilger 1956).

Characteristic compared	*C. fuscescens*	*C. guttatus*	*C. ustulatus*	*C. minimus*
Breeding range	Southernmost	More northerly	Boreal	Arctic
Wintering area	No. South America	So. United States	C. America to Argentina	No. South America
Breeding habitat	Bottomland woods with lush undergrowth	Coniferous woods mixed with deciduous	Mixed or pure tall coniferous forests	Stunted northern fir and spruce forests
Foraging	Ground and arboreal (forest interior)	Ground (inner forest edges)	Largely arboreal (forest interior)	Ground (forest interior)
Nest	Ground	Ground	Trees	Trees
Spotting on eggs	Rare	Rare	Always	Always
Relative wing length	Medium	Short	Very long	Medium
Hostile call	*veer* *pheu*	*chuck* *seeeep*	*peep* *chuck-burr*	*beer*
Song	Very distinct	Very distinct	Very distinct	Very distinct
Flight song	Absent	Absent	Absent	Present

their different foraging and migratory habits the four species differ from one another (and from other thrushes) in the relative length of wing and leg elements and in the shape of the bill. There are thus many small differences between these at first sight very similar species. Most important, no hybrids or intermediates among these four species have ever been found. Each is a separate genetic, behavioral, and ecological system, separated from the others by a complete biological discontinuity, a gap.

Difficulties in the Application of the Biological Species Concept

The practicing taxonomist often has difficulties when he endeavors to assign populations to the correct rank. Sometimes the difficulty is caused by a lack of information concerning the degree of variability of the species with which he is dealing. Helpful hints on the solution of such practical difficulties are given in the technical taxonomic literature (Mayr 1969).

More interesting to the evolutionist are the difficulties that are introduced when the dimensions of time and space are added. Most species taxa do not consist merely of a single local population but are an aggregate of numerous local populations that exchange genes with each other to a greater or lesser degree. The more distant two populations are from each other, the more likely they are to differ in a number of characteristics. It will be shown in Chapters 10 and 11 that some of these populations are incipient species, having acquired some but not all characteristics of species. One or another of the three most characteristic properties of species taxa—reproductive isolation, ecological difference, and morphological distinguishability—is in such cases only incompletely developed. The application of the species concept to such incompletely speciated populations raises considerable difficulties. There are six wholly different situations that may cause difficulties.

(1) *Evolutionary continuity in space and time.* Widespread species may have terminal populations that behave toward each other as distinct species even though they are connected by a chain of interbreeding populations. Cases of reproductive isolation among geographically distant populations of a single species are discussed in Chapter 16.

(2) *Acquisition of reproductive isolation without corresponding morphological change.* When the reconstruction of the genotype in an isolated population has resulted in the acquisition of reproductive isolation, such a population must be considered a biological species. If the correlated morphological change is very slight or unnoticeable, such a species is called a sibling species (Chapter 3).

(3) *Morphological differentiation without acquisition of reproductive isolation.* Isolated populations sometimes acquire a degree of morphological divergence one would ordinarily expect only in a different species. Yet some such populations, although as different morphologically as good species, interbreed indiscriminately where they come in contact. The West Indian snail genus *Cerion* illustrates this situation particularly well (Fig. 2.1).

(4) *Reproductive isolation dependent on habitat isolation.* Numerous cases have been described in the literature in which natural populations

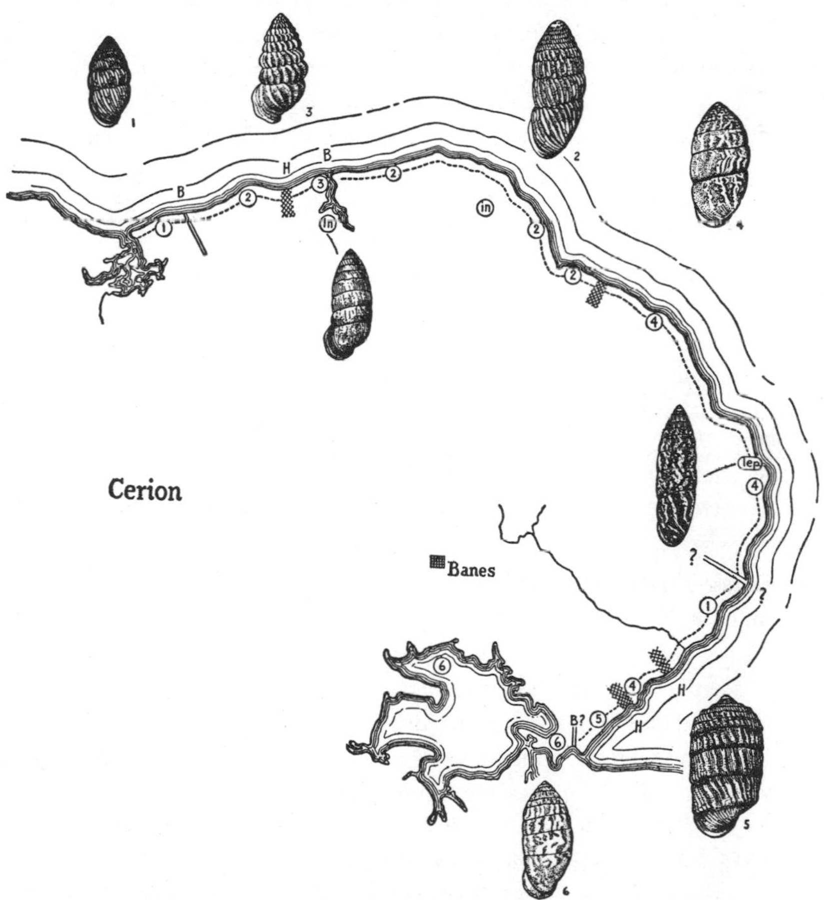

Fig. 2.1. The distribution pattern of populations of the halophilous land snail *Cerion* on the Banes Peninsula in eastern Cuba. Numbers refer to distinctive races or "species." Where two populations come in contact (with one exception) they hybridize (*H*), regardless of degree of difference. In other cases contact is prevented by a barrier (*B*). *In* = isolated inland population.

acted toward each other like good species (in areas of contact) as long as their habitats were undisturbed. Yet the reproductive isolation broke down as soon as the characteristics of these habitats were changed, usually by the interference of man. Such cases of secondary breakdown of isolation are discussed in Chapter 6.

(5) *Incompleteness of isolating mechanisms.* Very few isolating mechanisms are all-or-none devices (see Chapter 5). They are built up step by step, and most isolating mechanisms of an incipient species are imperfect and incomplete. Species level is reached when the process of speciation has become irreversible, even if some of the secondary isolating mechanisms have not yet reached perfection (see Chapter 17).

(6) *Attainment of different levels of speciation in different local populations.* The perfecting of isolating mechanisms may proceed in different populations of a polytypic species (one having several subspecies) at different rates. Two widely overlapping species may, as a consequence, be completely distinct at certain localities but may freely hybridize at others. Many cases of sympatric hybridization discussed in Chapter 6 fit this characterization (see Mayr 1969 for advice on handling such situations).

These six types of phenomena are consequences of the gradual nature of the ordinary process of speciation (excluding polyploidy, see p. 254). Determination of species status of a given population is difficult or arbitrary in many of these cases.

Difficulties Posed by Uniparental Reproduction

The task of assembling individuals into populations and species taxa is very difficult in most cases involving uniparental (asexual) reproduction. Self-fertilization, parthenogenesis, pseudogamy, and vegetative reproduction are forms of uniparental reproduction. The biological species concept, which is based on the presence or absence of interbreeding between natural populations, cannot be applied to groups with obligatory asexual reproduction because interbreeding of populations is nonexistent in these groups. The nature of this dilemma is discussed in more detail elsewhere (Mayr 1963, 1969). Fortunately, there seem to be rather well-defined discontinuities among most kinds of uniparentally reproducing organisms. These discontinuities are apparently produced by natural selection from the various mutations that occur in the asexual lines (clones). It is customary to utilize the existence of such discontinuities and the amount of morphological difference between them to delimit species among uniparentally reproducing types.

The Importance of a Nonarbitrary Definition of Species

The clarification of the species concept has led to a clarification of many evolutionary problems as well as, often, to a simplification of practical problems in taxonomy. The correct classification of the many different kinds of varieties (phena), of polymorphism (Chapter 7), of polytypic species (Chapter 12), and of biological races (Chapter 15) would be impossible without the arranging of natural populations and phenotypes into biological species. It was impossible to solve, indeed even to state precisely, the problem of the multiplication of species until the biological species concept had been developed. The genetics of speciation, the role of species in large-scale evolutionary trends, and other major evolutionary problems could not be discussed profitably until the species problem was settled. It is evident then that the species problem is of great importance in evolutionary biology and that the growing agreement on the concept of the biological species has resulted in a uniformity of standards and a precision that is beneficial for practical as well as theoretical reasons.

The Biological Meaning of Species

The fact that the organic world is organized into species seems so fundamental that one usually forgets to ask why there are species, what their meaning is in the scheme of things. There is no better way of answering these questions than to try to conceive of a world without species. Let us think, for instance, of a world in which there are only individuals, all belonging to a single interbreeding community. Each individual is in varying degrees different from every other one, and each individual is capable of mating with those others that are most similar to it. In such a world, each individual would be, so to speak, the center of a series of concentric rings of increasingly more different individuals. Any two mates would be on the average rather different from each other and would produce a vast array of genetically different types among their offspring. Now let us assume that one of these recombinations is particularly well adapted for one of the available niches. It is prosperous in this niche, but when the time for mating comes, this superior genotype will inevitably be broken up. There is no mechanism that would prevent such a destruction of superior gene combinations, and there is, therefore, no possibility of the gradual improvement of gene combinations. The significance of the species now becomes evident. The reproductive isolation of a species is a protective device that guards against the breaking up of its well-integrated, coadapted gene system. Organizing organic diversity into species creates a system that permits genetic diversification and the accumulation of favorable genes and gene combinations without the danger of destruction of the basic gene complex. There are definite limits to the amount of genetic variability that can be accommodated in a single gene pool without producing too high a proportion of inviable recombinants. Organizing genetic diversity into protected gene pools, that is, species, guarantees that these limits are not overstepped. This is the biological meaning of species.

3 · Morphological Species Characters and Sibling Species

The morphological species definition, which is based on the typological species concept, dominated animal taxonomy during the nineteenth and early twentieth centuries, but is steadily losing ground. Yet much contemporary thinking about species still rests on it. In view of the historical importance of the morphological concept and the correctness of some of its elements, it deserves more detailed discussion.

The argument of proponents of a morphologically defined species runs about as follows: "Natural populations considered by general consent to be species are morphologically distinct. Morphological distinctness is thus the decisive criterion of species rank. Consequently, any natural population that is morphologically distinct must be recognized as a separate species." This conclusion is fallacious, even though based on the correct observation of a frequent correlation between reproductive isolation and morphological difference. It is fallacious because it overlooks the strictly secondary role of morphological differences. The primary criterion of species rank of a natural population is reproductive isolation. The degree of morphological difference displayed by a natural population is a secondary by-product of the genetic divergence resulting from reproductive isolation.

Application of a purely morphological species concept to sexually reproducing species leads to insurmountable difficulties, owing to (1) the presence of conspicuous morphological differences among conspecific individuals and populations (intraspecific variation) and (2) the virtual absence of morphological differences among certain sympatric populations (sibling species) that otherwise have all the characteristics of good species (genetic difference and reproductive isolation).

Intraspecific Variation

There is often greater morphological difference between individuals of a single population or between conspecific populations than between related species. In the mallard the sexes are so different that Linnaeus originally described the male as *Anas boschas* and the female as *Anas platyrhynchos*. In many other groups of birds (birds of paradise, hummingbirds, tanagers, wood warblers, and so forth), the females may appear more different from males of their own species than from females of related species. Even greater are such sex differences among fishes (for example, female deep-sea fishes with attached dwarf males), insects (for example, army ants, mutillid wasps), and lower invertebrates (for example, the dwarf male of the echiurid worm *Bonellia viridis*).

The difference between immature stages and adults in many kinds of animals is astonishingly great. Age variation is marked in insects, lower invertebrates (free-swimming larvae of crustaceans, mollusks, echinoderms), and particularly in parasites. Sexual dimorphism and age variation are only two of the many kinds of individual variation found in organisms (see Chapter 7 and Mayr 1969). All these forms of variation demonstrate a formidable amount of intraspecific variability.

What attitude does the supporter of the morphological species concept take toward this variation? Curiously, he takes precisely the same attitude as the proponent of the biological species concept. No matter how different a morphological variant may be, as soon as it is revealed as a member of the same breeding population (sex and age differences, polymorphism) or as a local variant (genetic or nongenetic), he deprives it of its species status. It is quite evident, then, that even those who profess to hold a morphological species concept base their taxonomic decisions ultimately on the biological criterion of interbreeding. Degree of morphological difference is completely useless as a yardstick for species status unless it is applied in conjunction with such biological criteria as population, interbreeding, and reproductive isolation.

This conclusion is reinforced by a consideration of the reverse situation, the so-called sibling species. Again the vulnerability of a purely morphological criterion is evident.

Definition and Characteristics of Sibling Species

The naturalist occasionally encounters sympatric populations that are morphologically similar, if not identical, but are reproductively isolated.

What shall he do with such populations? Adherents of a purely morphological species concept will not classify them as species because for them species must be separable on the basis of ordinary preserved material. Natural populations that are not readily distinguishable but are nevertheless reproductively isolated have caused considerable difficulties in the biological and taxonomic literature. Such populations have sometimes been called "biological races," a term often used to encompass many unrelated problems and phenomena (Chapter 15). Adoption of the biological species concept makes it evident that most "biological races" are actually valid species, distinguished only by the slightness of the morphological difference. For such exceedingly similar species, the term *sibling species* was introduced (Mayr 1942) as a translation of the equivalent terms *espèces jumelles* in French and *Geschwisterarten* in German. Sibling species may be defined as *morphologically similar or identical natural populations that are reproductively isolated.*

Sibling species are of threefold importance in biology: (1) they permit us to test the validity of the biological versus the morphological species concept; (2) they are of great practical importance in applied biology, in agricultural pest control, and in medical entomology; (3) they are of historical importance in the study of speciation (Chapter 15), having been cited by some authors as evidence for a special kind of speciation.

The characteristics of sibling species are best revealed by describing in detail some of the better-known cases. (For a far more detailed account see Mayr 1963: 34–57.) In the genus *Drosophila* most of the species complexes contain groups of sibling species (Patterson and Stone 1952). The kinds of differences that may exist between two sibling species are typified by the extensively studied pair *D. pseudoobscura* Frolova and *D. persimilis* Dobzhansky and Epling. When originally discovered (1929), these flies were designated *D. pseudoobscura*, race A, and *D. pseudoobscura*, race B. Crosses between the "races" produced F_1 hybrids of which the females were fertile, the males sterile. The *Y*-chromosome of "race A" is J-shaped; that of "race B," V-shaped. From this modest beginning, the number of known differences has increased steadily, and when it was discovered that the two "races" coexist over wide areas without interbreeding, they were raised to the rank of full species. The salivary-gland chromosomes of the two species are quite different and fully diagnostic, even though the gene arrangements are variable within either species. At first it was thought that the two species were identical morphologically, but then it was found that the average number of teeth in the sex combs of the males is greater in *D. pseudoobscura* than in *D. persimilis* and that the wings of *D. persimilis* are on the average larger than those of *D. pseudoobscura.* Finally, the species were found to exhibit clear-cut differences in the shape of the male genitalia.

What were once considered two morphologically indistinguishable "biological races" are now accepted as two similar species, distinguished by diagnostic characters in salivary-gland chromosomes, male genitalia, sex combs, and relative wing size. This development is typical of our knowledge of most sibling species. When first discovered, they are believed to be morphologically identical or nearly so, but on closer study one morphological difference after another is discovered. These are fortified by ecological, physiological, and other differences. *Drosophila persimilis,* for instance, has a more northerly distribution than does *D. pseudoobscura,* and is more frequently found at higher altitudes, showing a preference for lower temperatures. The two species also differ in their diurnal activity rhythm, in their reaction to light, and in other ecological-physiological characteristics. Females of *D. pseudoobscura* reach sexual maturity at an age of 32–36 hours after hatching; those of *D. persimilis,* at an age of 44–48 hours. There may be differences in their scents, as manifested by interspecific mating preferences. In short, under scrutiny these species are found to differ, slightly or conspicuously, in almost every property. In spite of their superficial similarity, the two species represent two very different gene complexes.

Perhaps the most celebrated case of sibling species is that of the malaria-mosquito complex in Europe. According to the older literature, malaria in Europe is caused by *the* malaria mosquito, *Anopheles maculipennis.* A study of the distribution and ecology of this mosquito revealed all sorts of puzzling irregularities. *Anopheles* mosquitoes were found to be quite common in certain parts of Europe where malaria was absent. In some districts they fed only on domestic animals; in others they preferred man. In some districts they were associated with fresh water; in others, with brackish water. Understanding dawned only when Falleroni discovered constant differences in the eggs of mosquitoes differing in biological characteristics. Finally it was shown that *the* malaria mosquito of Europe was actually a group of six sibling species (Table 3.1).

After 30 years of search the most reliable morphological differences to be found among these species of *Anopheles* are still those of the eggs. However, additional differences have been found. Number and branching of the hairs in larval appendages are usually diagnostic for each species at a given locality. There are differences among some of the species in size and shape of the wing scales. Finally, there are constant differences among the species in gene arrangements, as indicated by the banding pattern of the salivary chromosomes (Table 3.1) (Kitzmiller et al. 1967).

Occurrence of Sibling Species

Sibling species occur in all groups of animals, yet they seem to be much more common in some groups, such as insects, than in others, such as vertebrates. It would be interesting to have exact numerical data, but this will not be possible until the taxonomy of these groups is better known. Several surveys of sibling species have been published (Mayr 1942, 1963; Walker 1964; Niethammer and Kramer 1966). The examples discussed in the following pages have been chosen almost at random from the enormous number of such cases in the taxonomic literature. There is hardly a taxonomic monograph or revision that does not give new instances of sibling species.

Vertebrates

Among mammals sibling species are rare, but a few exist among the shrews, and a few more among the rodents, where chromosomal analysis has recently revealed several previously overlooked species (Matthey 1964). Among birds they are even rarer, being limited to a few genera, for instance the honey eaters (*Meliphaga*), the cave swiftlets (*Collocalia*), the honey guides (*Indicator*), and several genera of tyrant flycatchers (*Empidonax*, *Elaenia*, and *Myiarchus*) where analysis of calls and songs has greatly clarified the taxonomy.

Sibling species are frequent among frogs and toads of all continents, as analysis of vocalizations has revealed. Even the well-known leopard frog (*Rana pipiens*), so widely used in embryological, physiological, and genetic studies, turns out to be a complex of sibling species (Littlejohn and Oldham 1968).

Exceedingly similar species occur in many genera of fishes. Their discrimination is particularly difficult among fresh-water fishes, owing to the plasticity of the phenotype, as for instance in *Salmo* and *Coregonus*. Sometimes the same lake or river system is invaded repeatedly by colonists from the same source population, each group of colonists being reproductively isolated from the others and usually spawning at different seasons or water temperatures. Some of these species are very similar to each other, and the difficulty of distinction is aggravated if there is occasional hybridization, as is known to occur between sympatric species of *Coregonus*.

Table 3.1. Biological differences among members of *Anopheles maculipennis* group of mosquitoes.

Characteristic compared	A. *melanoon and subalpinus*	A. *messeae*	A. *maculipennis*	A. *atroparvus*	A. *labranchiae*	A. *sacharowi* (*elutus*)
Egg color	All black or (*subalpinus*) with dark cross bars	Transverse bars part of a diffuse dark pattern	Two black cross bars on light background	Dappled or with wedge-shaped black spots	Similar to *atroparvus* but paler, dark spots smaller	Gray without pattern
Egg float	Large and smooth	Large and rough	Large and rough	Small and smooth	Very small and rough	None
X-chromosome	Standard	Extensive rearrangement	Standard	Standard	Standard	Small inversion
Third chromosome	Inversion in right arm	Inversion in right arm	Inversion in right arm	Standard	Standard	Inversion in left arm
Habitat	Often rice fields	Cool, standing fresh water	Cool, running fresh water	Cool, slightly brackish water	Mostly warm, brackish water	Shallow, standing water, often brackish
Hibernation	No	Yes	Complete	No	No	No
Feeding on man	?	Rarely	No	Yes	Yes, with preference	Almost exclusively
Malaria carrier	No	No (rarely)	No	Slightly	Very dangerous	Very dangerous
Range	Mediterranean	Continental and northern Europe	Mountains of Europe	Northern Europe	Chiefly southern Europe	Eastern Mediterranean and Near East

Insects

Sibling species appear to be especially common among insects. Their discovery has been relatively slow, since even conspicuously different species remain undescribed in most families of insects. On the whole, sibling species are discovered only when the group to which they belong is of economic, medical, or behavioral importance. In plant-feeding insects, different sibling species usually live on different host plants. Prior to the abandonment of the morphological species definition these plant-feeders were usually described as host races or biological races (see Chapter 15).

Lepidoptera. Sibling species are particularly frequent in the order Lepidoptera (butterflies and moths). A well-analyzed case is that of the budworms of the genus *Choristoneura* (Tortricidae). One species (*fumiferana*) feeds on spruce and fir, another one (*pinus*) on jackpine. The two species, although exceedingly similar, differ not only in food-plant preference, but also in average size, coloration, wing expanse, wing pattern, male genitalia, and season of flight. Equally similar species occur on other conifers (Freeman et al. 1953:121–151; 1967:449–506). Most of the sibling species of Lepidoptera were discovered through differences in their preferred food plants and through morphological differences in the male genital armatures. Other species-specific characters were usually found as soon as the "host races" became recognized as different species.

Diptera. Sibling species are widespread among the Diptera (flies, etc.) and on the whole are well studied, because many of the species are important medically (*Anopheles, Aedes, Simulium*), genetically (*Drosophila*), or cytologically (*Sciara, Chironomus*). Sibling species in *Drosophila* and in the *Anopheles maculipennis* complex have been discussed above.

Beetles. Groups of sibling species occur in almost every family of beetles; they are particularly common in Curculionidae, Chrysomelidae (Brown 1959), and Lamellicornia. There is an interesting group of about twenty-five species of leaf beetles of the genus *Calligrapha* (Chrysomelidae), many of which cannot be separated by morphological characters. Feeding and breeding tests, and the study of the beetles in nature, indicate that these are valid species. The analysis is impeded by the fact that some of these "species" are parthenogenetic (in at least five "species" no males are known). The genus *Chrysomela* is also rich in sibling species.

Orthoptera. Sibling species are very common among the Orthoptera (grasshoppers, crickets, etc.) but, until a few years ago, the specialists who discovered these different forms described them as races. Not until Mayr (1948) had pointed out that they have all the characteristics of good species were they recognized as sibling species. In all cases they were discovered through differences in their calls (see below).

Hymenoptera. Sibling species seem common among bees, wasps, and other groups of hymenopterans. Ants are particularly rich in sibling species. This fact has been the cause of much confusion and was in part responsible for a peculiar polynomial nomenclature that long characterized ant taxonomy. The application of the modern species concept to North American ants led to a dramatic simplification of their taxonomy.

Scattered cases of sibling species have been reported for many other groups of insects, such as aphids and Collembola. Only rarely has the classification matured sufficiently to permit a quantitative analysis of the frequency of sibling species. Kontkanen (1953) found that about 40 percent of the 292 species of leafhoppers (Homoptera) of Finland are sibling species, with as many as 13 in a single genus.

Other arthropods. Sibling species are undoubtedly common among mites, but the study of them has hardly begun. They occur in many genera of spiders, for instance in the poisonous black-widow spiders (*Latrodectus*).

Several genera of crustaceans are noted for the occurrence of sibling-species complexes. Kinne (1954) showed that a number of similar forms of the genus *Gammarus* first described as "races" or "subspecies" are full species: *G. zaddachi, G. salinus,* and *G. oceanicus.* Sibling species also occur in *Artemia.* The number of species of *Daphnia* in North America has been much debated, estimates ranging as low as four, with numerous "subspecies" and varieties. Brooks (1957a) found that no fewer than 13 species can be distinguished quite clearly, but nongenetic variation (cyclomorphosis and so forth) and occasional hybridization may make delimitation difficult.

Other Invertebrates

Sibling species are common among mollusks, and some are pointed out in almost every major taxonomic revision. The difficulty in recognizing them is increased by the great individual variation in many genera. Intraspecific variants are frequently more different than are separate species.

In view of the relative frequency of sibling species in such morphologically elaborate organisms as Diptera, Hymenoptera, Coleoptera, Lepidoptera, and Crustacea, one would expect an even greater frequency in the lower invertebrates with their simplified external morphology. However, the difficulty of discovering sibling species is correspondingly increased, particularly in groups like the nemerteans, turbellarians, and nematodes, in which most taxonomic characters are internal. Yet even in these groups a close study of ecological and physiological characteristics has led to the recognition of previously overlooked complexes of sibling species. Once recognized, such species often are found to be well defined morphologically. For instance, the nemertean *Lineus ruber,* at first believed to be a single species with two "varieties," was found to consist of four species that differ in color, size, proportions, size and position of the eyes, and color of the cerebral ganglia. Some of these species can contract, others roll up in a bundle; two species reproduce sexually, the others reproduce asexually; gamete formation and larval development are species specific. The asexual species regenerate extremely well, but the sexual ones do not (Gontcharoff 1951). The nematode genus *Rhabditis* appeared to be rich in sibling species, but Osche's (1952) painstaking analysis indicates that there are more morphological differences than was formerly suspected and most of these species cannot be considered sibling species.

The simpler a group of organisms is morphologically, the more difficult it should be to distinguish species. On the basis of this consideration, one would expect the highest number of sibling species to occur in the morphologically simple protozoans. Among them sibling species were first discovered (1938) in *Paramecium aurelia* and in *P. bursaria.* Although usually recorded as "varieties," they are reproductively isolated and correspond to the species of higher organisms, as is evident from Sonneborn's (1957) analysis. Such sibling species have also been found in other species groups of *Paramecium* and *Euplotes.* The recognition of sibling species in the class Protozoa depends mainly on experiment. In the most intensively studied species group, *P. aurelia,* no fewer than 16 "varieties" are now known. In *P. caudatum,* similarly, 16 "varieties" are known, and in *P. bursaria,* five. All are clearly sibling species. Even more difficult is the problem of determining sibling species in parasitic and partially asexual protozoans. In many of them, so-called biological races are known, but it is not known to what extent they represent reproductively isolated species. The widespread occurrence of sibling species in these organisms is highly probable on the basis of much indirect evidence.

The Recognition of Sibling Species

Because of their superficial morphological similarity, sibling species are normally discovered through various differences in habits, ecology, or physiology. Some examples of biological attributes that distinguish sibling species and that may aid in their discovery are recorded in the following paragraphs. Others were mentioned in the preceding section.

Biometric differences. Although qualitative structural differences between sibling species may be absent, their distinctiveness can sometimes be substantiated by biometric studies. Such an analysis need not be based exclusively on metric characters, but may involve any type of multiple-character analysis, as was demonstrated for butterflies of the genus *Boloria* and for lycaenid butterflies of the genus *Everes*.

Breeding tests. In many cases, the genus *Paramecium* for example, sibling species were first discovered when strains were crossed in the laboratory. Patterson and Stone (1952) gave a detailed account of such discoveries in the case of *Drosophila*. Crossing is particularly important when sibling species are not sympatric. Moore (1954), when crossing frogs of the genus *Crinia* of eastern and western Australia (previously classified as *C. signifera* and morphologically indistinguishable), found that these crosses were sterile. The western Australian population was actually a sibling species, *C. insignifera*. Such breeding tests are the only way in which the existence of allopatric (occupying mutually exclusive areas) sibling species can be substantiated, although differences in call notes and other behavioral traits may provide suggestive clues.

Habits. Nest structure is helpful in species recognition among cave swiftlets (*Collocalia*), termites, ants, and certain wasps. Eighteen "varieties" of the wasp *Polistes fuscatus* were traditionally recognized, differing in color pattern but not in structure. Rau found that near Kirkwood, Missouri, three of these so-called "varieties" differed in many biological attributes (Table 3.2) and showed no evidence of interbreeding. The three forms are now universally recognized as full species.

A study of the light flashes produced by eastern North American fireflies led Barber (1951) to recognize 18 species of *Photuris* instead of the traditional two or three. Once recognized, these species were found to differ not only in the frequency, pattern, and color (yellow, green, or reddish) of their flashes, but also in breeding season, in preferred habitat, and in minor color differences. Even though there are no differences in the male genitalia, there is little doubt that good species are involved (Fig. 3.1). Similar species-specific flash patterns have been described for the genus *Photinus* (Lloyd 1966).

The breeding seasons are often different in sibling species, such as those of frogs, *Coregonus* (whitefish), *Gryllus* (crickets), *Photuris* (fireflies), the moth *Choristoneura*, and the sponge *Halisarca*, to mention only a few.

Vocalization. From the very beginning of natural history, differences in songs and call notes were used to discover morphologically similar species. By analyzing these, Gilbert White, the vicar of Selborne, recognized the

Table 3.2. Characteristics of three sibling species of the *Polistes fuscatus* group (after Rau 1946).

Characteristic	*P. metricus*	*P. variatus*	*P. rubiginosus*
Color	Somber black-brown body	Yellow bands on brown	Solid bright brick red
Nest	In well-lighted parts of man-made structures or in dense vegetation	In hollow places in the ground, such as old mouse holes	In total darkness in hollow trees or between walls of buildings, under the roof
Colony founding	One queen	One queen	Many queens
Average size of colony at end of summer	70–85 cells	120–140 cells	⩾140
Hibernation	In cracks of buildings	In cracks of buildings	In hollow logs
Guards at nest entrance	Absent	Absent	Present, also ventilating nest

Chiffchaff (*Phylloscopus collybita*) and Willow Warbler (*P. trochilus*) as distinct species 30 years before the formal description by Vieillot. Call differences often permit rapid field identification, even in mammals such as chipmunks of the genus *Eutamias*. The recent advances in electronic sound-recording instruments have led to a rapid expansion of this area. Use of the new techniques has advanced species discrimination in the avian genera *Sturnella*, *Empidonax*, and *Myiarchus*, in many amphibian genera, such as *Hyla*, *Crinia*, *Bufo*, and *Acris*, in the orthopteran genera *Nemobius* and *Gryllus*, and in *Magicicada* (Alexander and Moore 1962).

In certain groups of orthopterans the analysis of call notes has led to a veritable revolution in species recognition (Alexander 1968). Among the sound-producing ensiferans of the eastern United States, for instance, approximately 40 of 167 species (24 percent) were overlooked altogether or listed as doubtful races until the analysis of their songs led to a careful study of their life histories (Walker 1964). The common field cricket (*Gryllus*) of eastern North America turned out to be a complex of six sibling species, of great interest to the ecologist and the evolutionist. Modern taxonomic analysis has made this genus accessible for further biological research. A similar clarification has taken place among the crickets of the genus *Nemobius* (Table 3.3) (Alexander and Thomas 1959) and the tree crickets (*Oecanthus*). Some of these species can be crossed in the laboratory.

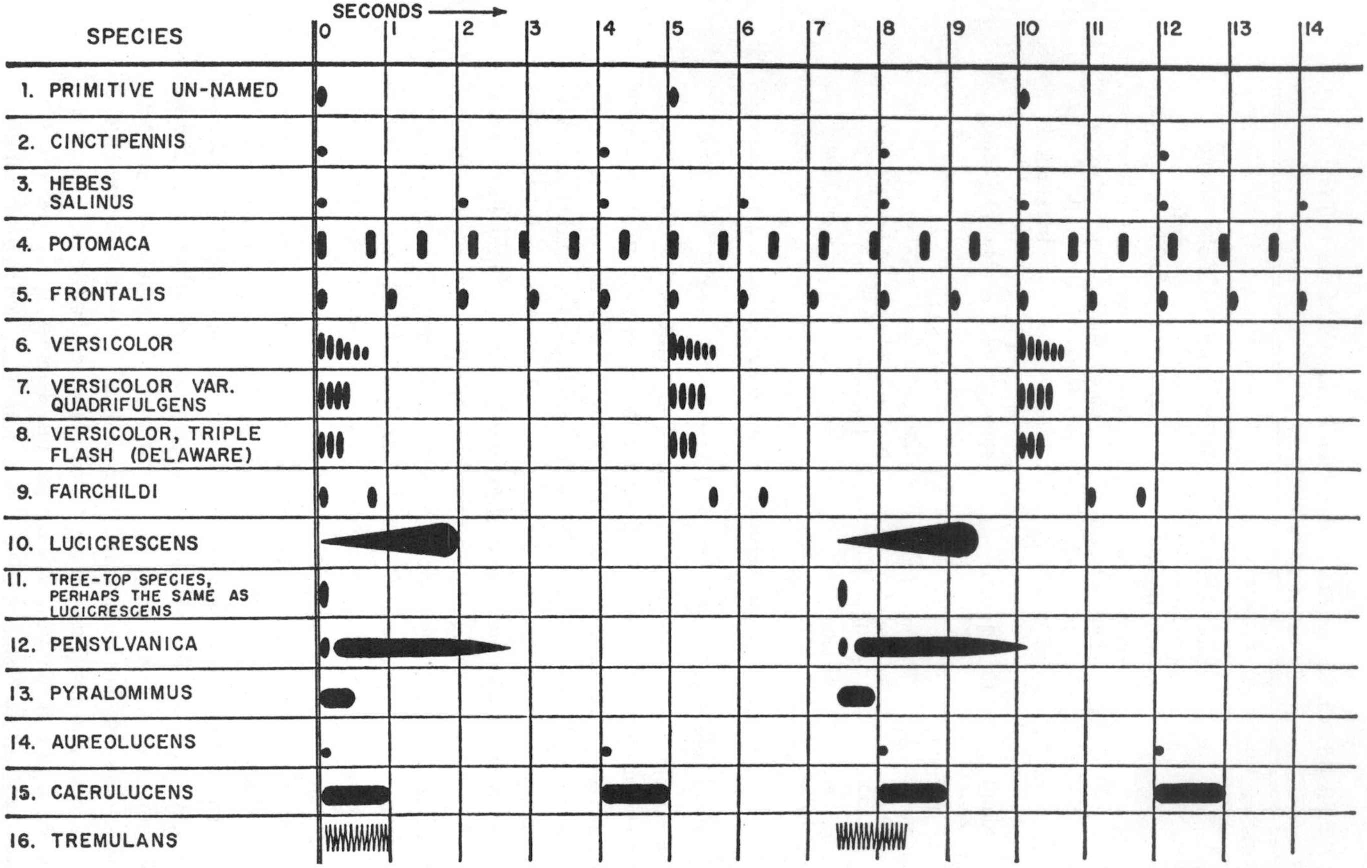

SPECIES

SECONDS →

0 1 2 3 4 5 6 7 8 9 10 11 12 13 14

1. PRIMITIVE UN-NAMED
2. CINCTIPENNIS
3. HEBES SALINUS
4. POTOMACA
5. FRONTALIS
6. VERSICOLOR
7. VERSICOLOR VAR. QUADRIFULGENS
8. VERSICOLOR, TRIPLE FLASH (DELAWARE)
9. FAIRCHILDI
10. LUCICRESCENS
11. TREE-TOP SPECIES, PERHAPS THE SAME AS LUCICRESCENS
12. PENSYLVANICA
13. PYRALOMIMUS
14. AUREOLUCENS
15. CAERULUCENS
16. TREMULANS

Fig. 3.1. Pattern of light flashes in North American fireflies (*Photuris*). Pattern, timing, and color of the flashes is distinctive for each species. The time scale gives the usual frequency and duration of the flashes. The height and length of the marks indicate the intensity and pattern of the flashes. (From Barber 1951.)

Table 3.3. Morphological differences of the sound file and characteristics of the song in the *Nemobius fasciatus* group of crickets (from Pierce 1948).

Measurement	*N. allardi*	*N. fasciatus*	*N. tinnulus*
Average number of teeth in file (right wing)	192 (165–220)	118 (101–126)	214 (196–218)
Average file length (mm)	1.438 (1.32–1.50)	0.992 (0.81–1.12)	1.600 (1.5–1.74)
Duration of pulse (sec)	0.002	0.006–0.010	0.02
Number of teeth struck per pulse	162 (84%)	56 (47.5%)	126 (58%)
Frequency (cy/sec)	7500	7740	6300
Number of pulses or chirps per second	14–20 pulses per second, lasting 8 sec	4–12 pulses per chirp, 1.4–5.0 chirps per second	5–10 single-pulse chirps per second
Nature of song	Series of separate and distinct pulses	Series of discrete chirps or trains of pulses	High-pitched bell-like note

The fact that the characteristic song of the hybrids is not encountered in nature, where these species overlap, indicates the efficiency of the behavioral isolating mechanisms.

Host preference. Sibling species that feed on plants or are parasitic are often discovered through differences in host specificity. Many sibling species of moths, beetles, and leafhoppers have been identified through such differences. Sibling species have been found even in the much-studied butterflies when individuals raised from different host plants were compared (see Chapter 15, under Biological races).

Parasites, commensals, and symbionts. Sibling species often differ in the number or the kind of parasites they carry. A pair of sibling species of *Octopus* in California was distinguished when it was found that some octopuses were parasitized by the mesozoan *Dicyemennea abelis* and others by *D. californica.* The *Octopus* individuals parasitized by *D. abelis* live in deep water on rocky bottoms, and have minute eggs (1.8 to 4 mm long) with long stalks, attached in festoons. In the adult the arms are relatively longer, the suckers larger, and the hectocotylized arms relatively shorter. These were distinguished as *Octopus bimaculatus.* The individuals parasitized by *D. californica* live in shallower water where rocks rest on soft bottom, and have large eggs (9.5 to 17.5 mm) with shorter stalks, attached in small clusters. This species was given the name *O. bimaculoides.* Males of one species will not mate with females of the other.

Cytology. The study of chromosomal patterns has led to the discovery of numerous sibling species, or at least it has established differences between stocks and strains that were suspected of belonging to different species because they were difficult to cross in the laboratory. Among a number of such cases described for the genus *Drosophila* (Patterson and Stone 1952; White 1954), the case of *D. pseudoobscura* and *D. persimilis* is best known. A study of the banding patterns of salivary-gland chromosomes has helped to remove difficulties also in the taxonomically difficult genus *Chironomus.* The same is true for black flies (simuliids), where numerous sibling species have been described, particularly in *Prosimulium.*

Cytological analysis revealed several sibling species of mammals, for instance in the genera *Mus* (*Leggada*), *Spalax*, and *Sorex*. Diagnostic chromosomal differences between sibling species exist in a number of genera of crustaceans, such as *Cyclops* and *Jaera*. These are merely a few selected examples of conspicuous chromosomal differences between species that are morphologically exceedingly similar.

Biochemical analysis. It is to be expected that sensitive biochemical methods, such as electrophoresis, chromatography, and other methods of protein analysis, will be employed to an increasing extent to confirm suspected species differences.

Special Kinds of Sibling Species

Although most sibling species differ from other species in no way except the slightness of their morphological differences, there are exceptions.

A special class is represented by the "parthenogenetic species," which are widespread in some groups of insects and lower invertebrates and occur also in fishes, amphibians, and reptiles (Maslin 1968). Among the sibling species of chrysomelid beetles described by Brown (1945), many are known only in the female sex and apparently reproduce strictly parthenogenetically. The same is true of the so-called species of white-fringed weevils, of some psychid moths of the genus *Solenobia*, and of isopods of the genus *Trichoniscus*. Whether or not to list such parthenogenetic clones as sibling species depends on the criteria adopted for designating "species" in asexual organisms (see Chapter 15).

Parthenogenetic animals tend to develop polyploidy (Chapter 15), and this makes the task of determining species status even more difficult. The polyploid "races" of weevils (Suomalainen 1950) and *Solenobia* (Seiler 1961) are well-analyzed cases; others are found in *Trichoniscus* and in additional genera discussed by White (1954). Most of these polyploids are reproductively isolated from the parental diploid species even in the cases in which no morphological differences are visible. Since they are definitely not "biological races," they are best considered sibling species.

The Significance of Sibling Species

Speciation among sibling species is in no way different from that in other species. Aside from a few cases of autopolyploidy and of parthenogenesis (see above), geographic speciation is the normal process by which sibling species originate. This is very evident in the taxonomically better-analyzed groups. Among the anopheline mosquitoes, for instance, subspecies are frequent and many species are still essentially allopatric: *Anopheles atroparvus* and *labranchiae,* or *occidentalis, freeborni,* and *aztecus.* Reid (1953) concluded that the *A. hyrcanus* group of sibling species "seems to conform to the classical pattern of speciation by geographical isolation." The most distinct forms of this group occur in the Philippines, which constitute the most isolated portion of its range. Patterson and Stone (1952) came to the same conclusion with respect to speciation in the sibling species of *Drosophila.* The occurrence of entirely or essentially allopatric sibling species in many genera strengthens the evidence for geographic speciation (Chapter 16).

A much more difficult question concerns the evolutionary significance of sibling species. Why are some closely related species very different, others morphologically indistinguishable?

Up to the 1940's the misconception was widespread that sibling species—which at that time were usually called varieties or biological races—were species *in statu nascendi.* They were believed to have acquired only the "biological" properties of species, not the genetic ones. At that time it was vital to stress that sibling species are perfectly good species, as distinct in their reproductive isolation as morphologically differentiated species (Mayr 1948). They belong in the category of species, not in that of races. Now that this point is universally accepted, we can ask how sibling species differ from other species. Presumably within a species group those species that are morphologically most similar are most closely related and hence genetically most similar. Yet it is quite possible, for instance, that sibling species of *Drosophila* are genetically more different from each other than closely related but morphologically distinct species of cichlid fishes. Developmental homeostasis, which prevents the manifestation of genetic change in the visible phenotype, is much stronger in some groups of organisms than in others. Considering the incomplete correlation between phenotypic and genotypic change it would be a great mistake to place sibling species in a different class altogether from species that are morphologically distinct from each other.

Conclusions Concerning Sibling Species

(1) There is no sharp division between ordinary species and sibling species. The latter are merely near the invisible end of a broad spectrum of increasingly diminishing morphological differences between species. The occurrence of natural populations with all the biological attributes of good species but with little or no morphological difference confirms the vulnerability of a purely morphological species concept.

(2) Sibling species, when subjected to a thorough analysis, are usually shown to differ in a whole series of minor morphological characters. Like ordinary species, they are separated from each other by distinct gaps.

(3) Sibling species are apparently particularly common in those kinds of animals in which chemical senses (olfactory and so on) are more highly developed than the sense of vision. Although indistinguishable to the eye of man, these sibling species are evidently dissimilar to each other, as is shown by cross-mating experiments. Sibling species are apparently rarest in organisms, such as birds, in which epigamic characters that function as visual stimuli are most frequent.

(4) There is no indication that sibling species arise by a process of speciation different from that which gives rise to other species.

(5) Sibling species may be, on the average, genetically more similar to each other than closely related morphologically different species. However, degree of morphological similarity in sibling species is an indication not merely of genetic similarity, but also of developmental homeostasis. A reconstruction of the genotype, resulting in the reproductive isolation of two species, can take place without visible effect on the morphology of the phenotype.

4 · Biological Properties of Species

When first introduced into biology, the term "species" designated primarily a morphological-systematic unit, or, worse, merely a Latin binomen. As the study of species was taken up by field naturalists on one hand, and by laboratory biologists on the other, it became increasingly clear that the morphological distinctness of each species indicates the existence of an entirely distinct biological system.

Biological Species Characteristics

The taxonomic literature stresses "species characters." This term in general refers to any attribute of a species that differentiates it from other species (and is therefore "diagnostic"), and that is reasonably constant (invariable), so that a species can at once be recognized by it. Morphological characters are most useful for diagnostic purposes and with preserved material; this is the reason for their high esteem in the taxonomic literature. More detailed comparisons, however, always show that species may differ from each other not only in aspects of external morphology, but also in one or several of the following traits: size, color, internal structure, physiological characters, cell structures, chemical constituents (particularly proteins), ecological requirements, and behavior. Whenever a particular aspect of animals receives special attention, numerous previously unsuspected species differences are discovered, for instance in physiology or behavior.

Darwin was convinced that species differences are, in the last analysis, always adaptive, and most naturalists share this view. Others have maintained the opposite: "A survey of the characters which differentiate species (and to a less extent genera) reveals that in the vast majority of cases the specific characters have no known adaptive significance" (Robson and Richards 1936:314). A renewed analysis of species characters reveals that this conflict of opinions can largely be resolved. Every species is the product of a long history of selection, and its genotype, as a whole, is the product of this selection. The phenotype that is produced by this genotype is bound to have properties which are "well adapted," that is, which favor survival in the given environment of the species. However, each component of the phenotype is the product of the interactions of the genotype as a whole; consequently, some visible components of the phenotype may be adaptively irrelevant, as long as other components produced by the same genotype are favored by selection. Since the atomistic one-gene–one-character hypothesis is only rarely true in higher organisms, it is not to be expected that every component of the phenotype is the result of an ad hoc selection. It is principally the pleiotropy of genes (the capacity of a gene to affect several characters of the phenotype) that is responsible for selectively neutral aspects of the phenotype (see Chapter 7).

This chapter and the next are devoted to a discussion of three sets of biological attributes of species, those that (1) adapt species to their physical environment, (2) enable species to coexist with potential competitors, and (3) permit species to maintain reproductive isolation from one another (Chapter 5).

Adaptation to the Physical Environment

The statement that every species is adapted to its environment is a self-evident platitude. In continental areas without physical barriers the border of the species range indicates the line beyond which the species is no longer adapted, and the very existence of such borders is tangible proof of the limitations of this adaptation. Some factors that contribute to the adaptation are obvious, particularly those expressed in the visible phenotype. The white color of many Arctic birds and mammals and the sandy coloration of desert species are among these evident adaptive characters. Less easily observed, but far more important, are various physiological regulatory mechanisms that not only permit survival in the breeding range but also secure a rate of reproduction adequate to maintain the size of the breeding population at a more or less steady level. It was shown in the last chapter that even very closely related and similar species may differ from each other in various nonmorphological characters relating to physiology, ecology, and behavior. Many such species differences are cited in the vast physiological literature (for instance, Prosser and Brown 1961) and in the ecological literature (for example, Allee et al. 1949; Hesse, Allee, and Schmidt 1951). All these comparisons of closely related species indicate that each species is a separate biological system with species-specific tolerances to heat, cold, humidity, and other factors of the physical environment, with a species-specific habitat preference, productivity, and rate of population turnover, and with numerous other species-specific biological attributes. Related species may overlap in these properties, yet each species is characterized by well-defined mean values. Physiological steady states are maintained by complex homeostatic systems whose essential properties are shared not only by members of a single local population but also by all members of the species, a certain amount of geographic variation notwithstanding. The dispersal of individuals through smaller or larger portions of the species range, resulting in the mixing of genotypes, puts a premium on the existence of species-wide homeostatic mechanisms. It is probable that generalized adaptive genes and, even more important, balanced allelic and epistatic gene combinations are involved in the optimal functioning of these species-specific homeostatic mechanisms. They serve as the element that gives a species its unity (Chapter 10), but they are at the same time largely responsible for the existence and location of *species borders*. Such homeostatic devices have definite limits of tolerance beyond which they can no longer adjust to external conditions. The result is that every species has an optimal environment, presumably somewhere near the center of its range, and definite limits of tolerance with respect to latitude and altitude.

A few examples will illustrate such physiological differences between species. Wallgren (1954) has shown that two northern European species of buntings differ in temperature preference and tolerance (Fig. 4.1). The more northerly species, *Emberiza citrinella*, is less resistant to high temperatures and particularly to prolonged spells of heat. *Emberiza hortulana*, a more southerly species, tolerates much greater heat, but shows greater heat loss at low temperatures. It has a very different fat metabolism, correlated with a strong migratory drive. Far more conspicuous is the temperature dependence of aquatic cold-blooded vertebrates, as shown, for instance, by Moore (1949) for frogs of the genus *Rana* (Table 4.1).

Among species differences in terrestrial animals, temperature tolerance and humidity requirements seem to determine the location of species borders most frequently. Six species of ants of the genus *Formica* occurring in the Chicago region differ markedly in their temperature and humidity tolerances. At high humidity *incerta* is superior to *montana* and *subsericea* to *ulkei*, at low humidity the situation is reversed. The differences are correlated with habitat preferences and geographic range (Table 4.2). Sim-

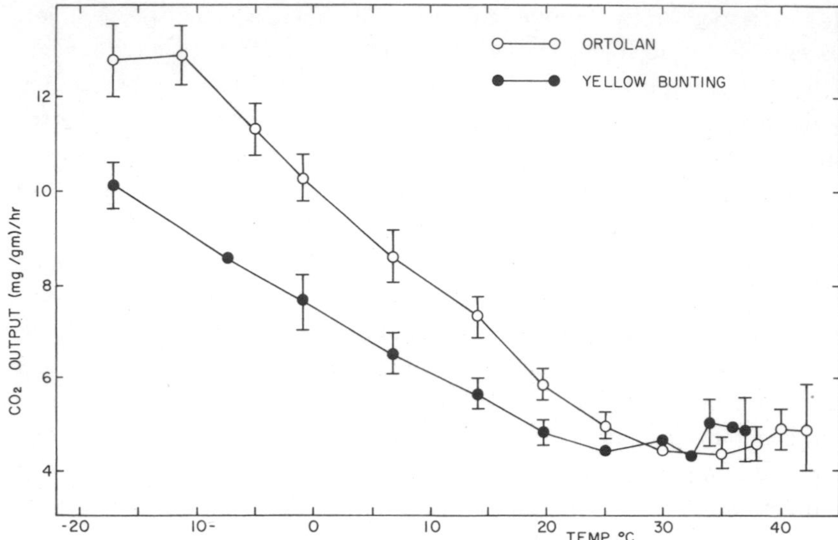

Fig. 4.1. Carbon dioxide output of two species of buntings at various environmental temperatures. Ortolan = *Emberiza hortulana,* Yellow Bunting = *E. citrinella.* (From Wallgren 1954.)

ilar differences have now been established for many species of snails, insects, and vertebrates.

Students of population dynamics have found that variations in aspects of productivity, such as fertility, fecundity, life expectancy, and, in social animals, colony size, may have a strong genetic component and may differ from species to species (Table 4.3). How much difference there can be in all sorts of biological characteristics has been demonstrated by Sonneborn (1957) and his co-workers for 16 sibling species of the *Paramecium aurelia* complex. These species may differ not only in temperature preference, size, and rate of maturation, but also in mating-type system, length of mature period, serotype system, and the form of fertilization in senility (autogamy versus selfing).

These examples could be extended ad libitum. Every recent investigation has confirmed the widespread conviction that each species is a unique biological system. No matter how similar two species may be morphologically, a detailed analysis will reveal numerous differences in their physiological preferences and tolerances and in many aspects of their life cycle from the egg stage to death. The precise description of such differences and their relation to the particular niche of a species, its geographic range, and its evolutionary history, however, has hardly been begun.

Table 4.1. The relation between geographic distribution, breeding habits, and adaptive embryological characteristics in four species of North American frogs, *Rana* (from Moore 1949).

Characteristic (mean values)	Species			
	R. sylvatica	*R. palustris*	*R. clamitans*	*R. catesbeiana*
Limiting latitude:				
Northern	67°	51–55°	50°	47°
Southern	34°	32°	28°	23°
Beginning of breeding season	Mid-March	Mid-April	May	June
Limiting embryonic temperature [°C]:				
Lower	2.5	7	12	15
Upper	24	30	32	32
Time (hr) between stages 3 and 20 at 20°C	72	105	114	134
Egg diameter (mm)	1.9	1.8	1.4	1.3
Type of jelly mass	Globular, submerged	Globular, submerged	Film, at surface	Film, at surface

Adaptations That Make the Coexistence of Species Possible

Among the numerous biological characteristics of species, two sets are of particular significance in the process of speciation (Mayr 1948). They are the special properties that permit closely related species to occupy the same area: (1) mechanisms that guarantee reproductive isolation (see Chapter 5), and (2) the ability to avoid competition from other species that utilize the same or similar resources of the environment.

Table 4.2. Average survival times (hours) of ants (*Formica*) at different temperatures and relative humidities (from Talbot 1934).

Species	Temperature (°C) and relative humidity (percent)		
	30.4 ± 0.34°, 44.3 ± 1.2	30.6 ± 0.3°, 12.1 ± 2.7	41.2 ± 0.8°, 8.3 ± 0.1
F. obscuriventris	—	14.06	1.93
montana	25.35	9.50	1.57
incerta	54.49	8.12	1.43
subintegra	—	6.62	1.04
ulkei	18.16	6.18	0.84
subsericea	41.64	2.81	0.92

Table 4.3. Biological differences (average values) of three species of mice (*Peromyscus*) (from McCabe and Blanchard 1950).

Characteristic	*P. maniculatus*	*P. truei*	*P. californicus*
Number of litters per season	4.00	3.40	3.25
Number of young per litter	5.00	3.43	1.91
Number of offspring per breeding female per season	20.00	11.66	6.21
Known life span (days)	152	190	275

The remainder of this chapter is devoted to a study of these ecological interactions among closely related species, the nature of such interactions, and their evolutionary effects.

Competition

Darwin (1859:76) strongly stressed the evolutionary significance of competition and its contribution to natural selection:

> As the species of the same genus have usually, though by no means invariably, some similarity in habits and constitution, and always in structure, the struggle will generally be more severe between them, when they come into competition with each other, than between species of distinct genera.

Unfortunately, he referred to competition in such drastic terms as "severe struggle" and "the great battle of life." To certain authors ever since then, competition has meant physical combat, and, conversely, the absence of physical combat has been taken as an indication of the absence of competition. Such a view is erroneous. To be sure, actual combat or aggression among competitors does occur, particularly among territory-holding higher vertebrates (see p. 50). Among sessile marine invertebrates, where the amount of available substrate may be the limiting factor, the fastest-growing species may overwhelm and literally kill slower-growing competitors. Connell (1959) has described such a situation among competing species of intertidal barnacles. The Tricolored Blackbirds (*Agelaius tricolor*) of California, when settling by the thousands in the territories of Redwinged

Blackbirds (*A. phoeniceus*), reduce the breeding success of the latter species to zero, though through disturbance rather than aggression. In most cases, however, dramatic competition occurs only where two species come newly into contact or where a radical change of the environment upsets the previously existing dynamic balance. As Brown and Wilson (1956) correctly point out, such an acute phase of competition is often "a relatively evanescent stage in the relationship of animal species" and will be replaced by a new balance in which severe competition is avoided. Thus the relative rarity of overt manifestations of competition in nature is proof not of the insignificance of competition, as asserted by some authors, but, on the contrary, of the high premium natural selection pays for the development of habits or preferences that reduce the severity of competition.

No unanimity has yet been achieved concerning the precise definition of *competition*, but it always *means that two species seek simultaneously an essential resource of the environment* (such as food, or a place to live, to hide, or to breed) *that is in limited supply*. Consequently, competition becomes more acute as the population of either species increases. Any factor whose effect becomes more severe as the density of the population increases (and this may be true for all causes of mortality and fecundity) is called a *density-dependent*, or *controlling*, factor.

The result of competition between two ecologically similar species in the same locality is either (1) the two species are so similar in their needs and their ability to fulfill these needs that one of the two species becomes extinct, either (1a) because it is "competitively inferior," that is, it has a smaller capacity to increase, or (1b) because even though competitively equivalent it had an initial numerical disadvantage; or (2) there is a sufficiently large zone of ecological nonoverlap (area of reduced or absent competition) to permit the two species to coexist indefinitely. In sum: *Two species cannot indefinitely coexist in the same locality if they have identical ecological requirements*. This theorem is sometimes referred to as the Gause principle, after the Russian biologist Gause, who was the first to substantiate it experimentally. Yet, as Hardin (1960) and others have pointed out, the principle was known long before Gause. Darwin discussed it at length in his *Origin of Species*, and Grinnell and other naturalists have referred to it frequently in the ensuing 100 years. Instead of associating the principle with the name of one of its many independent discoverers, Hardin suggests calling it the *competitive-exclusion principle*.

The exclusion principle has great heuristic value. Attempts to prove or disprove it have stimulated numerous comparisons of the ecologies of closely related species and analyses of the factors permitting them to coexist. Yet a complete analysis of the process of competition has proved difficult.

The operation of competition between individuals of two species is best understood if one compares it with the fate of two "competing" genes (alleles) in a single gene pool (disregarding the complications introduced by heterosis and the occasional advantage of rare genes; see Chapters 8 and 9). Let us assume that gene A is superior to its allele a in a population and that the heterozygote Aa is intermediate in viability between the two homozygotes. Then, even if the difference between A and a is very slight, a will inevitably disappear from the population, as calculated by Fisher (1930) and first demonstrated experimentally by L'Héritier and Teissier (1937). Into a population cage containing a pure population of more than 3000 *Drosophila melanogaster* with the Bar gene they introduced a few flies with the wild-type allele of Bar. The frequency of the Bar allele dropped rapidly at first, then more slowly, until, 600 days later, it had a frequency of less than 1 percent. Many similar experiments have since been conducted with different genes and gene arrangements; the outcome except in cases of heterozygote superiority, has always been the same. Two species are rarely as similar as genotypes of a single species. If, however, two species were to depend entirely on the same resources of the environment, one or the other would prove superior in the utilization of these resources. This superiority would lead to an exclusion of the other species from the zone of potential ecological overlap. To avoid such fatal competition, the two species must utilize the resources of the environment in a somewhat different way. This is why ecological compatibility with potential competitors is one of the most important species characteristics. In order to survive, each species must be supreme master in its own niche.

Observed Exclusion

The avoidance of competition is achieved by either conspicuous or subtle factors. As conspicuous factors, I classify geographical separation, occurrence in different habitats, and temporal isolation of potentially competing stages of the life cycle. As subtle factors, I include all niche differences in the same habitat, that is, all differences in the utilization of the habitat. In an analysis of the British songbird fauna, Lack (1944) found that, where a genus had more than one British species, in 21 cases there was conspicuous exclusion (geographical 3, habitat 18), while in 11 to 13 cases exclusion was more subtle. Whenever two species share the same habitat, a more refined analysis usually reveals differences. For instance, the cormorant (*Phalacrocorax carbo*) and the shag (*P. aristotelis*) seem to have identical ecologies. However, the cormorant feeds in shallow water and takes fish that live on or near the bottom, such as pleuronectids and gobies, while the shag feeds in the open sea on free-swimming fish (clupeoids, *Ammodytes*) (Lack 1945). Recent volumes of biological journals, such as *Evolution, American Naturalist,* and *Ecology,* describe literally hundreds of cases of such exclusion. Many cases are cited in Mayr 1963:69–88. For a more recent summary see Selander 1969.

A detailed analysis of niche exclusion is that made by MacArthur (1958) of North American wood warblers (*Dendroica*) in coniferous forests. Several species that at first sight seemed to have identical niches actually occupied different parts of a tree or fed on the outer or inner parts of a branch. Coexistence in the same habitat is also facilitated by differences in vocalization that reduce interspecific antagonism.

The study of exclusion in the oceans is only beginning. I know of no good evidence, for instance, demonstrating competition or exclusion among plankton-feeding pelagic fishes. The more specialized a species is, the less the probability that it would share its niche with another species. On these premises one would expect rather rigorous exclusion among parasites. This is not entirely borne out by the parasitological literature, which frequently reports three or four species, for instance of tape worms, coexisting in the intestinal tract of a single host. Further studies are needed to determine, on the basis of statistically significant data, how often true coexistence occurs and how often the potential competitors are segregated in different portions of the intestinal tract or in different individuals of the host species and how often they utilize different components of the host's food.

Evidence for Competition

The mere fact of exclusion does not prove that it is the result of selection for the avoidance of competition. Other explanations may apply to certain cases. There are, however, situations that are difficult to explain except as the result of competition.

Exclusion in impoverished habitats. Species that habitually coexist in a rich, diversified habitat may exclude each other in more homogeneous or ecologically marginal habitats. Since islands are generally far less diversified ecologically than are neighboring mainlands, related species of birds often replace each other on different islands while they are able to live side by side on the mainland. Many documentations of this phenomenon have been published in the ornithological literature. Lizards of the genus *Lacerta* were found by Radovanović (1959) on 46 small islands off the Dalmatian coast. On 28 islands the genus was represented by *L. melisellensis* and on 18 islands by *L. sicula*. On no island did both species of lizards coexist. *L. sicula* is apparently the more recent arrival and once it establishes itself on an island it displaces *L. melisellensis* (Radovanović 1965). A similar exclusion has been observed in *Anolis* and other genera of island lizards. The island principle is also illustrated by isolated mountain forests in the African savannas. Each mountain island is occupied by only one species of woodpeckers, but it may be a *Dendropicos*, a *Campethera*, a *Mesopicos*, or a *Yungipicus*. There is apparently neither sufficient niche diversity nor total available habitat to permit occupation by a second species of woodpeckers. The numerical advantage of the first colonist is decisive.

Exclusion limited to area of overlap. Widespread species sometimes display a sharp reduction in habitat utilization in areas where their range is overlapped by related species, or, conversely, a marked broadening of tolerance in a zone of nonoverlap. For instance, in the Philippines, where only three species of the *Anopheles hyrcanus* group occur, the two species *A. lesteri* and *A. peditaeniatus* are found both on the coast and inland. On the Malay peninsula, however, where there are seven species of this group, *lesteri* is on the whole restricted to the coast near or in brackish water, while *peditaeniatus* is an inland species (Reid 1953). The Chaffinch (*Fringilla coelebs*) occurs throughout its European range in both broad-leaved and coniferous woods. On the islands of Gran Canaria and Tenerife, where a second species occurs, the Chaffinch is restricted to deciduous woods, being replaced in the pine woods by the Blue Chaffinch (*F. teydea*) (Lack and Southern 1949).

The impact of invasions. The effect of competition in nature is best demonstrated by the impact of an invading species on species already established in the area. For instance, the introduced wasps *Aphytis lignanensis* and *A. melinus*, parasitic on the red citrus scale, displaced the previously established *A. chrysomphali* in southern California between 1948 and 1961 (De Bach and Sundby 1963). Island faunas are particularly vulnerable to new competitors. The Moth Skink (*Lygosoma noctuum*) disappeared from the Hawaiian Islands, where it was once common, soon after its close relative *L. metallicum* was introduced (Oliver and Shaw 1953). The Red Squirrel (*Sciurus vulgaris*) disappeared from many parts of the British Isles after colonization of the particular district by the introduced American

Gray Squirrel (*S. carolinensis.*) Competition, in the strict sense of the word, is only one of several mechanisms leading to replacement of an indigenous species by an invader. In other cases the new arrivals were the vectors (carriers) of disease organisms that caused the extinction of the native species. For example, species of birds in the Hawaiian Islands succumbed to diseases carried by introduced species (Warner 1968).

Particularly dramatic are the competitive encounters of entire faunas. The fossil record as well as the analysis of existing faunas graphically documents the effects of faunal exchanges across the Bering Straits bridge and the pervasive changes resulting from the invasion of the Australian region by Asian elements and of South America by North American types after the closing of the Panamanian water gap. Many South American mammals, like marsupial carnivores, were supplanted by the competitively superior northern types; however, the large-scale extinction of peculiar South American families of edentates, ungulate-like types, autochthonous rodents, and so forth, during the late Tertiary had begun before the invasion of North American competitors (Patterson and Pascual 1969).

There are essentially two ways in which enrichment of a fauna can be achieved. In one, the newcomer (let us say a termite-feeding lizard) may utilize the same niche ("termites") as an incumbent species, but in a somewhat different habitat. There will be a local exclusion of the two species paralleling the distribution of the respective habitats. In the other, the newcomer succeeds in entering the same habitat as the incumbent and forces him to yield part of a heterogeneous niche in a well-diversified habitat. The coexistence in the same piece of woodland of five or six species of *Parus* in Eurasia or of *Dendroica* in North America is made possible by such a specialization of their niche requirements. The richer and more diversified a habitat, the more easily such niche partitioning can develop.

As far as the overall evolutionary picture is concerned, it should be mentioned that every new arrival in an area tends to add to the total diversity and to enrich thereby the opportunities of other organisms except the most immediate competitors. The new food pyramid that the evolution of the angiosperm flora made possible is a striking illustration of this principle.

Experimental competition. The validity of the exclusion principle has been tested in numerous recent laboratory experiments in which mixed populations of two species were established in a uniform environment. In virtually every case, one of the two species was eliminated sooner or later. The situation is different in experiments where the environment is not homogeneous. A culture bottle of *Drosophila* may have enough differential between the dry crust and the moist interior to allow a long-continued coexistence of two species. When two competing species of grain beetles are raised in whole wheat kernels rather than in sifted wheat flour, the two species can usually coexist indefinitely. These experiments in heterogeneous environments indicate how subtle the differences between the ecological niches of two closely related species can be.

Difficulties of the Exclusion Principle

Naturalists have discovered numerous cases in which two or more related species appear to occupy the same niche. Ross (1957) described the coexistence in the Mississippi Valley of six species of leafhoppers (*Erythroneura*) on the same food plant (*Platanus occidentalis*). Cooper (1953) found that the three most common species of eumenine wasps in New York state frequent the same flowers, hunt the same caterpillars on the same trees, and nest in 6-mm burrows in the same places at the same time. They appear to be competing for all the important elements of their existence. Even more perplexing is the extraordinary species diversity in the deep-sea benthos. As many as 200 to 400 different species were found in a single dredge haul operating for one hour (Hessler and Sanders 1967). With the uniformity of the deep-sea substrate, temperature, and seasons and the absence of light not many different niches would seem possible.

In all these cases (and they are only a small selection from an extensive literature), it has so far been impossible to determine the factor or factors that permit coexistence of several species in what seems to be the same niche. This difficulty is one of several reasons why the niche concept has been critically examined (Hutchinson 1965) and indeed rejected by some recent authors. The "niche" of a species is actually nothing more than the outward projection of its needs, and since these needs presumably can be filled in a number of different ways, the niche characteristics of a species are influenced to a greater or lesser degree by its physical and biotic environment. Identity of the niches of two species cannot be claimed until it is shown that the needs of the two species are at all times the same. How careful one must be in judging the "identity" of needs is nicely illustrated by the case of *Drosophila mulleri* and *D. aldrichi* (Wagner 1944). The larvae of these two closely related species occur simultaneously in the ripe fruit of the cactus *Opuntia lindheimeri*, feeding on the microflora of the fermenting pulp. Competition between the larvae of the two species would appear to be complete. Yet analysis of the intestinal contents shows that these two species of *Drosophila* tend to feed on different species of yeast and bacteria and that there is a certain amount of nonoverlap in their requirements.

Recent studies indicate that natural environments are far more heterogeneous than they appear at first sight and that competition may be severe only during unusually severe seasons. The respective superiority of the competing types may shift with environmental conditions and with population density. Competition is further reduced by differences in habitat selection resulting from differences in the physiology of the species. The diets of two European species of sticklebacks (*Gasterosteus aculeatus* and *G. pungitius*) appear to be virtually identical. Competition is apparently mitigated through a preference for different breeding habitats in which the level of population density is regulated by territorial habits. Two species of parasitic wasps continued to coexist for 70 generations in a population of the azuki bean weevil (*Callosobruchus chinensis*) without either being able to eliminate the other. It appears that one species of parasites is more successful at low, the other at high densities of the host species (Utida 1957).

Possible Absence of Exclusion

The frequency of demonstrable exclusion is so great that its importance can no longer be questioned. We are now ready to ask, without denying the principle as such, whether there are factors or constellations of factors that may reduce or altogether eliminate the operation of exclusion. Competition is always greatest at highest population densities and there should be a selective premium on negative feedbacks that would counteract too great a build-up of populations. Such mechanisms do exist, although many of them have been misinterpreted in the past as evidence of conscious population control by species of animals or as evidence for the population as the unit of selection (Wynne-Edwards 1962).

Territoriality is one factor that is occasionally able to reduce the impact of interspecific competition. Marshall (1960), for instance, found no noticeable difference between two species of towhees (*Pipilo aberti* and *P. fuscus*) with respect to feeding and nesting. The two species coexisted in the same habitat, seemingly completely tolerant of each other, even though they seemed to use exactly the same resources of the environment. The explanation is apparently that the population density of one of the species is set at such a low level, owing to intraspecific intolerance, that its individuals are not competing with individuals of the other species, whose populations, owing to territorial intolerance, are also widely spaced.

The foregoing is presumably an exceptional situation. There are far more observations documenting aggressive interference between species with similar ecological requirements, for instance, hummingbirds, titmice, woodpeckers, grackles, and other birds (Mayr 1963:87, Orians and Willson 1964). Noncompetitive interactions among invertebrates usually take the form of predation or interference rather than of interspecific aggression.

Apparently selection strongly favors any mechanism that builds up a negative feedback to counteract too-violent population fluctuations, such as restlessness at high population densities (leading to emigration), regulation of fertility, age at maturity, total length of life span, and so forth. Not much is known about the effect of such intraspecific mechanisms on the operation of interspecific exclusion.

Evolutionary Results of Competition

Whenever the ranges of two closely related species overlap, there are five possible results (Lack 1949):

(1) One species is superior to such a degree that it eliminates the other one. This is the situation, discussed above, that results from introductions and invasions. It can be proved only during the actual period of displacement.

(2) In part of the geographic area one species is competitively superior; in another part, the other species is superior. As a consequence the two species will replace each other along a sharp but frequently fluctuating line of balance (parapatric distribution). Range expansion by either species is prevented by its competitor.

(3) Where two species come into contact along a similarly well-defined border, expansion is prevented not by the competing species but by the ecological unsuitability of the terrain beyond the borderline.

The choice between |2| and |3| is not always simple. Allopatric species pairs are very common, for instance among birds. In most cases they meet along a sharp climatic or vegetational break, and it appears probable that differences in adaptation rather than competition are responsible for the geographic replacement. This is evidently the case where such conspicuously different habitats come into contact as savannas and the Congo rain forest in Africa. The interpretation must, however, remain tentative where more elusive climatic or biotic factors are involved. The fact that the ranges of two widespread Asiatic bee-eaters, *Merops superciliosus* and *M. philippinus*, are largely allopatric has, for instance, been ascribed to competition. Yet these two species coexist in some border districts, and there is a strong probability that the essentially allopatric distribution pattern is due to differences in physiological adaptation. *Merops superciliosus* appears to be superior where the rainfall is less than 20 inches per year, *M. philippinus* where it exceeds this amount. Whether either species could invade the range of the other species if the latter were not present is the big question. I do not believe, for instance, that competition has been involved in the many cases of a northward retreat of northern animals during the recent amelioration of the climate and the corresponding northward expansion of southern species.

Although most cases of allopatric species will probably have to be listed under |3|, there are also some clear-cut instances of |2|. The essential geographic replacement of *Gammarus pulex* and *G. duebeni* in the British Isles, for instance, is due to the reproductive superiority of *pulex* in fresh water and of *duebeni* in brackish waters.

(4) One species is superior in some habitats, the other in other habitats. The result is geographic overlap combined with habitat exclusion. The case of the chaffinches (*Fringilla*) on the Canary Islands is a typical illustration. The altitudinal ranges of mountain species are often very much compressed where they encounter a competitor.

(5) Both species enter the same habitat but occupy different niches. Most of the previous discussion was devoted to such cases.

The meeting of two similar or closely related species along a line of contact or in a zone of overlap may thus have very different consequences.

Sympatric Character Divergence

Competition in an area of geographic overlap of two similar species exerts strong selection pressure. Most favored by selection should be those individuals of either species that have the least need for the resources jointly utilized by both species. Darwin (1859:111–126) devoted an entire section to this "Divergence of Character," saying, "The principle which I have designated by this term is of high importance . . . Natural Selection . . . leads to divergence of character; for more living beings can be supported on the same area the more they diverge in structure, habits, and constitution." Although such divergence will at first be strictly ecological, consisting of a different utilization of the environment (see above, p. 45), it will subsequently be reinforced by the selection of such morphological differences as facilitate the ecological divergence. Among modern authors, Lack (1947) was the first to give an example of such character divergence (Table 4.4). The ecological niches on the outlying islands of the Galapagos may be filled by two or three finches of the genus *Geospiza* whose bill sizes are adjusted to the available food niche. On Tower Island three strongly differentiated species divide the habitat among themselves. On each of three other islands only two species are present, one of which occupies two potential niches and is able to adjust its bill size owing to the lowered selection pressure. Several other cases are discussed by Mayr (1963:83) (Fig. 4.2). Brown and Wilson (1956) have properly stressed the evolutionary importance of Darwin's character divergence (which they call "character displacement) and have cited additional examples. Selander (1966) has found indications that increased sexual dimorphism of birds, particularly in bill size, may be the result of selection for differences in food utilization. Laboratory studies have clearly demonstrated the effect one species can have on the genetic composition of coexisting and competing species (Ayala 1968).

There is not yet complete certainty concerning the importance of character displacement, owing to the existence of numerous exceptions. Cases are known, for instance, in which several species are more similar to each other in their area of overlap than elsewhere (see, for example, Moynihan

Table 4.4. Character divergence in *Geospiza* on the Galapagos: bill depth (mm) and niche occupation (from Lack 1947).

<table>
<tr><th>Niche</th><th>Tower Island</th><th>Hood Island</th><th>Wenman Island</th><th>Culpepper Island</th></tr>
<tr><td>Large ground finch</td><td>magnirostris
21.2</td><td rowspan="2">conirostris
16.0</td><td>magnirostris
20.4</td><td>conirostris
16.5</td></tr>
<tr><td>Cactus feeder</td><td>conirostris
13.0</td><td rowspan="2">difficilis
9.0</td><td rowspan="2">difficilis
9.0</td></tr>
<tr><td>Small ground finch</td><td>difficilis
7.9</td><td>fuliginosa
8.3</td></tr>
</table>

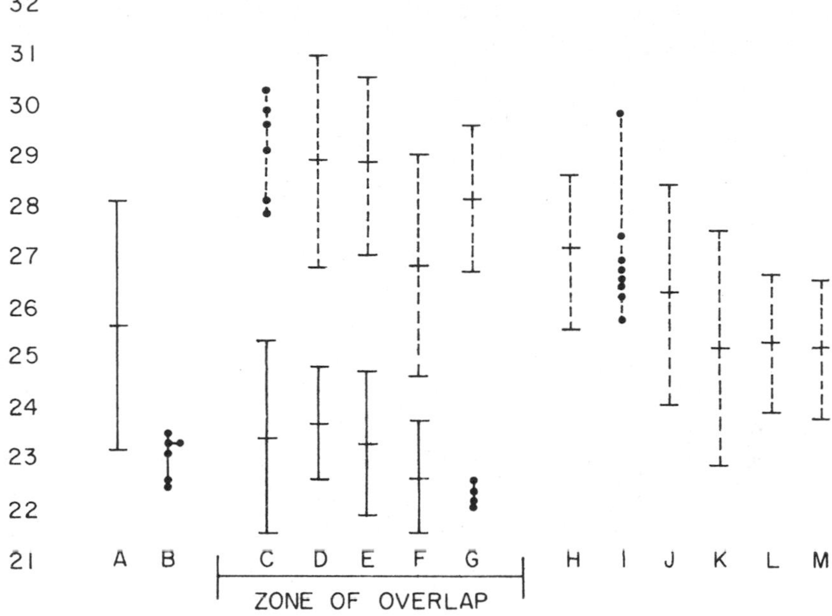

Fig. 4.2. Geographic variation of bill length (in mm) in two partially sympatric species of rock nuthatches, *Sitta neumayer* (solid lines) and *S. tephronota* (broken lines). Each letter (*A*-*M*) designates a geographic region. Westernmost population (Dalmatia, Greece) on left (*A*), easternmost population (Tian Shan) on right (*M*). Complete character divergence in the zone of sympatry in Iran (*C*-*G*). (From Vaurie 1951.)

1968). Furthermore, there is a large class of cases where the overlap seems to have had no effect whatsoever on the phenotype of the sympatric populations. There is need for an unbiased statistical analysis of unselected cases of overlapping species to permit the establishment of valid generalizations.

Open Problems

Ecologically similar species that coexist in the same area interact with each other in numerous ways. Selection tends to increase the ecological difference between such species and to reduce the zone of ecological overlap. Competition, therefore, has many and important evolutionary consequences. Yet our ignorance of the nature, the mechanisms, and the amount of competition is vast. In recent years the study of competition has become one of the most active branches of ecology, and it promises to yield large returns for many years to come. Some of the problems that invite further study are the following:

Which observed cases of geographical vicariance (allopatry) of species are due to ecological incompatibility and which cases are simply the result of physiological adaptations that would make the unoccupied area unsuitable even if it were not inhabited by the potential competitor?

Which of the differences among closely related, sympatric species are the result of character divergence after establishment of sympatry and which had already been acquired as by-products of the preceding geographic speciation?

Are there cases of sympatric species that do not compete with each other even though their essential needs are the same? If so, what is it that permits coexistence? In answering this question it is particularly important to avoid circular reasoning. It is very tempting to say that the mere coexistence of species A and B proves ecological exclusion and, therefore, that whatever differences are discovered between these species are responsible for the coexistence. Such a conclusion is not necessarily valid.

What differences, if any, exist between the central and the peripheral parts of the species range with respect to competition? Haldane (1956) suggests that population size near the periphery of the species range is largely controlled by density-independent factors (for example, weather). Competition with other species would then be less drastic, and a shift into a new niche facilitated. Mayr (1954) presents evidence indicating that shifts into new niches are particularly frequent in peripherally isolated populations.

Conclusions

Most resources of the environment are limited, and competition for them, wherever it occurs, has various evolutionary consequences. It makes speciation more difficult, because an incipient species cannot invade the range of a sister or parent species until it has acquired ecological compatibility with it, that is, until its niche has become sufficiently distinct to permit exclusion (rather than extinction!).

The premium put by natural selection on increased ecological differences between competing species is a powerful centrifugal force. It favors the entry into new niches and, more generally, adaptive radiation. Competition thus is an element in speciation and an important cause of evolutionary divergence.

5 • Isolating Mechanisms

The various devices that prevent the interbreeding of a species with others are called isolating mechanisms. That there is a whole set of special devices by which the gap between species is maintained was not realized until fairly recently. Darwin, thinking that species were rather arbitrarily delimited, neglected the problem of the nature and the origin of the species gap. The typologists, on the other hand, accepted the gap as an inherent property of organic nature. It was not until Dobzhansky coined the term *isolating mechanisms* and devoted a whole chapter to them in his classic *Genetics and the Origin of Species* (1937) that the great importance of these devices was recognized by evolutionists. Since that time a truly staggering amount of literature has accumulated. (For details see Mayr 1963:89–109; also Grant 1963:chap. 13 and Littlejohn 1969).

The active study of isolating mechanisms has led to the development of an entirely new branch of biology that combines the study of behavior, ecology, and genetics. These studies have led to a deeper insight into the quality and quantity of species differences and to the precise experimental testing of various assumptions.

Before discussing the various mechanisms, two subjects require clarification: sterility and geographic isolation.

Sterility. In spite of everything that has been written in the last 30 years there are still some authors who seem to think that reproductive isolation and sterility are synonymous terms. This is not the case. An ever-increasing number of very distinct, reproductively isolated sympatric species have become known that are not isolated from each other by a sterility barrier. For instance, the mallard, *Anas platyrhynchos*, and the pintail, *Anas acuta*, are perhaps the two most common fresh-water ducks of the Northern Hemisphere. The total world population of these two species may well exceed 100,000,000 individuals. Phillips (1915) showed that in captivity these two species are fully fertile with each other and that there seems to be no reduction of fertility in the F_1, F_2, F_3 hybrids or in any of the backcrosses.

One would therefore expect a complete interbreeding of these species in nature, as their world breeding ranges largely coincide. In northern Europe, Asia, and North America they nest side by side on literally millions of ponds, sloughs, and creeks; yet the number of hybrids found among the many birds shot every year is on the order of one in many thousands. Nor is there evidence of backcrossing between these hybrids and the parent species. Obviously, then, the two species are being kept apart not by the sterility barrier but by some other factors. High, if not complete, fertility is known for many species crosses, not only among ducks and other families of birds, but throughout the animal kingdom. The idea that hybrids always are sterile "mules" is quite erroneous.

More decisive evidence for the importance of isolating mechanisms in animals other than sterility is the observation by field naturalists that males and females of a species are brought together by sensory stimuli, that these stimuli are ineffective between the males of one species and the females of another, and that it is therefore extremely rare for the male of one species to copulate with the female of another species. *The sterility barrier, even when present, is only rarely tested.*

Geographic isolation. It is quite impossible to discuss intelligently the nature and the origin of isolating mechanisms if one includes with them so extraneous an element as geographic isolation. The term "isolation" refers to two very different phenomena, spatial isolation and reproductive isolation (Mayr 1959a). The term "isolating mechanisms" refers only to those properties of species populations that serve to safeguard reproductive isolation. In order to exclude extraneous phenomena from the category of isolating mechanisms, it is important to define them precisely: *isolating mechanisms are biological properties of individuals which prevent the interbreeding of populations that are actually or potentially sympatric.* This definition clearly excludes geographic or any other purely extrinsic isolation. The walls of a penitentiary are not an isolating mechanism, nor is a mountain or a stream that separates two populations that are otherwise able to interbreed. Isolating mechanisms always have a partially genetic basis, although some components of behavioral isolation may be reinforced by conditioning or imprinting.

Classification of Isolating Mechanisms

The classification of isolating mechanisms that I have adopted (Table 5.1) arranges them in the sequence in which the barriers have to be overcome. I avoid the terms "genetic" or "physiological" because genetic and

Table 5.1. Classification of isolating mechanisms.

1. Mechanisms that prevent interspecific crosses (premating mechanisms)
   (a) Potential mates do not meet (seasonal and habitat isolation)
   (b) Potential mates meet but do not mate (ethological isolation)
   (c) Copulation attempted but no transfer of sperm takes place (mechanical isolation)
2. Mechanisms that reduce full success of interspecific crosses (postmating mechanisms)
   (a) Sperm transfer takes place but egg is not fertilized (gametic mortality)
   (b) Egg is fertilized but zygote dies (zygotic mortality)
   (c) Zygote produces an F_1 hybrid of reduced viability (hybrid inviability)
   (d) F_1 hybrid zygote is fully viable but partially or completely sterile, or produces deficient F_2 (hybrid sterility)

physiological factors play a role in every one of the seven major subdivisions.

For the vast majority of animals, it is still not known which particular isolating mechanisms prevent closely related species from interbreeding. The naturalist, the cytologist, the geneticist, all have unlimited opportunity for discoveries in this area.

Isolating mechanisms clearly fall into two very different classes (Mayr 1948): "premating mechanisms" and "postmating mechanisms" (Mecham 1961). There is a fundamental difference between the two types: premating mechanisms prevent wastage of gametes (germ cells) and so are highly susceptible to improvement by natural selection; postmating mechanisms do not prevent wastage of gametes, and their improvement by natural selection is indirect. A clear understanding of this difference is of the greatest importance for an understanding of the origin of isolating mechanisms (Chapter 17).

Interspecific Crosses Prevented (Premating Mechanisms)

This category includes seasonal and habitat isolation, ethological isolation, and mechanical isolation.

Habitat Isolation

The less often two potential mates in breeding condition come into contact with each other, the less likely they are to interbreed. Habitat selection thus effectively reinforces other isolating mechanisms. Habitat exclusion among close relatives has been known to naturalists ever since Darwin's days.

Habitat segregation is often very slight in terms of distance. The situation of the spawning grounds of river fish is determined not only by the nature of the substrate (gravel, sand, mud), but also by the rate of water flow (and other factors). A sudden change of gradients in a creek bed may bring the normally well-separated spawning sites of two species so close together that sperm of one species is washed into the nests of the other species; the result is interspecific hybridization.

Habitat isolation is not a very effective isolating mechanism in mobile animals. Consequently, a breakdown of sharp borders between habitats, owing to human or other disturbances, has been the cause of most recorded cases of sympatric hybridization in animals (Chapter 6).

Seasonal Isolation

Differences in the breeding season can effectively prevent the meeting of individuals of different species. Seasonal isolation is common in plants and occurs frequently among insects and other invertebrates (see Chapter 15).

Seasonal barriers are particularly frequent among water animals, because water temperatures are more stable than air temperatures and embryonic development is more closely attuned to definite temperatures. These two factors combined permit a close regulation of the breeding season. Yet most closely related species of anurans (frogs and toads) have overlapping breeding seasons and isolation is provided by acoustic stimuli.

The actual contribution of seasonal isolation to the maintenance of reproductive isolation between species is largely unknown. In some cases in which the seasonal isolation broke down owing to unusual weather conditions, reproductive isolation was maintained by other factors. No natural hybridization occurred in western Canada between the spruce budworm (*Choristoneura fumiferana*) and the pine budworm (*C. pinus*) when unseasonably high temperatures one year resulted in simultaneous hatching. A combination of different host preferences and behavioral barriers maintained the isolation. Differences in breeding season seem to have arisen as a by-product of selection for optimal niche utilization, and they function only secondarily as isolating mechanisms.

Ethological (Behavioral) Barriers (Restriction of Random Mating)

Ethological barriers to random mating constitute the largest and most important class of isolating mechanisms in animals. The word "ethological" (derived from the Greek *ethos* = habit, custom) refers to behavior patterns.

Ethological isolating mechanisms, then, are barriers to mating due to incompatibilities in behavior. The males of every species have specific courtships or displays to which, on the whole, only females of the same species are receptive. The specific reaction of males and females to each other is often referred to loosely as "species recognition," but this term is somewhat misleading since it implies a mechanism of cognition not known in animals.

Ethological isolation is based on the production and reception of stimuli by the sex partners. What are these stimuli that assure the mating of conspecific individuals? The song of the nightingale belongs here and so does the strutting of the peacock and the light flashes of fireflies. The literature on behavioral stimuli exchanged by potential mates is very large. Some of it is summarized by Mayr (1963:97), Hinde (1966), Bastock (1967), Manning (1965), and Wickler (1967). It is convenient to classify these ethological devices on the basis of the principal sense organs involved.

Visual stimuli. It has long been known to naturalists that peculiarities in color, pattern, and form are among the most frequent species differences. That they play an important role as ethological isolating mechanisms has been recognized only recently. The study of visual displays, particularly among birds, fishes, many insects and spiders, and even squids, has contributed greatly to the development of ethology and to our understanding of isolating mechanisms. Rich as the literature of this field is, there is still much uncertainty about the relative importance of optical versus nonoptical (acoustical, chemical) signals and of color versus motion. Most conclusions are arrived at by inference. For instance, when it is found that the reproductive isolation between *Drosophila melanogaster* and *D. simulans* is the same in the light and in the dark, one tentatively concludes that optical stimuli are of minor importance. If two sympatric species of cichlid fishes of markedly different coloration have identical display movements, one concludes that the species-specific coloration is largely responsible for their reproductive isolation.

Analysis is particularly difficult because the signals that function as isolating mechanisms may also serve various other functions (Tinbergen 1954): (1) advertising the presence of a potential mate; (2) synchronizing mating activities; and (3) suppressing fleeing or attacking tendencies in the sex partner. The isolating function cannot be separated from the other three functions. Indeed, one can go so far as to say that only the appropriate, species-specific signals lead to full success of functions (1), (2), and (3) and that these signals thereby automatically function as isolating mechanisms.

Visual stimuli are usually reinforced by auditory, tactile, or chemical stimuli. It is true, however, that other stimuli tend to be less well developed when the visual stimuli are predominant. Naturalists have long stated that the most beautiful birds have undistinguished songs and vice versa. The number of exceptions, however, is too great for this to be accepted as a general rule. The bird that is perhaps the most gifted vocalist in the world, the Australian Lyrebird (*Menura superba*), has also conspicuous displays and wonderful plumes.

The light signals sent out by male fireflies (beetles of the family Lampyridae) are perhaps the most fascinating visual isolating mechanisms. They differ in timing (see Fig. 3.1) and in color, which may be clear white, bluish, greenish, yellow, orange, or reddish. Some species emit exceedingly bright flashes, others dull ones (Barber 1951; Lloyd 1966). The stimuli produced by the females are equally specific, as indicated by the behavior of the males.

Auditory stimuli. Songs, calls, and other acoustic signals have a precision and specificity that would seem to make them even more suitable as isolating mechanisms than visual stimuli. Recent work indicates that this is certainly true of the anurans and of most orthopterans and cicadas. Species-specific sounds play an important role also in the courtship of birds (Marler 1960) and certain diptera. It is often easier to identify a bird by its song than by the visual characters of its plumage. The female, during pair formation, is attracted to the stations of the singing males of her species. Observational evidence is accumulating that female anurans move toward conspecific males. Auditory stimuli are of particular importance among the Orthoptera: nearly every species has a species-specific song. Receptive females approach the male while he is stridulating but stop as soon as the male interrupts his song. They do not respond to the stridulation of a male of a different species. The importance of sound in the reproductive isolation of two sibling species of cicadas (*Magicicada*) was established by Alexander and Moore (1962). For acoustic communication in arthropods in general see Alexander (1967), for anurans, Mecham (1961) and Littlejohn (1969).

Technological progress in sound recording has greatly stimulated research in this field (Lanyon and Tavolga 1960; Busnel 1963; Greenewalt 1968). The ability to translate variable sound into visible sound "spectra" permits measurement and precise description. These are now available for anurans from America and many other parts of the world and they have led to the discovery of many sibling species.

Chemical stimuli. Differences in chemical signals, which act at a distance ("olfactory") or on contact, often serve as isolating mechanisms. They seem to be particularly important among mammals (except the higher primates) and in some groups of insects (Wilson and Bossert 1963). Every naturalist knows that a freshly hatched female moth may attract males for distances of many hundred yards, if not miles. Since the human nose is notoriously insensitive, the study of chemical isolating mechanisms is rather difficult. The experimental approach consists of eliminating visual, auditory, and other nonchemical stimuli and of then testing individuals that have been deprived of one sense organ after another. Amputation of the antennae, for instance, greatly reduces the ability of females of *Drosophila pseudoobscura* and *D. persimilis* to discriminate between their own and alien males. The same operation has little effect on females of *D. melanogaster* and *D. simulans,* which apparently discriminate through contact chemoreceptors, as do many other species of *Drosophila* and some species of roaches. Larger wasps and bees are apparently also attracted to females and are excited by various scents, but there is no good evidence yet for species specificity (Kullenberg 1961). Much work has been done on the chemical nature of sex-attractant scents in insects, but the evidence indicates that the males of other species are frequently attracted as well and that cross-fertilization is prevented by additional isolating mechanisms.

Chemical stimuli are apparently of paramount importance in the spawning of many marine organisms. In the polychaete worm *Grubea clavata* there is an interesting mutual stimulation. The mature females lose egg protein into the water, which induces some sperm ejection by males. In the sperm liquid is a sex stuff (which can be separated from the spermatozoa by centrifugation) that induces the females to spawn. The release of spawn is followed a minute later by massive sperm ejaculation. The sex stuff is genus-specific and probably also species-specific. Such species-specific sex stuffs are probably the most important isolating mechanisms in marine invertebrates with external fertilization, although there are also numerous devices guaranteeing close contact of conspecific males and females during spawning (Thorson 1950). Internal fertilization, however, is widespread in marine invertebrates, including all the higher prosobranchs, all opisthobranchs and cephalopods, most crustaceans, and some polychaetes (*Capitella*), and places a selective premium on the development of behavioral isolating mechanisms.

The functioning of ethological isolating mechanisms. Analysis of animal courtship by students of behavior (Tinbergen 1951; Spieth 1958) reveals courtship to be an exchange of stimuli between male and female continuing until both have reached a state of physiological readiness in which successful copulation can occur.

Among most animals it is the male that actively searches for a mate. He is usually rather easily stimulated to display to objects, sometimes quite inappropriately. When he does not receive adequate responses from his display partner, or is actively repulsed, his display drive eventually becomes exhausted. Consequently, if such a displaying male encounters an individual of a different species, or a male of his own species, he will break off his courtship sooner or later. If the male is displaying to a nonreceptive female, the same will happen, perhaps after a longer interval. However, if the male encounters a receptive female of his own species, he will be sufficiently stimulated by her to continue his displays until the female has passed the threshold of mating readiness. This threshold is on the average much higher in females than in males. "Species recognition," then, is simply the exchange of appropriate stimuli between male and female, which insures the mating of conspecific individuals and prevents hybridization of individuals belonging to different species. The difference in ease of arousal between the sexes is favored by natural selection because a large supply of sperm permits one male (in species without pair formation) to inseminate many females, while females have a limited supply of eggs and it is of great selective value that they be fertilized in such a manner that zygotes of optimal fitness are produced. The slower a female is in accepting a male, the greater the opportunity for discrimination and the smaller the danger of producing inviable offspring (Richards 1927). An extreme case occurs in flies of the genus *Sarcophaga,* in which males mount, almost indiscriminately, individuals (male or female) of their own species, of other species of the genus, and even of other genera gathered on the same source of food. It is apparently the female alone that determines the success of the mating. The lack of discrimination by males is by no means always as great as in *Sarcophaga;* in many species males display to females of related species only perfunctorily.

An isolating mechanism is rarely an all-or-none affair. It has been remarked by several students of *Drosophila* courtship, for instance, that the basic pattern of courtship is the same in closely related species. The differences are often quantitative rather than qualitative (Fig. 5.1); yet such quantitative differences between species are apparently sufficient to prevent the successful synchronization of readiness for mating. The quantitative nature of courtship differences has been recorded also for birds, and seems to be the rule for many related species (Hinde 1959).

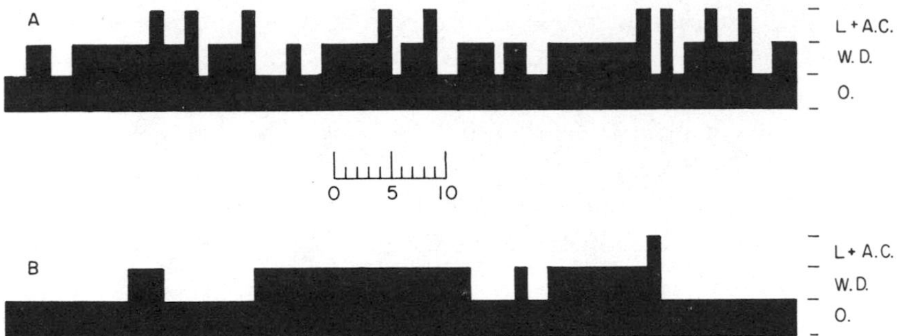

Fig. 5.1. Quantitative differences in the major courtship components of two sibling species of *Drosophila*. *A* = *D. melanogaster*, *B* = *D. simulans*. The sequence reads from left to right, the scale showing time units of 1½ seconds. The height of the black columns indicates the courtship element being performed. The lowest level is *O* (orientation), the middle level *W.D.* (wing display), and the highest level *L* + *A.C.* (licking and attempted copulation). (From Manning 1959.)

Ethological isolation is the result of an interaction between external stimuli and the totality of internal drives. If no appropriate sex partner is available, internal drives continue to build up until readiness for mating can be induced even by a highly inadequate stimulus.

Mechanical Isolation

Soon after the discovery of the manifold structural differences in the genital armatures of different species of insects, it was asserted by Dufour (1844) that these armatures act like lock and key, preventing hybridization between individuals of different species. There is occasional observational evidence to support this claim. Interspecific crosses in *Drosophila* may cause injury or even death to the participants, and the same is true in *Glossina* (tsetse fly).

The mechanical barrier formed by incompatibilities of the copulatory apparatus may well be an important isolating mechanism in the pulmonate snails. For instance, it has been shown that the complicated structure of the genitalia prevents interspecific mating in the subfamily Polygyrinae of the polygyrid snails. In the closely related subfamily Triodopsinae, the genitalia are of simpler structure, and in this group interspecific crosses have been reported.

Mechanical isolation was for a long time considered a most effective isolating mechanism, particularly when it was found how widespread genitalic differences are among various orders of insects. Karl Jordan (1905) concluded, however, on the basis of comprehensive detailed studies, that Dufour's hypothesis is not valid. Among 698 species of Sphingidae examined by him, 48 were not different in their genitalia from other species of the family, while in about 50 percent of the species with geographic variation in color there was also geographic variation in the structure of the genitalic armatures. Since that time much additional information has accumulated indicating the slight importance of the genitalic armatures as isolating mechanisms.

The true significance of the differences in the genitalia is presumably the following. The genitalic apparatus is a highly complicated structure, the pleiotropic by-product of very many genes of the species. Any change in the genetic constitution of the species may result in an incidental change in the structure of the genitalia. As long as this does not interfere with the efficiency of fertilization, it will not be selected against. On the whole, of course, the genitalic structures are as much subject to normalizing selection as any other vital structure of an organism (Chapter 10). In many groups of insects, the morphology of the genitalia does not change appreciably during speciation.

Interspecific Crosses Unsuccessful (Postmating Mechanisms)

The ecological and ethological barriers in animals are very efficient and prevent interspecific crosses in most cases. If, however, this first set of barriers should fail, a second set of barriers may prevent successful hybridization, although it cannot prevent wastage of gametes: the potential mates complete copulation but no offspring are produced, or the offspring produced have reduced viability or fertility. This second group of isolating mechanisms can be classified (in part after Patterson and Stone 1952) into four categories.

Gametic Mortality

The sperm may encounter an antigenic reaction in the genital tract of the female and be immobilized and killed before it has a chance to reach the eggs. Patterson found that an "insemination reaction" occurs in many species of *Drosophila* which leads to an enormous swelling of the walls of the vagina and the subsequent killing of the spermatozoa. Antigenic reactions may also occur when there is no evident insemination reaction. And even in the absence of antigenic reactions the sperm may die because it cannot penetrate the egg membrane of the alien species.

Zygotic Mortality

The development of the fertilized hybrid egg is often irregular, and development may cease at any stage between fertilization and adulthood. The details and causes of this incompatibility are thoroughly described in the textbooks of embryology, cytology, and genetics (for instance, Dobzhansky 1951).

Hybrid Inviability

Many naturally occurring animal hybrids have been observed to leave no offspring, even though they seem to have somatic hybrid vigor and are fully fertile, with normal eggs or spermatozoa. Hybrid inferiority has been very little studied (it is difficult to study!), but may well be the reason for the small amount of introgression in groups, like the ducks, in which hybrids are fully fertile.

The reason for the reproductive failure of fully fertile species hybrids is perhaps that they are less well adjusted to the available ecological niches than individuals of the parental species. Also, hybrids are usually less successful than individuals of pure species in courtship, when definite behavior patterns and species-specific stimuli play an important role. Ecological and ethological inferiority reduces their chances of leaving offspring.

Hybrid Sterility

Species hybrids have considerable or complete fertility in some groups of animals, but are more or less sterile in others. The cytological and genetic reasons for this are excellently discussed by White (1954) and Dobzhansky (1951). There is a detailed survey of hybrid sterility in *Drosophila* (Patterson and Stone 1952), but few investigators have systematically crossed all the available species of a genus and determined their cross fertility. However, such crossing is now being done for toads of the genus *Bufo* (W. F. Blair) and for ducks (P. Johnsgard).

Backcross individuals are likely to be even more strongly inferior, owing to various imbalances of their gene complexes. Incompatibility of genes of one of the parental species with the genome of the other species may lead (even in the F_1 hybrid) to severe or lethal physiological disturbances. A well-known case is that of the melanotic tumors that arise in certain crosses between the fish species *Xiphophorus maculatus* and *X. helleri.* Similar incompatibilities have been described for many species crosses. Other aspects of hybridization are discussed in Chapters 6 and 13.

The Coaction of Isolating Mechanisms

The interbreeding of closely related sympatric species of animals is usually prevented by a whole series of ecological, behavioral, and cytogenetic factors, usually several for each species pair. One factor is often dominant, such as acoustic stimuli in certain frogs, grasshoppers, cicadas, and mosquitoes, chemical contact stimuli in some insects, sex stuffs in Lepidoptera and certain marine organisms, and visual stimuli in certain birds, fishes, and insects. The sterility barrier may be strong or weak, but it is rarely tested except when the other isolating mechanisms break down.

That each of the categories of Table 5.1 is, furthermore, composite is evident in the case of ethological isolation; it has been shown also for sterility factors, and is indicated for habitat and temporal isolation. Consequently, the total isolation between two species is normally due to a great multitude of different isolating factors, each more or less independent of the others. In animals behavioral isolation is by far the most important of the various isolating mechanisms. In plants, on the other hand, sterility is the most important isolating factor. The ability of animals to search actively for potential mates is undoubtedly the reason for this striking difference between the two kingdoms (Grant 1963:chap. 13).

In only a few cases has an attempt been made to analyze all the factors that isolate two species from each other. And we still do not know in what sequence isolating mechanisms are acquired (see pp. 328–330). A study comparing the isolating mechanisms of groups with a low frequency of hybridization (birds, reptiles, spiders) with the mechanisms of groups with more frequent hybridization (fishes, ?mollusks) is highly desirable.

The isolating mechanisms are arranged like a series of hurdles; if one breaks down, another must be overcome. If the habitat barrier is broken, for example, individuals of the two species may still be separated from each other by behavior patterns. If these also fail, the mates may be unable to produce viable hybrids, or if hybrids are produced they may be sterile.

The Genetics of Isolating Mechanisms

From the multiplicity of isolating mechanisms that protect every species, one can conclude that a considerable number of genes are involved. Almost any gene that changes the adaptation of a population may have an incidental effect on the interaction between male and female. Conversely, a mating advantage of a male will become established in a population provided it does not seriously lower the fitness of his offspring.

An annually increasing number of investigators study the genetics of behavior. Among recent summaries are those of Fuller and Thompson (1960) and Hirsch (1967). Detailed analyses of the genetics of isolating mechanisms other than sterility exist only for the genus *Drosophila*. Many mutations in *D. melanogaster*, such as *yellow*, *white*, and *bar*, influence the mating success of the mutant individuals. However, degree of sexual isolation is certainly not solely governed by individual genes, since it often changes drastically when the mutants are transferred to different genetic backgrounds. It is now well established that different gene arrangements may differ greatly in their contribution to mating speed and success (Spiess et al. 1966). It is therefore quite evident that behavioral isolation has a highly complex genetic basis. In addition to genes controlling the specificity (and variation) of the signals exchanged between males and females, there are others that control mating drive and levels and rhythms of activity.

Selection continuously adjusts the mating propensity of females to the prevailing level of the mating drive of the males. When one of these two components varies geographically in a species, the other always varies correspondingly. An understanding of the extraordinarily sensitive adjustment of the drives contributing to sexual isolation, as well as of the complex genetic basis of the isolating mechanisms, is important for the explanation of the origin of reproductive isolation between species (Chapter 17).

There is evidence that genetic isolating mechanisms are occasionally modified or reinforced by conditioning. Young birds can sometimes be imprinted on a foster species when raised by foster parents. There is no danger that such conditioning will lead to hybridization, because in nature young birds are invariably raised by parents of their own species. In parasitic birds—cuckoos and cowbirds, for example—such conditioning is absent, and species recognition is rigid. Young cowbirds, for instance, leave the company of their foster parents and flock together as soon as they are independent. In the human species, however, conditioning is an important isolating mechanism. The free interbreeding of individuals coexisting in a geographical region is strongly influenced by religious, economic, and cultural barriers.

The Role of Isolating Mechanisms

One of the evident functions of isolating mechanisms is to increase the efficiency of mating. Where other closely related species do not occur, courtship signals can "afford" to be general, nonspecific, and variable. Where other related species coexist, however, nonspecificity of signals may lead to wasteful courtship and delays, even where no heterospecific hybridization occurs. Under these circumstances there will be a selective premium on precision and distinctiveness of signals.

This situation is perhaps best documented by cases in which a species moves from a continental area that has several congeneric species to an island where it is the only representative of its genus. No longer being exposed to stabilizing selection for precision and distinctiveness, it may lose the high specificity of the isolating mechanisms. Such a loss has been recorded not only for visual characters (Mayr 1942) but also for the song of birds on islands, for instance *Parus* on Tenerife and *Regulus* on the Azores (Marler 1957).

Each species is a delicately integrated genetic system that has been selected through many generations to fit into a definite niche in its environment. Hybridization usually leads to a breakdown of this system and results in the production of disharmonious types. It is the function of the isolating mechanisms to prevent such a breakdown and to protect the integrity of the genetic system of species. Any attribute of a species that would favor the production of inferior hybrids is selected against, since it results in wastage of gametes. Such selection maintains the efficiency of the isolating mechanisms and may indeed help to perfect them. Isolating mechanisms are among the most important biological properties of species.

6 • The Breakdown of Isolating Mechanisms (Hybridization)

Not all isolating mechanisms are perfect all the time. Occasionally they fail and permit the crossing of individuals that differ from each other genetically and taxonomically. Such interbreeding is called *hybridization.* This term is difficult to define precisely and has therefore been applied to very different phenomena.

Both the concept and the term "hybrid" were taken from the realm of animal and plant breeders. They referred originally to the product of a cross between two unlike individuals, usually members of two different species. Difficulties arise when one attempts to extend this typological concept from individuals to populations. Is it legitimate, for instance, to refer to the interbreeding of conspecific populations as hybridization? If so, under what circumstances? In most cases it would seem highly misleading to apply the term hybridization to ordinary gene exchange among conspecific populations. Yet when there is a secondary contact between previously long-isolated populations, the term "hybridization" is sometimes appropriate (see p. 220). Hybridization may be defined conveniently as *the crossing of individuals belonging to two unlike natural populations that have secondarily come into contact.* The disadvantage of a subjective determination of "unlike" in this definition is counterbalanced by the emphasis on populations and the avoidance of circular reasoning.

Genetic Aspects of Hybridization

In most treatises on hybridization the emphasis is placed primarily on hybrid sterility and its causes, secondarily on other genetic aspects of hybridization. For a detailed discussion of such genetic and cytological aspects of hybridization consult the authoritative treatments by Dobzhansky (1951), Stebbins (1950), Grant (1963), Cousin (1967), and White (1954). Only a few facts will be summarized here.

Hybrids differ from individuals of the parental species not only in morphology but usually also in fertility and viability. The inability of hybrids to produce the normal number of viable gametes is called *hybrid sterility.* This inability may range from slight to complete and may be caused by a variety of genetic factors. Such a reduction in fertility is not necessarily correlated with a reduction in viability. Indeed, hybrids often show a marked phenotypic "luxuriance," called *hybrid vigor.* The hybrid vigor of the mule, sterile product of the cross of horse and donkey, is proverbial, yet it does *not* constitute Darwinian fitness, which is measured in terms of the contribution to the gene pool of the next generation.

Hybridization as a Population Phenomenon

The present treatment emphasizes the natural history and particularly the population aspects of hybridization. When hybridization is regarded as a population phenomenon, various aspects ignored by the breeders become important. Does a given case of hybridization involve otherwise valid species, or allopatric populations of the same species? Does it consist of the occasional production of a hybrid individual, or is it a massive phenomenon resulting in a more or less complete breakdown of the barrier between species? These various kinds of hybridization differ in their evolutionary significance and it would be misleading to confuse them.

On the basis of the stated criteria, five kinds of hybridization might be distinguished, as far as the taxonomist and evolutionist are concerned:

(a) The occasional crossing of sympatric species resulting in the production of hybrid individuals that are ecologically or behaviorally inviable or sterile and therefore do not backcross with the parental species.

(b) The occasional or frequent production of more or less fertile hybrids between sympatric species, some of which backcross with one or both of the parental species.

(c) The formation of a secondary zone of contact and of partial interbreeding between two formerly isolated populations that had failed to acquire complete reproduction isolation during the preceding period of geographic isolation.

(d) The complete local breakdown of reproductive isolation between two sympatric species, resulting in the production of hybrid swarms that may include the total range of variability of the parental species.

(e) The production of a new specific entity as the result of hybridization and subsequent doubling of the chromosomes (allopolyploidy) (virtually restricted to plants).

The first four categories (*a–d*) grade into one another, and it is sometimes difficult to decide where a given case should be listed (Short 1969). Cases of hybridization involving "good species" are here arbitrarily included under *a* and *b* above, even when the distribution is largely allopatric. They do behave like sympatric species in the area of overlap.

In connection with a systematic survey of the occurrence of these five classes of hybridization in nature, it will be possible to consider some more general questions, such as the following:

(1) How common is hybridization in nature among different groups of animals?

(2) Is there a striking difference in the incidence of hybridization between plants and higher animals and, if so, why?

(3) What effect does hybridization have on species structure and intraspecific variability?

(4) What is the role of hybridization in speciation and evolution?

Occasional Hybridization

Hybridization is common among plants. In most groups of animals it is sufficiently exceptional to justify a report in the literature whenever a hybrid is discovered. The suggestion is sometimes made that this rarity of animal hybrids is more apparent than real; however, the frequency of hybrids is as low in groups where hybrids can be discovered easily (by cytological or other methods) as where the recognition of hybrids is difficult. Interspecific hybrids produced in nature are particularly rare in mammals, birds, and reptiles. To give a rough estimate of the order of magnitude of hybridization in these classes, one might say that about 1 in 50,000 individuals is a hybrid. Of most sympatric species combinations no wild hybrid has ever been found and recorded cases are frequent only in a few families, such as the ducks and gallinaceous birds (see Mayr 1963:114).

Although hybridization is evidently very rare among reptiles, it is comparatively frequent among the amphibians. Here its significance seems to change from genus to genus. Hybridization is rare in true frogs of the genus *Rana*, but widespread in toads of the genus *Bufo* (see below). Fertile hybrids between the frogs *Hyla cinerea* and *H. gratiosa* are not uncommon in the region of overlap. Hybridization has also been reported for urodele amphibians, for instance the genus *Triturus*.

The situation in fishes contrasts strongly with that in nearly all land vertebrates. Fertilization in fishes is usually external and for this reason (among others) hybridization is frequent. Excellent reviews by Hubbs (1955, 1961) summarize the previous literature and record the known data, particularly for the fresh-water fishes of North America (mostly based on the work of Hubbs and his students). In most cases only sterile F_1 individuals are found, with little or no evidence of backcrossing with the parental species. For instance, among 2000 *Catostomus commersoni* and *C. catostomus* caught in the Platte River there were 5 hybrids (0.25 percent); 7 percent of the specimens of *C. macrocheilus* and *C. syncheilus* caught in 9 lakes in British Columbia were F_1 hybrids (Nelson 1968). Every collected specimen was either an F_1 hybrid or a specimen of the parental species. The same is true of hybrid specimens found among marine fishes.

It has long been known that some hybridization occurs between kinds of three-spined sticklebacks (*Gasterosteus*). An armored species (*G. trachurus*) is essentially marine, but enters fresh water to spawn; a smooth-skinned species (*G. aculeatus*) is restricted to fresh water. Hybridization occurs in a narrow zone of overlap, apparently because neither an ethological nor a sterility barrier exists. There is, however, a powerful ecological isolation and probably hybrid inferiority away from the breeding area (Hagen 1967).

The frequency of natural hybridization in most groups of animals is difficult to determine because the information in the taxonomic literature is exceedingly scattered, and the primitive condition of the taxonomy does not permit comprehensive reviews by specialists. The general impression one gets from a perusal of the literature is that hybrids are rare. The best available data on invertebrates are those for some groups of Lepidoptera and for *Drosophila* (see below).

Introgression in Animals

The term *introgression* designates the incorporation of genes of one species into the gene complex of another species as a result of successful hybridization. This is not to be confused with *gene flow,* which designates the normal gene exchange among populations that belong to the same species. The term introgression is to be used only for gene exchange between semispecies or species. Although such a leakage of genes from one species to another is frequent among plants (Grant 1963), it is rather rare in animals. Among the better-known cases are the following. Where the Golden-winged Warbler (*Vermivora chrysoptera*) and the Blue-winged Warbler (*V. pinus*) meet in eastern North America, hybrids are not uncommon. They are distinctive in color and song ("Brewster's Warbler," "Lawrence's Warbler"), and numerous backcrosses have been studied by field ornithologists. This hybridization has been known for a century and presumably began some 200 years ago when the natural habitat barrier between the species was obliterated by deforestation and farming. Yet the delimitation of the two parental species is still quite sharp in most areas. There is no evidence of a blurring of the species border except in the zone of overlap.

Two yellow butterflies in North America, *Colias philodice* and *C. eurytheme,* hybridize regularly. This breakdown of reproductive isolation is in part caused by the spread of *eurytheme* over wide areas previously occupied only by *philodice,* following the increased planting of alfalfa, *eurytheme*'s food plant. Where the overlap is recent, only 1.0–1.5 percent of the population are hybrids. Elsewhere the percentage of hybrids seems to have risen to a level of about 10–12 percent, where it has stayed ever since this hybridization was first observed over 50 years ago. This stabilization is somewhat surprising since the mating between the two species is reported to be almost random, and F_1 individuals not only are fertile but have been shown to backcross to some extent with the two parental species. Presumably the hybrids are sufficiently inferior in general viability to prevent the development of a genuine hybrid population.

In the majority of cases where introgressive hybridization in animals has been reported, two species are involved that had been conspecific until recently and are still largely allopatric. They are *semispecies* (in the emended definition of Lorkovic 1953), showing some of the characteristics of species and some of subspecies. The fact that they hybridize to a greater or lesser extent proves that they did not acquire complete reproductive isolation during their geographic isolation. The success of the F_1 hybrids determines whether such occasional hybridization leads to a strengthening of isolating mechanisms or to the development of zones of secondary intergradation (see Chapter 13).

Sympatric Hybrid Swarms

The barrier between two sympatric species sometimes breaks down so completely, locally or over wide areas, that the two parental species are replaced by a hybrid swarm that forms a continuous bridge between the two parental extremes. A thorough knowledge of the taxonomy of the respective groups is a prerequisite for a sound analysis of such situations. Lack of this knowledge is presumably the reason that so few such cases have so far been described. Groups that are easily observed, like birds, or easily caught, like butterflies and fishes, supply the best-substantiated instances. In view of their great evolutionary interest, some of these cases will be described in detail.

Birds. Perhaps the most thoroughly analyzed case of the breakdown of isolation between two species of birds is that of two members of the bunting genus *Pipilo* in Mexico (Sibley 1954). The Red-eyed Towhee (*P. erythrophthalmus*) and the Collared Towhee (*P. ocai*) are more or less widespread as "pure" species (Fig. 6.1). *Pipilo ocai* occurs from Oaxaca to Jalisco. *Pipilo erythrophthalmus* is widespread in North America and extends south as far as Chiapas and Guatemala. In Oaxaca the two species live side by side without intermixing. In Puebla 16 percent of the 117 known specimens show evidence of hybridization. In the other states of the Mexican plateau, from northern Puebla through Nayarit and Michoacan to Jalisco, a series of introgressed hybrid populations is found, which in the east and north are similar to *P. erythrophthalmus* and toward the south and west are similar to *P. ocai.* If a hybrid index is designed which gives pure *erythrophthalmus* the value of 24 and pure *ocai* the value of 0, an east-west chain of populations is found with the mean value 22.4–19.8–16.9–15.8–13.5–7.8–4.0, and a north-south chain with the values 23.5–22.8–22.6–13.7–8.0–2.8–0.17. The variation within a local population is great but does not span the total range. In a population with a mean of 13.7 the index varied from 6 to 20 in 76 specimens; in another with a mean of 8.0 it varied from 3 to 16 among 58 specimens. If it were not for the pronounced differences between the species and their sympatry in Oaxaca, one might be tempted to consider them conspecific. Sibley's original papers much be consulted for many other interesting aspects of this hybridization. It is apparently a very recent event, caused by man's agricultural activities, and not dating back further than 300–500 years.

Similar cases have been described for sparrows (*Passer*), flycatchers (*Terpsiphone*), honeyeaters (*Melidectes*), bulbuls (*Pycnonotus*), and kingfishers (*Ceyx*). For details see Mayr (1963:119–123) and Short (1969).

Amphibians. An especially interesting case in the genus *Bufo* was analyzed in detail by A. P. Blair (1941). Fowler's Toad (*B. fowleri*), living primarily in grasslands, fields, and other open areas, and the American Toad (*B. americanus*), living primarily in wooded areas, interbreed extensively in many areas of their nearly completely coincident geographic ranges. The most interesting aspect of this hybridization is that the toads at the beginning of the breeding season (April) are almost pure *americanus* and at the end of the breeding season (June) almost pure *fowleri.* However, many samples captured at the middle of the breeding season (May) are intermediate between the two species. A study of rate of development and temperature tolerance of the embryos confirmed the hybrid nature of some populations. The amount of hybridizing is different from locality to locality,

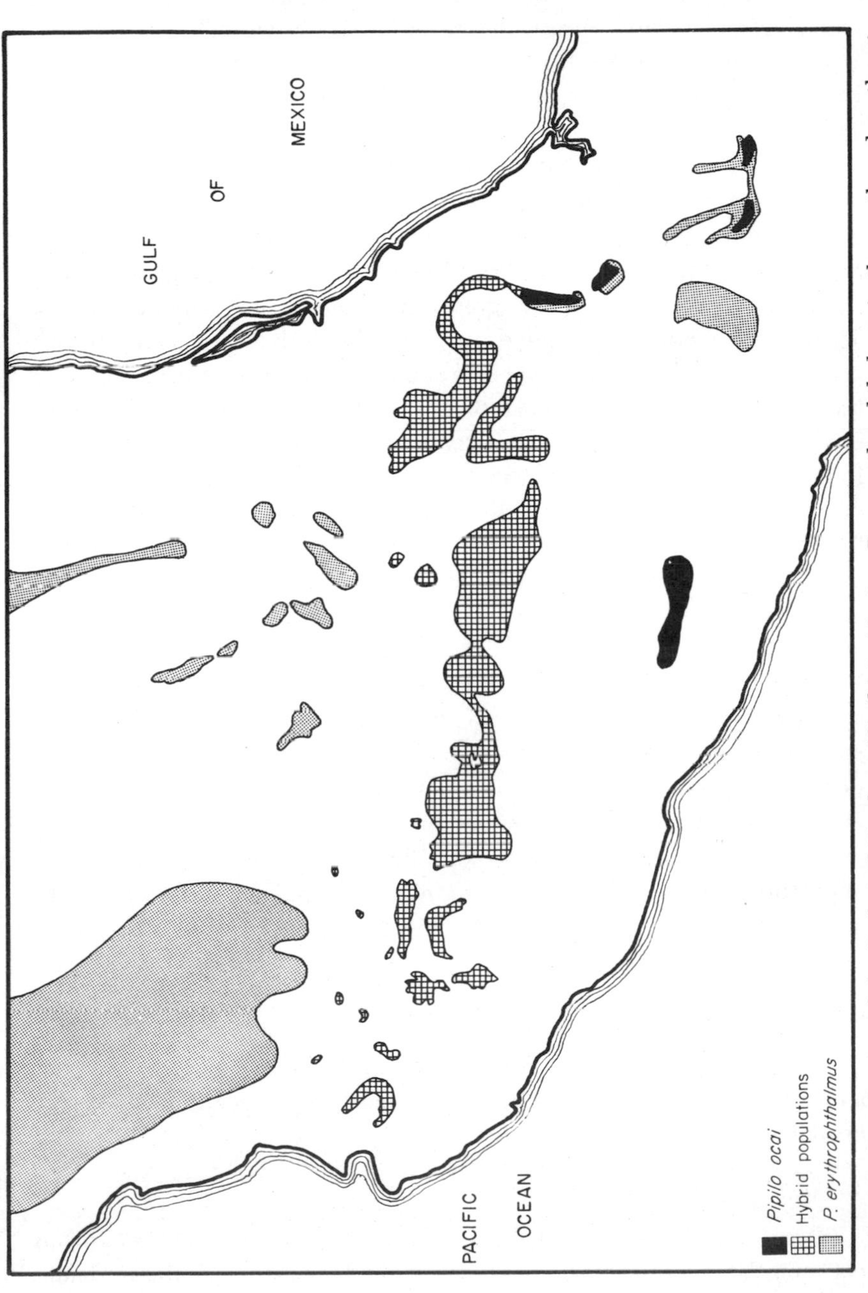

Fig. 6.1. Distribution of the Red-eyed Towhees (*Pipilo*) in Mexico. Pure *erythrophthalmus* in the north and southeast. Pure *ocai* in the south and southeast. Only hybrid populations exist in a vast area of central Mexico. These hybrid populations range in appearance from almost pure *erythrophthalmus* to almost pure *ocai*. Note the sympatry of the two species at several localities in the southeast. (After Sibley 1954).

and there is much evidence that hybridization is caused by the breakdown of ecological barriers between the species owing to agricultural activities. It is interesting to note that there are apparently "pure" populations of *americanus* and *fowleri* at localities near the hybrid areas. Various other species of *Bufo* also show evidence of introgressive hybridization.

Fishes. In spite of the high frequency of hybridization in fishes, particularly fresh-water fishes, the situation is only rarely auspicious for the occurrence of introgression. Although sunfishes (Centrarchidae) hybridize commonly, only three places have been found so far where the characters of the especially abundant hybrids grade into those of each parental species. These three hybrid swarms are exceptional also in that they do not consist predominantly of males. Another case of introgression occurs in the cyprinid fishes. In a stretch of the San Juan River in California two sympatric species of the genera *Hesperoleucus* and *Lavinia* have crossed to the extent of forming a hybrid swarm, with the borders between the species breaking down completely locally. Hybrids between *Notropis rubella* and *N. cornuta* have a normal sex ratio and seem to bridge the gap between the two species completely, thereby providing opportunity for introgression. Hybridization seems widespread in the genus *Coregonus* and occasionally involves introgression.

Invertebrates. Introgressing hybrid swarms are infrequently reported among insects and other invertebrates because taxonomic analysis has rarely reached the point at which hybridity of a variable or intermediate population can be proved unequivocally. Hybridization seems widespread in *Daphnia.* Brooks (1957b) ascribes the difficulty of identifying many populations of the *D. pulex* group to the fact that they are introgressing hybrid populations between *D. middendorfiana, D. schoedleri,* and *D. pulex.* In many areas, however, pure populations of these species can be found.

Causes of the Breakdown of Isolating Mechanisms

A study of the occurrence of natural hybridization shows that it is not a random phenomenon. There are certain factors that clearly facilitate the breakdown of isolating mechanisms.

Methods of Fertilization

Internal fertilization is preceded by a more or less extended courtship between the potential mates, and, as described in Chapter 5, this normally prevents interbreeding between individuals that are not conspecific. Similar courtships occur in aquatic animals with external fertilization. Yet if there is a simultaneous spawning of two related species and if water currents bring the sperm of one species in contact with unfertilized eggs of the other species, opportunity is provided for hybridization. External fertilization is unquestionably one of the reasons for the relatively high frequency of hybridization among fishes as compared with mammals, birds, and reptiles. In fishes with internal fertilization, like the viviparous cyprinodonts, hybridization is very rare.

However, the mode of fertilization is not the only determining factor in hybridization. Frogs (*Rana*) and toads (*Bufo*) have the identical mode of external fertilization, yet they differ considerably in frequency of hybridization. The more precisely the reproductive activity of an aquatic species is limited to a specific water temperature, the less danger of hybridization there is, each species tending to spawn at the species-specific temperature.

Nature of the Mating Bond

Even in species with internal fertilization, there are great differences in the frequency of hybridization, depending on the nature of the mating bond. In most species of birds, for instance, males and females form a definite pair during the mating season and share the duties of incubation and raising of the young. In such pair-forming species there is usually a more or less lengthy "engagement" period before copulation takes place. There are innumerable displays between the two mates during this period, and pairs not composed of conspecific individuals apparently break up at this stage. This is presumably the reason why hybrids are so rare in birds with this type of mating bond. Hybrids are more frequent among those groups of birds in which copulation is not preceded by pair formation and an "engagement" period. In most birds of paradise (Paradisaeidae), for example, one or several males perform on a display ground and a female appears there only when she is ready for fertilization. After this has been accomplished, she alone builds the nest, incubates the eggs, and raises the young. On rare occasions a female is attracted to the display ground of the wrong species and is fertilized. At least 14 kinds of such hybrids, involving 10 genera, are now known. A relatively high frequency of hybrids, many of them intergeneric, has been described in all the other families of birds in which pair formation is absent (at least in certain genera): hummingbirds (Trochilidae), grouse (Tetraonidae), and manakins (Pipridae). Even where such hybridization is common, as between Capercaillie (*Tetrao*) and Black

Cock (*Lyrurus*), it has not led to introgression. The frequency of hybrids in these families is easy to understand. The apparent brevity of contact between males and females leaves much room for error and little chance for correcting it.

Rarity of One Parental Species

The individuals that occur beyond the solid range of their species often have difficulty in finding a conspecific mate. In the absence of appropriate stimuli, that is, stimuli from conspecific individuals, they are apt to respond to inappropriate stimuli, that is, to stimuli emanating from individuals of a different species. Many of the known hybrids of animal species are found at the margin of the normal geographic range of one of the two parental species or even beyond it. Such hybrids have been reported for birds, fishes, and many other animals. This rule holds not only for geographic distribution in the narrow sense of the word but also for altitudinal and habitat distribution. It is not different in principle from situations in which a species hybridizes in captivity in the absence of conspecific mates.

Disturbance of Habitat

By far the most frequent cause of hybridization in animals is the breakdown of habitat barriers, mostly the result of human interference, particularly agricultural activities. Hybridization in the African Paradise Flycatchers (*Terpsiphone*), in the American toads, and in the yellow butterflies (*Colias*) is apparently due to man's disturbance of the natural habitats. The same is true of most other cases described above. Stebbins (1950) lists this as the most important single cause of hybridization in plants.

Not all habitat disturbances that lead to hybridization are man-made. Hubbs and Miller (1943) showed that the desiccation in the Western deserts forced certain species of fish together in the confined waters of springs and that as a result the isolating mechanisms between these species broke down. The same situation was found for the cyprinodont fishes of Asia Minor.

Difference in Hybridization Between Plants and Animals

It is evident that higher plants and animals differ considerably in frequency of hybridization. Contrary to the belief of some botanists, the low frequency of hybridization among animals is not an artifact produced by insufficient analysis of animal species. Rather, the striking difference between these two kingdoms is correlated with basic differences in their physiology, population structure, genetic constitution, and ecological needs.

Animals have mobility and can actively search both for a suitable mate and for a suitable habitat. These abilities allow much greater specificity and specialization than is normally found among plants. Speciation is made easy and occurs with great profusion: there exist perhaps five times as many species of animals as of plants. An increase in variability, such as is produced by hybridization, normally results in a lowered efficiency in the species-specific niche, especially if it is a narrow one. Selection will discriminate against genotypes with a propensity for hybridization.

Plants cannot move. A seed germinates where it falls and must succeed or die. In plants, therefore, natural selection favors great nongenetic (phenotypic) plasticity as well as genetic variability. Hybridization replenishes such variability and is therefore favored. Frequent hybridization is inevitable because the transfer of pollen from one plant to another is accomplished by extrinsic agents such as wind or insects. This situation permits many "mistakes" that lead to hybridization. Thus, both the ecological needs of plants and their reproductive mechanisms favor hybridization. Because there is much hybridization and because hybrids are as viable under many conditions as the parental types, there is selection pressure in favor of an increased ability to cope with the consequence of hybridization. This leads to further increase in hybridism, and, conversely, reduces the frequency of speciation, counteracts too narrow a specialization, and puts a premium on instantaneous speciation.

The Evolutionary Role of Hybridization

Some authors have placed heavy emphasis on hybridization in their evolutionary theories, others have virtually ignored this factor. The various aspects of the evolutionary significance of hybridization that have been stressed in the past can be classified under three headings.

(1) *Perfection of isolating mechanisms.* It has been suggested that the occurrence of hybridization may occasionally lead to the perfecting of isolating mechanisms, provided hybrids are sufficiently rare and inviable. The validity of this suggestion will be investigated in Chapter 17.

(2) *Source of new species.* If hybrids between two species were to form a third species coexisting with the two parental species, the result would

be a process of speciation through hybridization. The possibility of such an occurrence will be discussed in Chapter 15.

(3) *Increase of genetic variability.* The claim has been made that species owe much of their genetic variability to introgressive hybridization. However, all the evidence contradicts this conclusion so far as animals is concerned. Not only are F_1 hybrids between good species very rare, but where they occur the hybrids (even when not sterile) are demonstrably of inferior viability. The few genes that occasionally introgress into the parental species are not coadapted (Chapter 10) and are selected against. Introgressive hybridization seems to be a negligible source of genetic variation in animals.

The total weight of the available evidence contradicts the assumption that hybridization plays a major evolutionary role among higher animals. To begin with, hybrids among them are very rare, except in a few groups with external fertilization. The majority of such hybrids are totally sterile, even where they display "hybrid vigor." Even those hybrids that produce normal gametes in one or both sexes are nevertheless unsuccessful in most cases and do not participate in reproduction. Finally, when they do backcross with the parental species, they normally produce genotypes of inferior viability that are eliminated by natural selection. Successful hybridization is indeed a rare phenomenon among animals. For further discussion see Mayr 1963:130–133.

Summary

Occasional breakdown of isolating mechanisms has been found to occur in most taxonomically well-known groups of animals. Relatively most frequent among the various forms of hybridism is the occurrence of occasional sterile, or at least nonreproducing, species hybrids. Evidence of backcrossing with one or both of the parental species is found much more rarely, and rarer still is the complete breakdown of the barrier between species resulting in hybrid swarms.

The evolutionary importance of hybridization seems small in the better-known groups of animals. Even when fertile hybrids are produced, genetic unbalance of the hybrids results in markedly lowered ecological and ethological adaptedness, and there is little or no introgression. The contribution to the genetic variability of a population made by noneliminated genes, remaining as a residue of introgression, can be considered negligible in comparison with the contribution made by mutation and regular gene flow from adjacent conspecific populations.

In view of the rarity of allopolyploidy and successful hybridization, it is evident that reticulate evolution above the species level plays virtually no role in the higher animals. The systematics of lower animals is too little known to permit generalizations on the evolutionary role of hybridization in them. The possibility that these animals are more similar to plants than to higher animals can be neither supported nor excluded by the available evidence.

7 · The Population, Its Variation and Genetics

Between the individual and the species is a level of integration of particular importance to the evolutionist, the level indicated by the word *population.* It is proper that the study of natural populations has become a major preoccupation of several branches of biology: genetics, ecology, and systematics. The term "population" is used in several different ways and recourse to a dictionary is of little help. Ecologists may speak of the plankton population of a lake, including in it the individuals of several species. It is customary to speak of the human population, referring to the totality of individuals of a single species, the human species. These are legitimate uses of the term. Under the impact of modern systematics and population genetics, a usage is spreading in biology that restricts the term "population" to the *local population,* the community of potentially interbreeding individuals at a given locality. All members of a local population share in a single gene pool, and such a population may be defined also as "a group of individuals so situated that any two of them have equal probability of mating with each other and producing offspring," provided, of course, that they are sexually mature, of opposite sex, and equivalent with respect to sexual selection. The local population is by definition and ideally a panmictic (randomly interbreeding) unit. An actual local population will, of course, always deviate more or less from the stated ideal. A species in time and space is composed of numerous such local populations, each one intercommunicating and intergrading with the others.

In view of the diverse meanings of the word "population," it would be useful to have a technical term for the local population as defined above. The term *deme* was proposed by Gilmour and Gregor (1939); they, however, defined it vaguely, without clearly avoiding the ambiguities of the term *population.* Later, several zoologists, for instance Simpson (1953), gave "deme" a more specific meaning, definitely restricting the term to the local population, the interbreeding community, as defined above. Since there is no other technical term for this evolutionary unit, "deme" has been widely accepted for it, even though the meaning specified here is not the meaning originally proposed for the term.

The individual is only a temporary vessel, holding a small portion of the gene pool for a short time. It may, through mutation, contribute one or two new genes. It may, if it has a particularly viable and productive combination of genes, somewhat increase the frequency of certain genes in the gene pool, yet in sum its contribution will be very small indeed compared to the total contents of the gene pool. It is the entire effective population that is the temporary incarnation and visible manifestation of the gene pool. It is in the population that the genes interact in numerous combinations. Here is the proving ground of new genes and of novel gene combinations. The continued interaction of the genes in a gene pool provides a degree of integration that permits the population to act as a major unit of evolution.

A population has the capacity to change in time. This dynamic aspect of the population is of even greater biological significance than its role at a given moment. Evolution is sometimes defined as "a change in the genetic composition of populations" (Dobzhansky 1951). Such genetic changes manifest themselves in the phenotype in various ways, and the study of these phenotypic changes has been the basis of most evolutionary theories and hypotheses of speciation. A study of the variation of populations is a prerequisite for the understanding of these theories.

Kinds of Variation

From the evolutionary point of view one can distinguish two kinds of biological variation: *group variation*, referring to differences among populations (which is discussed in Chapters 11–13), and *individual variation*, referring to differences among individuals of a single population (which is the main theme of Chapters 7–10). The nature, source, maintenance, and biological functions of this individual variation, and the factors that account for the genetic changes of populations, deserve detailed consideration.

Individual Variation

No two individuals of a sexual species are completely alike. To say that The Robin or The Wolf has such and such characters is a generalization, and not necessarily always a correct one. A treatment of the species as a unit phenomenon was adopted in the early chapters of this book only for its obvious didactic advantages. It is now necessary to correct this oversimplification by showing how variable populations and species are.

Individual variation has long been of practical interest to the descriptive morphologist, and even more so to the taxonomist who wants to find out which of the described "species" are merely variants of previously known ones. A detailed discussion of this aspect of variation can be found in volumes on taxonomy (Mayr 1969; Simpson 1961). What interests the evolutionist is not the mere occurrence of variation, but rather its significance. On the basis of the criterion of heritability, all manifestations of intrapopulation variation can be divided into nongenetic and genetic variation. Broadly speaking, it can be said that *nongenetic variation adapts the individual, while genetic variation adapts the population.* They are alternative strategies of population adaptation, one of them sacrificing the individual, the other not.

Nongenetic Variation

Modifications of the phenotype that do not involve genetic changes have long been considered, as a reaction to Lamarckian concepts, to lack evolutionary significance. This view is not correct. Nongenetic* variation is usually adaptive and controlled by natural selection, since genetic factors determine the amount and the direction of the permissible flexibility of the phenotype. There are numerous ways in which the phenotype may vary in response to different demands in time and space (Table 7.1). For a more detailed treatment consult Mayr (1963).

Age variation. Every naturalist and biologist is familiar with the difference between immatures or larvae and adults. An individual is exposed to selection at every stage of its life cycle. Divergent adaptation of the larval stages of related species to different niches may result in their becoming more different from one another than are the adults. This difference is accentuated where intraspecific competition between adults and immatures is high. Such ontogenetic differences as those between caterpillars and butterflies are of selective advantage for the species when the immatures occupy a different niche from that of the adults. These differences are

*It must not be overlooked that the capacity of a genotype to produce several phenotypes is controlled by genetic factors. "Nongenetic" in the present discussion means simply that the differences as such between the modified phenotypes are not caused by genetic differences.

Table 7.1. Noninherited variation.

1. Individual variation in time
   (a) Age variation
   (b) Seasonal variation of an individual
   (c) Seasonal variation of generations
2. Social variation (insect castes)
3. Ecological variation
   (a) Habitat variation (ecophenotypic)
   (b) Variation induced by temporary environmental conditions
   (c) Host-determined variation[a]
   (d) Density-dependent variation[a]
   (e) Allometric variation[a]
   (f) Neurogenic color variation
4. Traumatic variation
   (a) Parasite induced[a]
   (b) Accidental and teratological variation[a]

[a]For details, see Mayr (1963, 1969).

intensified because each age group must be adapted for the specific role it plays in the life cycle in the performance of special tasks, such as dispersal (Chapter 18), growth, and reproduction.

Seasonal variation. Adult individuals of certain species of animals are subject to seasonal changes of the phenotype. For instance, mammals in the temperate and cold regions may molt in the fall into a winter fur. The snow-white winter dress of certain species of weasels (*Mustela*) and hares (*Lepus*) is well known. A seasonal change of plumage is very frequent in birds. The winter plumage of ptarmigans (*Lagopus*) is white, like the winter fur of ermines. Many other kinds of birds that have a simple eclipse plumage during part of the year develop a bright nuptial plumage before the beginning of the breeding season. A brightening of colors and a change of certain epidermal structures during the breeding season occur in many species of fishes and in some invertebrates. The function of this seasonal variation is to adapt individuals to the changes of the environment through the seasons and to the stages of their life cycles.

Generations. In some organisms with a rapid sequence of generations, seasonal variation involves generations rather than individuals. In well-known cases in insects, summer generations differ from spring generations, or dry-season individuals from those of the rainy season. The most frequent difference is one of coloration, but in some insects a long-winged generation may alternate with a wingless or short-winged one. Such differences can usually be shown not to have a genetic basis, particularly where only two generations per year are involved.

Habitat variation. Everyone is familiar with the difference between two plants, one planted in good, the other in poor soil. The direct effect of the physical environment on the phenotype is rarely as pronounced in animals as in plants. Among animals, it is perhaps most evident in sessile marine invertebrates (such as sponges and corals) and in some mollusks (for example, oysters and some fresh-water bivalves). Several factors of the physical environment—whether such animals grow in quiet waters or in the surf, in clear water or in water rich in plankton or silt, in an environment rich or poor in calcium (lime)—may greatly affect the appearance of an individual. Phenotypes resulting from modification by edaphic (substrate) or other ecological conditions, and not from changes in the genotype, are sometimes called *ecophenotypes* (Fig. 7.1).

Genetic or Nongenetic Variability?

Of phenotypic variability observed in nature, it is never possible to tell, except by careful breeding experiments, what part should be ascribed to nongenetic modification and what part to genetic factors. The identical environmental conditions may give rise to indistinguishable phenotypes that are produced either by alteration of the genotype through natural selection or by nongenetic, somatic modification. The short mammal tail favored by selection in a cool climate may also occur as a developmental response: mice raised at 26.3°C had an average tail length of 93.1 mm, whereas those raised at 6.2°C had tails that averaged only 75.9 mm long. Harrison (1959) reviews the role of the environment in the determination of the phenotype.

The respective contributions of environment and genotype to the final phenotype have been studied particularly well for fishes. Raising fish at cool temperatures leads to an increase in the number of vertebrae; however, populations from cool waters have a genetic tendency toward an increased number of vertebrae (Hubbs 1922; Tåning 1952; Fig. 7.2). The results of these and similar studies are of considerable practical importance, since different "races" or "stocks" of commercially important food fishes are often marked by differences in meristic (countable) characters, such as the number of vertebrae or fin rays.

The Significance of Nongenetic Variation

The ability of the phenotype to be modified by environmental factors has been interpreted in many different ways. Some Lamarckian naturalists

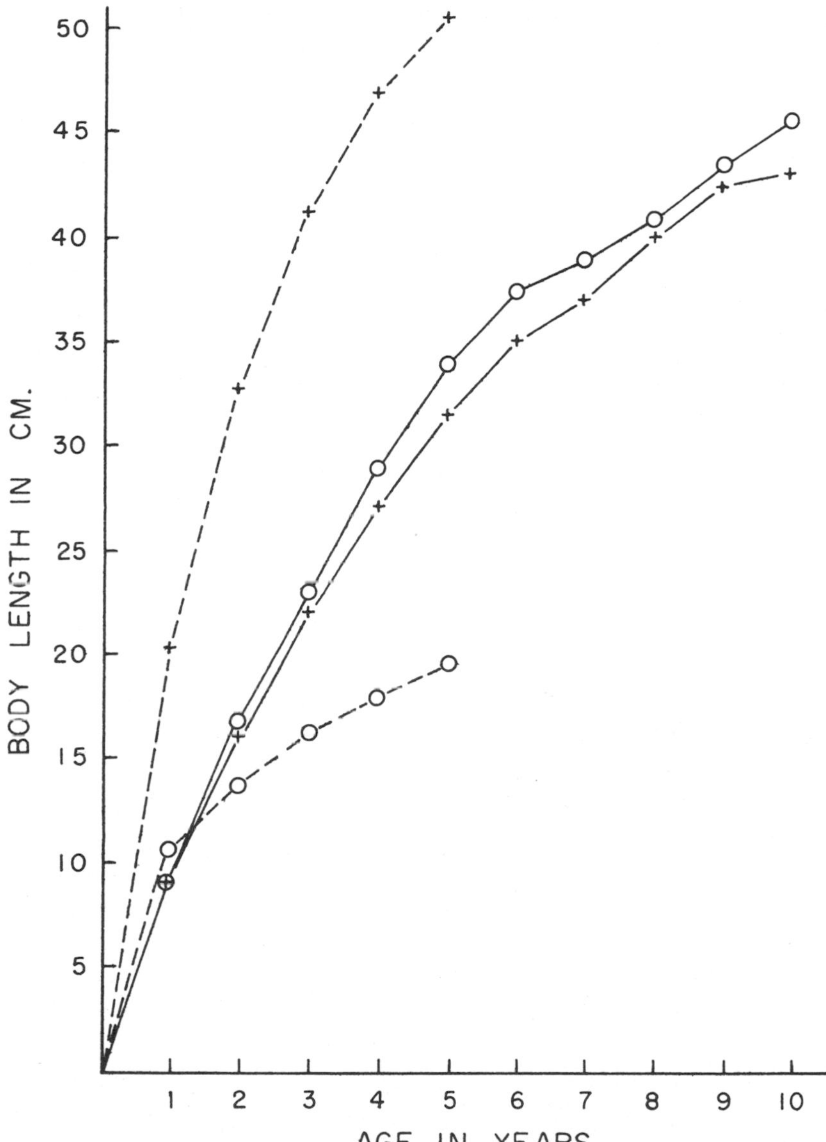

Fig. 7.1. Growth rates of two *Coregonus* species, *storsik* (circles) and *aspsik* (crosses) in different Swedish lakes. In Lake Uddjaur (solid line) the two species have almost identical growth; in Lake Kalarne (broken line) the growth of *aspsik* is accelerated, that of *storsik* retarded. The species are represented in each lake by a different phenotype. (From Svärdson 1950.)

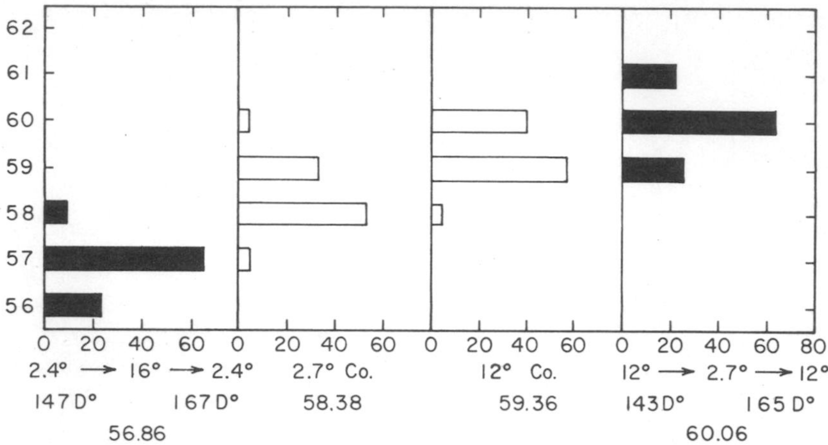

Fig. 7.2. Numbers of vertebrae (ordinate) in offspring of four samples of sea trout (*Salmo trutta*). Experimentals black, controls white. Heat shock (16°) during the supersensitive period at left, cold shock (2.7°) at right; *D*°, day-degrees since fertilization. Untreated controls at constant temperatures of 2.7° or 12°. (From Taning 1952.)

thought mistakenly that such nongenetic modifications could be transformed into genetic changes. This view has never been substantiated and the nature of the molecular basis of inheritance makes such a transformation quite impossible. This subject, which now has only historical interest, is treated in detail by Mayr (1963:145–148).

Genetic Variation

No two individuals in sexually reproducing species are genetically identical, presumably not even most monozygotic twins (owing to somatic mutations). Much of this genetic variation contributes to the variation of the phenotype. Consequently almost any feature of an animal may vary individually, be it a morphological character, a physiological attribute, a cytological structure (such as number, pattern, and form of the chromosomes), or whatever else. Morphological variation may be classified in various ways. It may concern *meristic* characters that can be counted, like numbers of vertebrae or scales, *quantitative* characters that can be measured, like dimensions or weight, or *qualitative* characters, like the presence or absence of spots. This is merely a distinction of convenience; the genetic basis for these classes of characters is the same. Variation can also be classified as to whether it is *continuous* (larger vs. smaller, darker vs. lighter) or *discontinuous* (blue eyes vs. brown eyes, with white spots vs. without white spots). The principal genetic difference between continuous and discontinuous variation is the number of genes affecting the character in question. When only one or a few genes is involved, phenotypic variation will be discontinuous (polymorphism); the more genes are involved (together with nongenetic modifications of the phenotype), the more continuous the variation of the character will appear (Mather 1943). Continuous variation, reflecting an interaction of genes, is dealt with in Chapter 10.

Polymorphism

Discontinuous phenotypes within a species occur commonly in the animal kingdom. They are either evident freaks, such as albinos, or normal components of a species population, such as redheads in the white human population.

The term "polymorphism" always refers to variability *within* a population. A simplified definition is "the occurrence of several strikingly different discontinuous phenotypes within a single interbreeding population." The term "polymorphic" must be strictly distinguished from "polytypic," a term applied to composite taxonomic categories (Chapter 9). A species is polytypic if it is composed of several subspecies, a genus is polytypic if it is composed of several species. Any human population, for instance, is polymorphic, but the human species as a whole is polytypic.

The term "polymorphism" is often used rather sweepingly for any kind of discontinuous variation. Ants and termites are called polymorphic because their castes differ from one another, and the Little Blue Heron (*Florida caerulea*) is called polymorphic because the young are white and the adults blue. There has been a growing tendency to refer to such nongenetic discontinuous variation by a different term (polyphasy or polyphenism, for instance). For the sake of precision, this distinction is here preserved, and "polymorphism" is used only to designate discontinuous genetic variation.

Polymorph variants ("morphs") are sometimes so strikingly different from the "normal" type of the population that they have been mistakenly described as separate species (Chapter 3). This they are not, nor are these intrapopulation variants subspecies or races. It must be emphasized strongly that a phenotype within a population does not constitute a race. The silver foxes, the black hamsters, the blondes of a local European population, and the Rh-negative individuals are not races. One must remember that polymorphism is an intrapopulation phenomenon, while race is an interpopulation phenomenon. The study of polymorphism is a convenient approach to an understanding of the genetics of populations. The genes involved in polymorphism have, in general, conspicuous discontinuous effects, and different genotypes (except some heterozygotes) can be distinguished phenotypically. Such genes are therefore much easier to analyze than the vast majority of the genes in a population that are more or less cryptic (not easily distinguished).

Polymorphism is exceedingly widespread. It has been recorded in virtually every class of animals from protozoans to vertebrates. In birds more than 100 cases are known in which a morph was originally described as a separate species. It is even more widespread in certain families of insects, crustaceans, and mollusks. For summaries of the literature see Mayr (1963:152) and Ford (1964, 1965).

Polymorph characters. The suffix "-morphism" suggests a limitation of the phenomenon to structural characters. Actually the term polymorphism covers any phenotypic character, be it morphological, physiological, or behavioral, provided it is genetically controlled and more or less discontinuous in its phenotypic expression. Polymorphism in color, being so conspicuous a character, is most often described, but presence or absence of certain tooth structures in mammals, wing veins or entire wings in insects, winding in snails (dextral vs. sinistral), and asymmetry in flatfishes are other well-known cases of polymorphism. A very large and increasingly important group of characters are the blood-group genes in vertebrates, demonstrated to be polymorphic not only in man, but also in other mammals and in birds. A similar polymorphism in wild populations is shown by a very large number of enzymes and other proteins.

The genetic basis. Polymorphism results from the simultaneous occurrence in a population of several genetic factors (alleles or gene arrangements) with discontinuous phenotypic effects. Very often there are merely two alternative types ("dimorphism"), such as male and female phenotypes, or white and colored *Colias* butterflies. In other cases more than two morphs occur, sometimes as many as a dozen, a score, or more.

The genetic analysis has been completed only in relatively few cases. In a number of cases, such as spotting genes in coccinellid beetles, pattern genes in grasshoppers, and blood groups in cattle, polymorphism is controlled by a large series of multiple alleles (or by pseudoalleles at closely linked loci). The reason for such large series of alleles is not at all clear. Reduction of the frequency of homozygotes is one possible selective advantage.

The most frequent phenotype in a polymorph population is by no means necessarily caused by the "dominant" gene. Genetic dominance does not equal numerical predominance. Indeed, recessives outnumber at many localities the dominant alleles in the banded snail *Cepaea nemoralis* and in other species. Why the most common allele in a series of multiple alleles is so often the universal recessive is not at all clear. A selective advantage of numerical equivalence of various phenotypes may be partly responsible. It must also be remembered that morph genes are selected in most cases for their physiological effect, and that a gene that is recessive for its contribution to the visible components of the phenotype may be dominant with respect to its physiological phenotype.

Cytological polymorphs. Various forms of cytological polymorphism in animals have been treated by White (1954) in masterly fashion. Polymorphism in chromosome number is most often produced by a (sub)terminal fusion of two rod-shaped (acrocentric) chromosomes into a single metacentric (V-shaped) chromosome (see also p. 313).

Various changes in the structure of chromosomes constitute a special class of polymorph characters. This includes supernumeraries, fusions, translocations, and inversions. Genotype and phenotype coincide in this type of polymorphism; indeed, the genetic material itself is polymorphic in its arrangement and this revelation of the genotype by the phenotype permits a complete genetic analysis. A particularly informative type of chromosomal polymorphism is that caused by the inversion of chromosome sections. In view of the outstanding interest of this phenomenon, a few words must be said about it (for a full discussion refer to genetic and cytological textbooks, Dobzhansky 1951, White 1954, and Grant 1964, and to the current literature). Genes are arranged along the chromosomes in linear sequence. If a chromosome breaks in two places and the middle piece is turned around ("inverted"), the gene sequence *ABCDEF* may be changed to *AEDCBF* (Fig. 7.3). This is called an *inversion*. In *Drosophila* and some

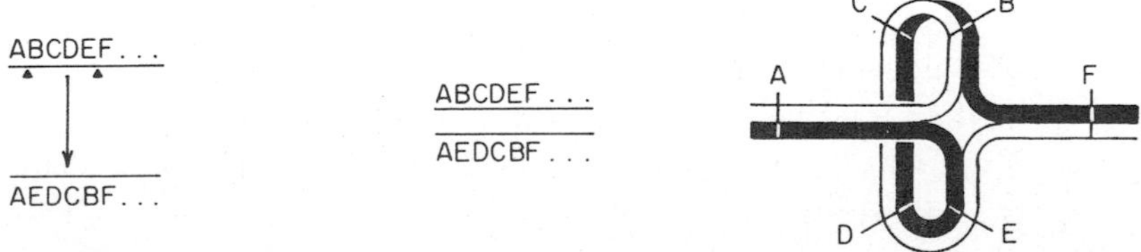

Fig. 7.3. Change of gene sequence by single inversion of the chromosome segment *BCDE*. The looping permits the pairing of homologous loci in the original and the inverted chromosomes. (From Dobzhansky 1951.)

other genera of dipterans, such as *Chironomus*, *Simulium*, and *Anopheles*, the giant salivary-gland chromosomes permit the direct study of gene sequences. Different parts of these chromosomes stain differently and give highly individualized patterns of light and dark bands, which permit identification of the various gene arrangements. Dubinin, Sturtevant, and particularly Dobzhansky, Patterson, and their associates have shown that several alternative gene arrangements may coexist in a population, and that such polymorphism of gene arrangements is very frequent in natural populations of most species of *Drosophila*. The third chromosome of *D. pseudoobscura*, for instance, has at least 16 different known gene arrangements, and *D. willistoni* has more than 50. The experimental study of gene arrangements in population cages has contributed greatly to our understanding of polymorphism and of the variability of populations in general.

Homologous polymorph series. Certain types of polymorphism are characteristic of whole genera or families of animals. Polymorphism in banding is exceedingly widespread among gastropod mollusks (snails), spotting polymorphism is widespread in coccinellid beetles, albinism in pierid butterflies, blood groups in mammals and birds, chromosomal inversions in *Drosophila* and other dipterans, and supernumerary chromosomes in grasshoppers, to mention merely a few conspicuous examples. The wide distribution of a single type of polymorphism in entire families and orders suggests not only that it has an important selective significance, but also that it is of considerable phylogenetic antiquity. Presumably at some time in the history of the phyletic branch a particularly valuable heterotic mechanism became linked with phenotypic polymorphism and was subsequently maintained in the entire group.

The Genetic Basis of Characters

The study of polymorphism has brought us much nearer to an understanding of the relation between gene and character, between genotype and phenotype. It used to be said that much of polymorphism is evolutionarily irrelevant because no selective difference was discernible in the visible differences between the morphs. Of what possible selective advantage could it be for a yellow snail, it was said, to have three black bands instead of five (or vice versa), or for a ladybird beetle (coccinellid) to have two spots on the elytra instead of three (or vice versa)? Such reasoning was based on the obsolete "one-gene–one-character" hypothesis of early Mendelism, which assumed that a gene does nothing but produce a single character and that the fate of a gene in a population depends on the "value" of this character. This view overlooked the fact that the relation between genes and the components of the phenotype is diverse and highly complex. This relation is discussed in depth in Chapter 10; here I shall mention only some aspects that are responsible for the maintenance of polymorphism in populations.

Multiple gene effects (*pleiotropy*). The capacity of a gene to affect several different aspects of the phenotype is called *pleiotropy* or *polypheny*. Every gene that has been studied intensively has been found to be pleiotropic to a greater or lesser extent. It is obviously naive to regard a gene as a mold into which a character is poured. There is no such mold for the antlers of a stag or the coloration of a bird of paradise. The diversity of manifestations of a gene is often surprising. For instance, nearly every known "coat-color" gene of the house mouse seems to have some effect on body size. Of 17 X-ray–induced eye-color mutations in *Drosophila melanogaster*, 14 showed definite effects on such an apparently altogether unrelated character as the shape of the sclerotic spermatheca of the female.

More important is the fact that pleiotropic genes often affect the very characters that are of the greatest evolutionary importance, such as fertility, fecundity, sexual vigor, longevity, and tolerance of environmental extremes. In well-studied organisms such as *Drosophila* and *Ephestia* it has been found that nearly every known mutation affects one or another component of fitness, usually several simultaneously.

A gene (*a*) affecting the pigmentation of the eye in the flour moth *Ephestia kuehniella* has ten other known morphological and physiological effects (Caspari 1949). Further analysis showed that the basic effect of this gene is its inability to synthesize kynurenine, a precursor of tryptophan. The pleiotropic action of this gene appears to be due to its effects on every biochemical process in the body in which the kynurenine-tryptophan reaction plays a role. This illustration may be used as a general model of the action of genes, although almost any level of differentiation may be involved.

The human blood-group genes have in the past been held up as an exemplary case of "neutral genes," that is, genes of no selective significance. This assumption has now been thoroughly disproved for the ABO blood groups. There is an excess of O in patients with duodenal ulcers (about 17 percent) or gastric ulcers (about 10 percent); there is an excess of A in patients with pernicious anemia (about 13 percent), diabetes (8 percent), and several other diseases. Far more important are the incompatibility reactions between A fetus and B or O mother, B fetus and A or O mother, and

AB fetus and A or B mother, which lead to a continuous loss of the genes for A and B. For the level of these genes to be maintained superiority of the heterozygotes must be postulated and Brues (1963), making appropriate assumptions, has shown with the help of computer simulation that the following selection coefficients would maintain observed frequencies: OO = 0.79, AA = 0.74, BB = 0.66, OA = 0.89, OB = 0.86, and AB = 1.00. Errors of sampling as well as the geographic variation of systematic selection pressures would account for the observed differences between human populations. The many observed effects of blood-group genes make them a particularly apt illustration of pleiotropic genes.

Neutral polymorphism. This term has been used to designate cases of polymorphism, such as the degree of banding in snails or of spotting in coccinellid beetles, in which there is no apparent selective difference between the alternative visible phenotypes. Such neutral polymorphism, it was once claimed, was maintained by "accident." Now that the cryptic physiological effects of "neutral" genes have been discovered, it is evident that such genes are anything but selectively neutral. It is altogether unlikely that two genes would have identical selective values under all the conditions in which they may coexist in a population. The strong geographic variation of most cases of polymorphism (Chapter 11), often closely paralleling climatic gradients, is further evidence of the existence of correlated physiological effects of polymorphic genes. The explanation of balanced polymorphism (Chapter 9) is likewise based on cryptic physiological differences of genotypes. Cases of neutral polymorphism do not exist, as has long been maintained by Ford (1945).

Pleiotropic gene action is the key to the solution of many other puzzling phenomena. The survival of a gene in a gene pool depends on its total contribution to "fitness" and not on the contribution to fitness made by the visible phenotype. Color, pattern, or some structural detail may be merely an incidental by-product of a gene maintained in the gene pool for other physiological properties. The curious evolutionary success of seemingly insignificant characters now appears in a new light.

Sources and Maintenance of Intrapopulation Variation

Every local population is adapted, through natural selection, to the specific environment in which it lives. It is sometimes concluded from this fact that a particular genotype will have optimum survival value at a given locality and that therefore natural selection should favor individuals with this genotype until every population is genetically uniform for such individuals. This conclusion, no matter how logical it seems to be, conflicts with the observed variation of populations (see p. 155), and disregards as well the obvious advantage to a population of storing genetic variability to serve as material for evolutionary responses to changing conditions. This contradiction worried Darwin and the early Darwinians considerably. Since they considered inheritance to be blending, they had to assume that, owing to blending, one half of the total variability would be lost in each generation. And yet, at the same time, the availability of an inexhaustible supply of genetic variation was one of the cornerstones of Darwin's theory of evolution. These two assumptions were completely in opposition to each other. There seemed to be only one answer to this conundrum, a truly colossal rate of mutation. Furthermore, in order to maintain adaptedness at all times in spite of this avalanche of mutations, genetic changes had to be appropriate. This, in turn, seemed possible only if the genetic changes were induced as an appropriate answer to a particular set of environmental demands, that is, if they were adaptive. Darwinian considerations of natural selection on the basis of the genetic premises of blending inheritance thus led "logically" to a Lamarckian interpretation of the induction of genetic changes. It is not surprising then that Darwin himself, as well as some of the best-informed students of evolution around the turn of the century, accepted certain Lamarckian assumptions.

It is one of the many achievements of the science of genetics to have shown that several of the major premises of this "logical" chain of arguments are wrong (Weismann had already undermined others). It is not true that a uniform genotype is the highest pinnacle of possible adaptation. Rather, a specific amount and very definite type of genetic variation may actually enhance the adaptedness and adaptability of a population.

The advantages of a store of genetic variation are evident: the greater the number of genetic types within a population, the greater the probability that the population will include genotypes that will survive seasonal and other temporal changes, particularly those of a violent nature. If there are drought-resistant genotypes in a population that normally lives in a humid environment, the population will have a chance to survive an abnormal period of drought during which the humidity-dependent genotypes are lost. Genetic variability will also permit a greater utilization of the environment because it will make it possible to colonize marginal habitats and various subniches. It will counteract too great a specialization and will give elasticity.

Yet too much genetic variability will inevitably result in the wasteful production of many locally inferior genotypes. Inviable hybrids illustrate this risk well. Extreme genetic variability is as undesirable as extreme genetic uniformity. How then does the population avoid either extreme? As so often in the realm of evolution, the solution is a dynamic equilibrium between two opposing forces. Since adaptation is maintained through natural selection, which always involves an elimination of genotypes, the most important question is: how can a population retain an adequate supply of genetic variability when this is continuously depleted by natural selection? This question can be answered only after a careful analysis of all the factors influencing the amount of genetic variation in a population.

A glance at Table 7.2 shows that one can group these factors essentially into three classes. The first class (*A*) consists of the mechanisms that produce new variation directly (mutation) or indirectly (recombination, gene flow) and that make it possible to maintain the integrity of these factors from generation to generation (particulate inheritance). The second class (*B*) consists of the forces that tend to reduce the amount of genetic variation either by natural selection or by chance elimination (Chapter 8). The third class (*C*) consists of the numerous phenomena and devices that protect genetic variation from the eroding effects of natural selection and of chance elimination (Chapter 9). The relative importance of these factors and their interplay deserve detailed discussion.

Sources of Genetic Variation

Integrity of Genetic Factors: Particulate Inheritance

The naive view of inheritance assumes that the genetic potencies of father and mother merge in reproduction and appear in the offspring as a new blend of the paternal and maternal characters, analogous to the mixing of two colored fluids. Virtually all genetic theories of the nineteenth century, with the spectacular exception of that devised by Mendel, were based on this concept of blending inheritance. Curiously, this is true even of those genetic theories that postulated particles as the carriers of inheritance. It was Mendel's great contribution to realize from his experiments that the genetic factors of father and mother, when they combine in a zygote, do not lose their identity, but reassort in the next generation. The full story of the Mendelian theory of particulate inheritance is so well told in various textbooks of genetics that I shall not repeat it here.

The hereditary material remains unchanged from generation to genera-

Table 7.2. Factors influencing the amount of genetic variation in a population.

A. Sources of genetic variation (Chapter 7)
   1. Integrity of genetic factors: particulate inheritance (Hardy-Weinberg)
   2. Occurrence of new genetic factors
      (a) Mutation
      (b) Gene flow from other populations
   3. Occurrence of new genotypes through recombination
B. Factors eroding variation (Chapter 8)
   1. Natural selection
   2. Chance and accident
C. Protection of genetic variation against elimination by selection (see Table 9.1)
   1. Cytophysiological devices
   2. Ecological factors

tion, unless altered by mutation. A white-eyed *Drosophila melanogaster* may be backcrossed with heterozygous red-eyed flies for hundreds of generations without either the white becoming pinkish or the red becoming diluted by white. This fact, now accepted in genetics as axiomatic, has the most profound evolutionary consequences. Particulate inheritance insures that the variability of a (large) population remains the same from generation to generation under conditions of random mating if the genes are of equal selective value, an enormous contrast to the theory of blending inheritance.

This basic law of particulate inheritance can be expressed in simple mathematical terms. Let us begin with the hypothetical case of two alleles A and a, of equal selective value, occurring together in a population. If the frequency of gene A is q, then that of a is $1 - q$, if A and a are the only alleles at this locus. If we assume random mating, then male gametes $qA + (1 - q)a$ fertilize eggs $qA + (1 - q)a$, producing offspring (zygotes) $[qA + (1 - q)a] \times [qA + (1 - q)a] = q^2AA + 2q(1 - q)Aa + (1 - q)^2aa$ according to the well-known binomial law. This formula gives the total frequency of zygotes (genotypes) in the population. How to calculate the frequency of the genes from the observed frequency of the genotypes is explained in the textbooks of genetics.

The law expressed in this formula is called the Hardy-Weinberg law, stating in mathematical terms the fact that owing to particulate inheritance *the frequency of genes in a population remains constant in the absence of selection, of nonrandom mating, and of accidents of sampling.*

The full understanding of the meaning of particulate inheritance had to await another major conceptual advance. The students of development had always assumed that everything in the fertilized egg (including the genetic material) participated directly in development. Weismann was the first to suspect that the "germ line" stayed outside the developmental process, but it was not until after 1944 that molecular genetics provided the real answer: the deoxyribonucleic acid (DNA) of the genetic material is merely a blueprint, so to speak, which gives instructions for the guidance of the ontogeny. The genetic material itself remains untouched during this process. Many of the potentialities of this blueprint are never activated, owing to recessiveness, epistatic suppression, and other processes described below. They remain available for the needs of future generations.

Particulate inheritance thus explains how genetic factors may remain indefinitely in a population. What the original source of these genetic variants is, is the next question that needs to be answered.

Occurrence of New Genetic Factors

Mutation. Evolution means change. To be more precise, it means the consecutive replacement of one genetic factor by another. The ultimate source of genetic novelties was the great unsolved problem in Darwin's theory of evolution. It has become customary in genetics to refer to the origin of a new genetic factor as a *mutation,* which one defined, with convenient vagueness, as "a discontinuous change with a genetic effect." The recent advances in molecular genetics now permit a far more precise definition.

The application of the term mutation in evolutionary biology has had an unhappy history (Mayr 1963:168). Most early Mendelians, particularly De Vries and Bateson, were essentialists. For them a mutation was a drastic reorganization of the type, capable of making a new species in a single step. They also applied the term indiscriminately to changes of genotype and phenotype. The naturalists could not accept this interpretation since evolution obviously proceeded through the gradual accumulation of small genetic changes.

The full understanding of the nature of mutation did not come until the chemistry of the genetic material was elucidated. Any change in the DNA molecule is de facto a mutation; when localized it is a "point" or "gene" mutation. Any change in the structure of the chromosomes is a chromosomal mutation.

The nature of the gene. The classical concept of the gene was that of a corpuscle on the chromosome with three characteristics: a specific function, a capacity for mutating, and being the smallest unit of recombination. Studies of *Drosophila* and maize showed, however, that the units of function, mutation, and recombination did not necessarily coincide (Stephens et al. 1955). Work on microorganisms further extended these findings. Why the three identifying characteristics of the gene do not coincide became clear when the chemistry of the gene was elucidated.

We owe to Avery, MacLeod, and McCarty (1944) the fundamental discovery that nucleic acid rather than protein (as formerly believed) is the genetic material. A double-stranded coil of DNA is, in most organisms (including all animals), the carrier of the genetic information (Watson and Crick 1953). A definite sequence of two kinds of nucleotide pairs (adenine-thymine, guanine-cytosine) permits the formation of a linear code. This is translated into a corresponding messenger, RNA (ribonucleic acid), which has the same base pairs except that thymine is replaced by the closely related pyrimidine uracil. The four bases A, U, G, C can form 64 different triplets (*codons*) and these are the letters of the genetic code. Each triplet, let us say UCU or AGA, translates into one of the 20 common amino acids. The discrepancy between 64 letters (triplets) and only 20 amino acids is caused by redundancy (several different letters may code for the same amino acid) and by the fact that some codons have different functions. A gene, to put it in very simple terms, consists of a series of letters (codons) that control the putting together of a polypeptide chain. For further detail see a modern textbook on molecular genetics (for instance Watson 1965).

It is now evident that the classical concept of the gene as an entity that serves simultaneously as unit of mutation, unit of recombination, and unit of function is not correct. A change in even a single nucleotide pair constitutes a mutation, yet a functional unit (*cistron*) may consist of a thousand nucleotide pairs. And as far as the unit of recombination is concerned, it seems that, at least in microorganisms, crossing over may occur between alleles.

The new insight into the chemical nature of the genetic material permits a more meaningful discussion of gene mutation. A gene mutation can be considered an error in the replication of genetic material prior to chromosomal duplication. Such an error is usually confined to a single nucleotide pair. Not all gene mutations are errors of replication; some may be failures of replication, resulting in small deficiencies in the chromosomes.

Curiously, the improved understanding of the nature of genes and mutation has so far not added much to the understanding of evolutionary phenomena. For most problems of evolutionary biology, particularly in higher organisms, it is legitimate to continue using the classical terminology of genes (alleles) and loci.

Mutability. The determination of the mutability of a given locus or of a species is very difficult. Mutability seems to be different at different loci, some loci appearing to be very stable. The subject is becoming more complex the longer it is studied. It is possible that the AT base pair with two hydrogen bonds mutates more easily than the CG with three bonds, but what effect this would have on a large gene is uncertain. One would expect that a gene would mutate more frequently the more base pairs it has, but again not much information is available. Estimates of an average mutation rate of 1 in 50,000 per gene per individual per generation are near the higher limit, while others range as low as 1 in 1,000,000 or less. Considering all these uncertainties it would seem premature to speculate whether and how selection adjusts the rate of mutation at individual loci.

Even if we take the lowest estimates we come to the conclusion that a species consisting of several million individuals is bound to have a couple of mutations per locus in every generation except at the most inert loci. However, mutations producing drastic changes of the phenotype are rare and any tendency to produce such mutations would certainly be selected against very strongly. The mutations that do occur on the thousands of mutable loci usually have very slight effects, which can be discovered only by special methods. This conclusion differs radically from the opinion of early authors, who considered mutation a dramatic but rare phenomenon. There is now abundant evidence, derived from the study of many different kinds of organisms (*Nicotiana, Drosophila,* microorganisms), that such comparatively high rates of phenotypically slight mutations actually do occur in most organisms.

Estimates vary as to the number of loci in a higher organism such as *Drosophila* or man. Calculations based on crossing-over tests suggest a minimum of 5,000 to 10,000 loci; calculations based on the amount of DNA per nucleus suggest the possibility of several millions of cistrons. Taking these estimates in conjunction with the data on mutation rates, one comes to the conclusion that every individual presumably differs from every other individual on the average by at least one newly mutated site. This estimate, even if ten times too high, indicates that mutation provides an enormous and steadily recurring source of genetic variation.

The fact that estimates as different as 10,000 loci or 5 million loci for *Drosophila* or the mouse can be equally well defended indicates how far we still are from understanding the genetic structure of the chromosomes of higher organisms. Evidence is now beginning to accumulate that not all DNA of the nucleus has the same function (to act as different genes) (Britten and Kohne 1968; Walker 1968). Future findings in this field may help to interpret certain observations of the evolutionist that up to now have remained unexplained.

The effect of mutation. It is frequently said that the deleterious nature of virtually all phenotypically conspicuous mutations is evidence that mutations are of little significance in evolution. This view is essentially correct as far as conspicuous mutations are concerned, for two reasons. The first is that the number of possible changes at each gene locus is limited, and it is therefore probable that a mutation giving the greatest fitness in the particular physical, biotic, and genetic environment has already occurred previously and is now the prevailing allele. A second reason is that any mutation so drastic as to short-circuit the normal developmental feedbacks and greatly affect the visible phenotype is likely to be selected against. It can hardly be questioned that most visible mutations are deleterious.

On the other hand, the genetic literature, and particularly the experiences of animal and plant breeders, provide abundant evidence of the occurrence of beneficial mutations. More important, the typological approach of classifying mutations into good and bad ones is misleading. New mutations invariably are first tested in heterozygous condition and might add to fitness even if they were deleterious as homozygotes.

The selective value of a gene, its "goodness," is determined by a complex constellation of factors in the external and internal environment. Some mutations lead to a breakdown of important metabolic processes and so are always deleterious. Others are deleterious under certain conditions but beneficial under others. In *Drosophila,* for instance, many genes are superior to the normal allele at high temperatures but inferior at low temperatures. A gene may add to viability on one genetic background but have the characteristics of a lethal on another genetic background. Such considerations (discussed in more detail in Chapter 10) make evident why mutation and its effects on fitness are largely unpredictable.

Mutation as an evolutionary force. In the early days of genetics it was believed that evolutionary trends are directed by mutation, or, as Dobzhansky (1959) recently phrased this earlier view, "that evolution is due to occasional lucky mutants which happen to be useful rather than harmful." In contrast, it is held by contemporary geneticists that mutation pressure as such is of small immediate evolutionary consequence in sexual organisms, in view of the relatively far greater contribution of recombination and gene flow to the production of new genotypes and of the overwhelming role of selection in determining the change in the genetic composition of populations from generation to generation (see also Chapter 8, 9, and 17). Yet it must not be forgotten that mutation is the ultimate source of all genetic variation found in natural populations and the only raw material available for natural selection to work on.

Randomness of mutations. To sharpen the contrast with Lamarckian ideas of the environmental induction of evolutionary changes, evolutionists stress the "randomness" of mutations. Since this term has often been misunderstood, it must be emphasized that it merely means (*a*) that the locus of the next mutation cannot be predicted and (*b*) that there is no known correlation between a particular set of environmental conditions and the particular allele among many potentially possible ones to which a gene will mutate. It does not question that the probability of mutation is much higher at some loci than at others and that the number of possible mutations at any given locus is severely limited by the other mutational sites of the cistron and indeed by the total epigenotype. The unity of the genotype places well-defined limits on the potential for variation (Chapter 10).

Gene flow. All populations studied in the genetic laboratory are closed populations. There is no input and inbreeding is high. Whenever a genetic novelty occurs in such a population it is due to a new mutation. The average wild population is an open population and thus a totally different system. Among the individuals composing a deme, many are immigrants from the outside. Depending on the species, as many as 30 to 50 percent of the members of a deme may be such newcomers in every generation. What we do not know is how different genetically these individuals are from the native population. One must assume that most of the new arrivals come from adjacent populations and differ only slightly in their genetic content.

The really crucial question for the evolutionist is what percentage of the genetic novelties in a local population is contributed by new mutation and what percentage by immigration. No precise information is available, but one may guess that immigration contributes at least 90 percent, if not more than 99 percent, of the "new" genes in every local population. Several lines of evidence support such high estimates of the magnitude of gene flow. One is the phenotypic uniformity of contiguous populations, often over very wide areas. Such uniformity indicates (Mayr 1954) that there is a high premium on stabilizing genes to counteract the disturbing effect of alien genes. A second indication is provided by the observation that dispersal curves are usually strongly skewed. Dispersal is not a purely passive phenomenon, like the scattering of ashes by a volcano, but one in which the individuals concerned participate actively, various opposing tendencies balancing each other (see p. 336).

Gene flow, that is, the exchange of genes between neighboring populations, is a very important but rather neglected evolutionary factor. Its importance for the cohesion of the genotype is discussed in Chapter 10 and its role in speciation in Chapter 17. Gene flow is essentially a retarding element in evolution. It has far-reaching effects on geographic variation, ecotypic adaptation, speciation, and long-range evolution.

The Occurrence of New Genotypes through Recombination

It is not the "naked gene" that is exposed to natural selection, but the phenotype, the manifestation of the entire genotype. It is the individual as a whole that is the target of selection, and what the student of evolution must consider is the total genotype rather than single genes. Genotypes are combinations of genes and the number of possible combinations of even a small number of genes is staggering. A species (or population) with 1000 loci, each with 4 alleles, could produce 4^{1000} gametic types, resulting in 10^{1000} diploid genotypes. Even a single pair of parents, differing from each other, let us say, in 100 loci, could in the course of time give rise to 3^{100} genetically different descendants. Since a given gene on different genetic backgrounds is likely to have a wide range of contributions to viability, potentially ranging from lethal to highly fit, genotypic variation is of crucial importance in evolution. *Recombination is by far the most important source of genetic variation,* that is, of material for natural selection.

The production of different genotypes can be vastly increased by the simple device of permitting exchange of genetic material among sexual individuals. This is the biological significance of sexual reproduction (Chapter 14). Genuine sexual reproduction always involves the fusion of two haploid gametes resulting in the production of a diploid zygote. At some time between fertilization and gamete formation, *meiosis* takes place, a sequence of two cell divisions during which the chromosome number is halved. In the higher animals this reduction of chromosome number immediately precedes gamete formation. In many lower organisms it immediately follows fertilization. The complex and not yet fully understood details of meiosis are described in the cytological literature. For our purposes it is sufficient to point out that the paternal and maternal chromosomes tend to pair with each other during one stage of meiosis, break in several places, and exchange pieces. This process is called *crossing over.* The chromosomes of the resulting gametes thus are a recombination of the homologous chromosomes of the parents. The amount of genetic variability released by such crossing over is enormous. Through recombination a population can generate ample genotypic variability for many generations without any new genetic input (by mutation or gene flow) whatsoever.

There seems to be an inverse correlation between the number of generations per unit of time and the relative importance of recombination (as compared to mutation) as a source of new genotypes. In an organism with a slow sequence of generations, a new mutation can be tested on a different genetic background only at intervals of many years. In man, for instance, a new mutation may be tested once every 25 years; in a large tree in a mature forest, a new mutation may be tested only once every 100–200 years. Mutation pressure as a determinant of evolutionary change is of negligible importance in such organisms. Such types rely for evolutionary plasticity on the storage of genetic variation and the production of a great diversity of genotypes by sexual recombination.

At the other extreme are microorganisms. They owe their evolutionary plasticity largely to an ability to keep up with their changing environment by mutation. A bacterium that divides every 20 minutes would have given rise, theoretically, to 2^{72} ($\approx 10^{22}$) individuals by the end of the day. A favorable mutation, which in these haploid organisms affects the phenotype almost immediately, will spread with extreme rapidity. Considering the enormous population size in microorganisms and the high number of generations per unit of time, even a low mutation rate can provide an amount of variability that might offer all the needed material for selection in a slowly changing environment.

It is possible that mutation may in part take over in microorganisms the function performed by recombination in higher organisms. I make this statement in full awareness of the fact that various forms of recombination (haploid sexuality, transduction, transformation, heterokaryotic fusion) are far more frequent in these microorganisms than is generally acknowledged. Indeed, there are obvious limits to the adaptation by mutation. Yet the rapidity of the genetic shift, compared with the changes of the environment, and the shortness of the pathway from gene to phenotype permit microorganisms to do a great deal of adapting merely by mutating. Like Dougherty (1955) and Stebbins (1960), I believe that most currently existing forms of asexuality are secondary. Certain types of organisms would not have relinquished sexuality to such a large extent if it had not been advantageous to do so, and thus favored by natural selection.

Summary. Phenotypes are produced by genotypes interacting with the environment, and genotypes are the result of the recombination of genes found in the gene pool of a local population. This is why the Mendelian population is such a crucial link in the evolutionary chain. The study of variation is the study of populations.

Most of the genotypic variation found in a population is due to gene flow and recombination. All of it, however, ultimately originated through mutation. The diversifying power of these factors is so great that no two individuals in a sexually reproducing population are identical.

8 · Factors Reducing the Genetic Variation of Populations

If mutation, gene flow, and recombination were to operate unchecked, the genetic diversity of populations would soon be far greater than it is in nature. Actually, a continued increase in the genetic diversity of populations is prevented by factors that, generation after generation, erode part of the accumulated genetic diversity. These eroding factors can be grouped under two headings, natural selection and accidents of sampling (p. 120). Aspects of natural selection that serve to increase genetic variation are discussed in Chapters 9 and 10.

Natural Selection

The elimination of deleterious genetic variability, which is one of the results of natural selection, is a conservative process. This stabilizing power of selection, well known to Blyth and other pre-Darwinians, was long neglected in the post-Darwinian period, when biologists concentrated on the direction-giving role of natural selection. The few authors who were aware of the importance of normalizing selection, like McAtee (1937: "survival of the ordinary"), thought that this property of selection disqualified it as an evolutionary agent. These two aspects of selection cannot be neatly separated from each other, and both will be considered in our discussion of the influence of selection on the genetic variation of populations.

The concept of natural selection was the cornerstone of Darwin's theory of evolution, and anyone speaking today of Darwinism or Neo-Darwinism has in mind a theory of evolution in which natural selection plays a decisive role. Although some authors before Darwin formulated the concept of differences in the probability of survival and reproductive success among differently endowed individuals, it was unquestionably

Darwin who gave the concept general recognition. Its subsequent adoption was closely, but inversely, correlated with the prevalence of typological thinking. History shows that the typologist does not and cannot have any appreciation of natural selection. The more widely statistical or population thinking spread in biology, the more the tremendous significance of natural selection was appreciated.

What is meant by "natural selection"? Darwin had a perfectly clear concept of it. He emphasized again and again that all individuals of a population differ from one another in countless ways and that the nature of these differences has a decisive influence on the evolutionary potential of their bearers. An individual that may "vary however slightly in any manner profitable to itself under the complex and sometimes varying conditions of life, will have a better chance of surviving, and thus be naturally selected." Unfortunately, Darwin sometimes also used Spencer's slogan, "survival of the fittest," and has therefore been accused of tautological (circular) reasoning: "What will survive? The fittest. What are the fittest? Those that survive." To say that this is the essence of natural selection is nonsense! This is not at all Darwin's reasoning. For him, the probability of reproductive success of an individual is determined by its genetic constitution. At a given time in a given environment each genotype has a different fitness, that is, a different probability of reproductive success. The word fitness simply designates the fact that a superior genotype has a greater probability of leaving offspring than an inferior one. *Natural selection,* simply, *is the differential perpetuation of genotypes.* Most of the objections raised against natural selection and its role in evolution become invalid and irrelevant as soon as the typological formulation of natural selection is replaced by one based on the probability of reproductive success of an individual as a consequence of its genetic properties.

Natural selection operates through any factor that contributes to differential reproduction. These factors include all prereproductive mortality and also aspects of differential reproduction that are independent of mortality. As Darwin said so rightly, what pays off is "success in leaving progeny" (1859:62).

Natural selection is a statistical phenomenon; it means merely that the better genotype has "a better chance of surviving" (Darwin). A light-colored individual in a species of moth with industrial melanism may survive in a sooty area and reproduce, but its chances of doing so are far less than those of a blackish, cryptically colored individual. It happens not infrequently in nature that, for one reason or another, a superior individual fails to reproduce while an inferior one does so abundantly. But the greater the difference in viability of the two genotypes, the smaller is the statistical probability of such an unlikely event. Natural selection, being a statistical phenomenon, is not deterministic; its effects are not rigorously predictable, particularly in a changeable environment.

It is necessary to realize that natural selection favors (or discriminates against) genes or genotypes only indirectly through the phenotypes (individuals) that they produce. Where genotypic differences do not express themselves in the phenotype (for instance, in the case of concealed recessives), such differences are inaccessible to selection and consequently irrelevant. Most of the phenotypic variation on which natural selection works (in sexual species) is the result of recombination and not of new mutations. The fact that fitness is determined by the phenotype is the reason for the extraordinary evolutionary importance of the developmental processes that shape the phenotype, as is so eloquently emphasized by Waddington (1957). Any improvements in the "epigenotype," for instance any genes that buffer development better against fluctuations of the environment or metabolic errors, will contribute to fitness.

The modern attitude toward natural selection has two roots. One is mathematical analysis (R. A. Fisher, J. B. S. Haldane, Sewall Wright, and others), demonstrating conclusively that even very minute selective advantages eventually lead to an accumulation in the population of the genes responsible for these advantages. The other root is the overwhelming mass of material gathered by naturalists on the effect of the environment. This evidence was given a largely Lamarckian interpretation in the days when mutations were claimed to be saltational and cataclysmic. When small mutations were discovered, and when it was realized that all variation had ultimately a mutational origin, this evidence became a powerful source of documentation for the selectionist viewpoint.

Basically, the arguments of the antiselectionists rest on an inability to appreciate the statistical nature of selection. Consequently, all those objections are irrelevant that are based on imperfections of adaptations or on conflicting selection pressures. For instance, antiselectionists will point out that much of the mortality of young animals is purely accidental rather than selective, as in the case of plankton, scooped up indiscriminately by a large fish or a whale. This observation overlooks the fact that among the remaining individuals (and it is immaterial whether they constitute 50 percent or 0.01 percent of the population), selective factors largely determine reproductive success. A second type of false argument stems from the observation that even among protectively colored animals some will fall prey to predators, or that an insect protected by its cryptic coloration against a vertebrate predator may succumb to a hymenopteran parasite. Such arguments reveal typological thinking at its worst. No one claims that natural selection gives immortality! It merely determines the probability of survival and of relative reproductive success among the members of a population. Whatever increases this probability will be selected for, regardless of residual mortality factors.

Antiselectionists often question the selective value of genes that add only very slightly to fitness, overlooking the fact that selection works not on individual genes but on phenotypes. A lot of very small advantages when combined into a single individual may add appreciably to fitness. Selective values are cumulative. To illustrate this, let us imagine an arbitrary model of a population with 1000 unfixed polyallelic loci, each allele having an exceedingly slight individual effect on fitness. Depending on the fluctuating environment and other factors, an allele may have a positive or a negative selective value. Owing to recombination and the particular local and contemporary constellation of environmental factors, most individuals will have an average mixture of positive and negative factors. But some individuals are sure to be on the tails of the curve of variation. Those that have plus genes on most of the 1000 loci are bound to have a considerably increased chance to survive and to reproduce more successfully. Those that have mostly inferior genes at these loci will almost certainly die without leaving offspring. Thus in every generation the frequency of the negative genes will decline and that of the positive genes will rise. This model, although crude, helps to illustrate the additive effect of viability factors.

Reaction norm and selection. The phenotype at a given time reveals only part of the potential of the genotype. The same weasel that is brown in summer has the potential to be a white ermine in winter. The difference between the office worker with a soft palm and the laborer with a heavily callused hand is not genetic. This kind of interaction between genetic potential and developmental response is overlooked in a frequently cited antiselectionist argument. Many vertebrates are born with calluses where the bare skin touches the ground, as on the knees in the warthog or on the breast in the ostrich. These calluses first appear in embryos, long before they are useful. On this basis it is argued that, if calluses are the reaction to friction on the skin and if there was no such friction in the embryos, the appearance of calluses in an embryo must be a case of "inheritance of acquired characters." This argument overlooks a number of points. First of all, most organs must be laid down long before they are used, outstanding examples being the eye and many parts of the central nervous system. If the presence of calluses is advantageous for the young animal, selection will certainly favor their formation at an early stage so that they are available when needed soon after birth. In the absence of preformed calluses, friction will produce blisters and inflammation, possibly leading to serious infection or feeding inefficiency. Ease of callus formation is under partial genetic control, like all other components of the phenotype. The shift of a species into a new niche, in which there is suddenly a "demand" for calluses, will set up strong selection pressure in favor of callus formation. How a given individual responds depends on the frequency in its genotype of genes facilitating callus formation. Even though the presence of such genes may be revealed only by a modification of the phenotype under exceptional conditions, it is nevertheless these genotypes that will be favored by selection. As callus-facilitating genes accumulate in the gene pool, the probability of their early penetrance into the phenotype increases. Finally calluses appear at the embryonic stage. This is an evolutionary model that has been well established for other characters and organisms (Mayr 1965b).

Phenotypic response (phenocopy) and selection. The same genotype may produce different phenotypes under different environmental conditions. An extreme environment may bring out developmental potencies that are not expressed under more normal conditions; it permits genetic factors to manifest themselves that do not normally reach the threshold of phenotypic expression. This threshold effect explains recent selection experiments by Waddington (1957). When pupae of *Drosophila melanogaster*, aged 21–23 hours, are given a temperature shock (4 hours at 40°C), about 40 percent of the hatching flies are "crossveinless." Stocks *formed from such selected crossveinless flies* respond more readily to the treatment with each generation, until crossveinless flies emerge even from untreated pupae. The simplest interpretation is based on the observation that about 60 percent of the flies do *not* respond to the shock treatment, owing to a lack or insufficient number of genes capable of responding to the treatment. Only those flies respond that have a sufficient number of the many genes contributing to the crossveinless condition (Milkman 1961). The presence of crossveinless genes at low frequency is not sufficient under normal conditions to lift the phenotype above the threshold of visibility. The treatment, however, reveals the carriers of such genes, and their continued selection permits an increasing accumulation in the gene pool of genes contributing to crossveinlessness until they express themselves phenotypically even without the shock treatment. The term "genetic assimilation," which Waddington uses for such situations, seems to me poorly chosen, because it fails to bring out the essential point that the treatment merely reveals which among a number of individuals already carry polygenes or modifiers of the desired phenotype. What we really have is *threshold selection*. Many similar situations, previously interpreted in Lamarckian terms or attributed to the Baldwin effect, are presumably due to the same threshold-selection effect.

The Force of Selection

The classical calculations of the power of natural selection (Fisher 1930; Haldane 1932) were deliberately based on very slight differences in the selective value of competing genes, in order to demonstrate that evolutionary changes would occur even when a gene was superior to its allele by as little as 0.1 percent. Much evidence is now accumulating to indicate that the differences in selective values among genotypes occurring in natural populations may be as high as 30 or 50 percent. This has been shown for the black morph of the hamster (*Cricetus*), for industrial melanism in the Peppered Moth (see Table 9.3), and for gene arrangements in *Drosophila* (see also Chapter 9).

Particularly impressive is the extraordinary sensitivity of the selective response to slight changes in the environment. The literature contains numerous records of genes or gene arrangements that are highly advantageous in one environment, say at 25°C, but neutral or deleterious in another environment, say at 16°C (Dobzhansky 1951; Ayala 1968). The extreme sensitivity of the genotype to environmental conditions cannot be doubted even in the cases where the viability differences are not expressed in the visible phenotype.

The efficiency of natural selection bedevils man in many ways. One is the rapid development of resistant strains of insects whenever insecticides are applied. In agriculture and horticulture new pesticides have to be introduced continuously to cope with the problem of resistance. DDT was first used on a broad scale in 1944. Within 2–3 years resistant strains of house flies had developed independently in different parts of the world. By 1960 more than 120 species of insects, including 62 species important in public health, showed resistance to one or several pesticides. Microorganisms, similarly, have the capacity to develop strains resistant to antibiotics and other drugs. This resistance results from the selection of a few resistant mutations or gene combinations, exactly as in higher organisms.

Disease is another selective factor of utmost importance, as was brilliantly discussed by Haldane (1949b). Resistance to disease may give a decisive advantage over a competitor, individual or species, that is not immune. Different genotypes differ in their immunological properties. Highly host-specific diseases (and types of parasitism) have very different ecological effects (for example, making food resources available for competitors) than less specific diseases.

The Phenotype a Compromise

One of the reasons that the role of natural selection in evolution is sometimes questioned is, curiously, that it is considered not sufficiently effective. Why does a species not acquire greater running speed if this would aid in escape from predators? Why does it not produce more young? Such questions overlook the fact that the phenotype is a compromise of all selection pressures and that some of these are opposed to one another. The breeder knows how much the production of eggs in the chicken or of milk in the cow has been increased by strong selection, but he also knows that this success was achieved at a price. The high-yielding strains of domestic animals would not be able to survive in nature, exposed to the elements and in competition with other species. Fitness is a property of the total genotype. The phenotype is the product of a compromise necessitated by the need for balance. Many illustrations of this compromise have been reported in the recent literature. For instance, Lack (1954) has shown that clutch size in birds is regulated to produce a maximum number of offspring. If too many eggs are laid, the feeding efficiency of the parent birds drops and the number of birds ultimately raised becomes smaller (Table 8.1). Studies of starlings and thrushes have provided evidence for the validity of this thesis. In the tropics predators rather than food may be the crucial selective factor. In these latitudes adults have a high life expectancy and small clutches may be of selective advantage. The smaller the nest and the fewer the feedings per day, the greater the probability that the nest will not be discovered by the exceedingly numerous nest predators. The geographic variation of selection pressures, such as seasonal cycles or day length during the breeding season, results in a geographic variation of many aspects of the life history. An early interpretation of increased clutch size in cooler climates was that birds lay larger clutches because they "need" them to compensate for the greater mortality in the higher latitudes. The invalidity of this "explanation" is evident when one considers that birds cannot anticipate the percentage of eventual loss in their offspring, nor consciously control the number of eggs they lay. Local

Table 8.1. Survival of Swiss Starlings (*Sturnus vulgaris*) after leaving nest (from Lack 1954).

Number in brood	Number of young ringed	Number of recoveries more than 3 months old:	
		Per 100 young ringed	Per 100 broods ringed
		Early broods	
1	65	—	—
2	328	1.8	3.7
3	1,278	2.0	6.1
4	3,956	2.1	8.3
5	6,175	2.1	10.4
6	3,156	1.7	10.1
7	651	1.5	} 10.2 (7–8)
8	120	0.8	
9, 10	28	0.0	—
Total	15,757	1.94	—
		Late broods	
1	44	} 2.3 (1–3)	} 5.8 (1–3)
2	192		
3	762		
4	1,564	2.2	8.9
5	1,425	1.8	8.8
6	438	1.4	8.2
7	49	0.0	—
Total	4,474	1.99	—

populations of bird species are selected in such a way that an optimal balance exists between various selection pressures: predation, available food supply, fluctuating seasons, available feeding time, and so forth (Cody 1965). Litter size in mammals is likewise regulated by natural selection (Batten and Berry 1967).

The phenotype is determined in all cases of conflict by the relative force of the two opposing selection pressures. This seems obvious enough but is too often forgotten in discussions of selection. Let us take, for example, the famous case of the *fremddienliche Zweckmässigkeit* of gall-making insects and their hosts. Why, it was asked, should a plant make the gall such a perfect domicile for an insect that is its enemy? Actually we are dealing here with two selection pressures. Selection works on a population of gall insects and favors those whose gall-inducing chemicals stimulate the production of galls giving maximum protection to the young larvae. The production of such galls is obviously a matter of life or death for the gall insect and thus constitutes a powerful selection pressure. The opposing selection pressure on the plant is in most cases quite small because having a few galls will depress viability of the plant host only very slightly. The "compromise" in this case favors the gall insect. Too high a density of the gall insect is usually prevented by density-dependent factors not related to the plant host.

Evidence for natural selection is so universal in nature and experimental proof for it is so abundant that it seems curious, in retrospect, that its importance should ever have been denied. Today it is no longer necessary to accumulate more evidence for its universality. Rather, the time has come to take up several of the puzzling and partly unsolved problems associated with the phenomenon of natural selection. The two problems I would like to single out are of very different nature: one raises the question how selection operates; the other, how it affects the further course of evolution.

Is the Population a Unit of Selection?

Many populations have attributes that are of strong selective advantage to the population as a whole but not, at least not on first sight, to any one individual or genotype within the population. Such attributes include the lengthening of generation time, the lengthening of the immature stage, reduction in the number of eggs or zygotes produced, unbalanced sex ratios, heterogeneity in dispersal drives, and similar phenomena. In some of these cases it is not at all obvious how selection of individuals will establish the particular balance of genotypes that produces optimum results for the population.

Considerations such as these have induced some authors to take refuge in *group selection*, that is, to postulate that the population rather than the individual is the unit of selection. The species, then, is envisioned as an aggregate of competing populations, each with a fortuitously differing mixture of traits. Populations with lucky combinations of genes will prosper, those with "losing tickets" will die out. This hypothesis might be reasonable if species were such aggregates of isolated populations. In reality, there is such extensive gene exchange among populations, generation after generation, that an interpretation of population characteristics as the product of group selection would seem altogether untenable.

Wynne-Edwards, currently the leading champion of group selection, therefore stresses social organization in animals, claiming that "social grouping is essentially a localizing phenomenon, and an animal species is normally made up of countless local populations, all perpetuating themselves on their native soil" (1962). His hypothesis requires that natural selection discriminate between these groups, promoting those with more effective social organizations and eliminating the less effective ones (Wynne-Edwards 1965). Evolutionists have long been aware of the importance of mechanisms, such as territory, parental care, and dispersal, that damp the effects of fluctuations in food supply and of other density-controlling factors. But since these mechanisms can be interpreted without difficulty in terms of Darwinian selection, their existence cannot be used as proof for group selection. No one denies that some local populations are more successful (and contribute more genes to the gene pool of the next generation) than others, but this is due to the aggregate success of the individuals of which these populations are composed, the population as such not being the unit of selection. A form of interdeme selection occurs in exceptional cases of a reproductive advantage of an otherwise deleterious gene, the result being the extinction of populations with too high a frequency of this gene, as in the case of the *t*-alleles in the house mouse (Lewontin 1962; Dunn 1964). However, all attempts to establish group selection as a significant evolutionary process are unconvincing.

Far more successful have been recent endeavors to explain population attributes as the result of ordinary selection operating on individuals. The validity of such an explanation can be shown for a number of population characteristics.

Aberrant sex ratio. Fisher (1930) was the first to show that natural selection normally tends to equalize the parental expenditure devoted to the production of the two sexes. All subsequent calculations by Edwards, Kolman, Shaw, and Williams have confirmed this conclusion (see also Hamilton 1967). Deviations from a 1:1 sex ratio may be selected for under special conditions, and the production of males may be replaced by parthenogenesis or some other form of uniparental reproduction when a species maintains or expands its range through single founding colonists (Baker and Stebbins 1965). The selective advantage of being able to reproduce under these circumstances in the absence of a conspecific mate is self-evident. For further literature see Mayr (1963:197) and Williams (1966:146–157).

Dispersal rate. The opposing tendencies of sedentariness and its converse condition, a flow of individuals from population to population, have an optimum balance. Too much dispersion destroys ecotypic adaptation, too little induces inbreeding. At first glance, it would seem to be the population rather than the individual that benefits by the proper balance; yet this situation again can be interpreted in terms of opposing selection pressures. Let us assume that there is genetic variation (as there surely is) in every population with respect to the degree of restlessness and philopatry (see p. 339). When the population is crowded and high density-dependent mortality prevails, individuals with dispersing tendencies will have a greater opportunity of finding a suitable breeding station than sedentary individuals and will be favored by selection. The effectiveness of this dispersal tendency will be reinforced by a genetic potential to respond to crowding or food shortage with migratory restlessness. When population densities are low, selection pressures will be reversed. Wynne-Edwards (1962) has given numerous examples of such regulation of dispersion, even though his interpretation often differs from mine. Murray (1967) explains dispersal rates in terms of individual selection, without, however, duly allowing for the factors responsible for long-distance dispersal.

Altruistic traits. Haldane (1932:207) was perhaps the first author to call attention to the problem of what he described as "socially valuable but individually disadvantageous characters." He was particularly concerned with "altruistic traits," such as would be of importance to mankind. He found that in large populations "the biological advantages of altruistic conduct outweigh the disadvantages only if a substantial proportion of the tribe behaves altruistically." Haldane's problem, how to make the altruistic trait reach such a high level of frequency, has now been solved, or rather shown to be a spurious difficulty, through the introduction of the principle of *kinship selection.* Let us illustrate this with a simplified example. A bird, by possessing a gene that makes it utter a special warning call against a predator, may slightly increase its own vulnerability but will appreciably reduce that of all other members of the species within hearing. There is a considerable probability that many of these benefiting individuals are relatives: its own young, brothers and sisters, half-brothers and half-sisters, and more distant relatives. Being its relatives, they share a portion of the genotype appropriate to the degree of relationship. Thus the altruistic trait is being selected for through the advantage it gives to the kin of the bearer of the trait (Hamilton 1964a, b; Maynard Smith 1964; Williams 1966). Whether or not such a gene spreads in the population depends on the net gain between the saved kin and the risk to the altruist. Haldane (1955) illustrated this principle vividly in a hypothetical situation in which

> you carry a rare gene which affects your behavior so that you jump into a flooded river and save a child, but you have one chance in ten of being drowned . . . If the child is your own child or your brother or sister, there is an even chance [on Mendelian principles] that this child will also have this gene, so five such genes will be saved in children for one lost in each adult. If you save your grandchild or nephew the advantage is only 2½ to

> one. If you only save a first cousin, the effect is very slight. If you try to save your first cousin once removed, the population is more likely to lose this valuable gene than to gain it.

In the early hominid populations, small in size and polygamous, there was a high opportunity for the spread of such genes.

Reproductive pattern and rates. Many organisms have very low reproductive rates. Large vultures, eagles, penguins, and albatrosses lay only one egg per clutch and may nest only every second year. Some albatrosses may not reach breeding condition until they are six or more years old. It is sometimes argued (for example, by Wynne-Edwards 1962) that natural selection could not possibly favor such a reduction of fertility and that the low reproductivity must be due to the fact that these birds "do not need" a higher fertility. Lack (1954, 1966:299–312), Cody (1965), and others have pointed out the fallacy of this argument. Not only is there no mechanism by which birds can consciously regulate such components of fertility according to "need," but, more important, it can be shown that feeding conditions (Ashmole 1963; Amadon 1964) and other factors indeed favor such low reproductivity.

There are several other properties of populations that, when only superficially analyzed, appear to have no obvious selective advantage. However, the mere fact that such traits have become established makes it highly probable that they are the result of selection, that is, of an unequal success of different genotypes. This probability is strengthened by the fact that several such cases (see above) have thus been successfully explained in recent years.

Selection for Reproductive Success

Even selection, with its almost unbelievable efficiency and sensitivity, has a weakness in its armor. Fitness is measured in terms of the contribution made to the gene pool of the next generation, that is, in terms of reproductive success. This is, normally, the best way of determining overall fitness. High sexual vigor in males or females is normally an indication of high general viability, of a good nutritional condition, and of an optimal functioning of all physiological processes. The reproductive success that results from it thus rather accurately reflects general viability. Reproductive success in other cases is due to special properties that have little relation to general viability. In these, a genotype spreads in a population not for any reason of general superiority but because it is a superior reproducer. In those families of birds, for instance, in which there is no pair formation, the male that is most stimulating sexually will inseminate the greatest number of females and contribute the greatest number of genes to the gene pool of the next generation. The reproductive success of the gaudiest males is the reason for the almost absurd ornaments of the male birds of paradise and of males in other groups in which a single male may fertilize many females (Sibley 1957). The extreme conspicuousness of such males may maneuver the species into an exceedingly precarious position, and the appearance of a new and more efficient predator might be the doom of such a species. In most cases, however, the development of exaggerated sexual dimorphism is probably relatively innocuous.

Epigamic characters are, however, only one of many attributes that lead to reproductive success. Selection of high reproductive success may be one of the factors contributing to the enormous fluctuations in population size among certain species of rodents and indeed among any species subject to catastrophic population declines. The individuals surviving such a crash find themselves in a virtually vacant habitat, in which a high premium is placed on the genotypes with the highest reproductive potential. Aphids or *Daphnia* in spring answer this challenge by abandoning sexual reproduction in favor of parthenogenesis, voles of the genus *Microtus* by early maturity and a rapid succession of generations. A female field mouse, *Microtus arvalis,* may be fertilized at an age of about 13 days, before or immediately after weaning, and produce her first litter 20 days later. She copulates again almost immediately after parturition and may produce litters every 3 weeks under favorable circumstances. One female in captivity produced 24 litters in 20 months (Frank 1956). Even though this runaway type of reproduction often leads to a partial or almost complete destruction of the habitat, natural selection has evidently been unable to incorporate factors that would check the unhealthy fluctuations in the face of the high premium placed on sheer reproductive success.

Darwin had an inkling of the importance of reproductive success and discussed it in part under the heading *Sexual Selection.* Yet selection in favor of reproductive success has in many instances nothing to do with sexual selection. How much of Darwin's sexual selection falls under the heading of "mere reproductive success" can be determined only through further research. Many recent studies, particularly of *Drosophila,* indicate that different genes and gene arrangements may differ quite drastically in their contribution to mating success. This situation may lead to the maintenance or even the spread of genes that are otherwise deleterious. It is even possible that such "parasitic genes" contribute to the extinction of species. Natural selection is apparently defenseless against genotypes that are successful reproducers but do not add to the survival value of the species as a whole. This essential weakness of natural selection is a potential danger to every species, including mankind (Chapter 20).

Is Selection Destructive or Creative?

In the days when the validity of "Darwinism" was still vigorously disputed, one of the crucial questions was: "Is selection creative?" If evolutionary change resulted from the mutational appearance of entirely new types, the only possible function of selection could be the a posteriori one of acceptance or rejection. Anything new and valuable that appeared in evolution was, according to this concept, the accidental product of mutation. Selection was assigned the purely negative role of eliminating inferior deviations from the type. One who interpreted evolution in this manner had to answer the question "Can selection be creative?" with an emphatic "No!"

But this is a completely distorted representation of the process of evolution. Evolution is not an all-or-none process. Genetic variation is enormous and does *not* consist merely in the production of a few new types. Natural selection favors certain individuals owing to their genetic properties; it "selects" them precisely in the same way in which a breeder "selects" the founder individuals for the next generation of breeding. This is a thoroughly positive process; inferior zygotes are simply lost. We do not hesitate to call a sculptor creative, even though he discards chips of marble. As soon as selection is defined as differential reproduction, its creative aspects become evident. Characters are the developmental product of an intricate interaction of genes, and since it is selection that "supervises" the bringing together of these genes, one is justified in asserting that selection creates superior new gene combinations. This viewpoint has been ably presented by Dobzhansky (1954), Lerner (1959), and virtually every recent biologist familiar with the newer findings of population genetics.

Natural selection sets severe standards and constitutes a sieve through which only a minority can pass. There is nothing accidental, nothing blind about its outcome. Since it achieves its object through differential reproduction, it tests a zygote again and again until the end of its reproductive cycle, in good times and in bad. The theory of natural selection escapes the fatal weakness of all vitalistic theories whose "improvements-of-the-type" factors lead to a uniform type unable to respond rapidly to drastic changes of the environment. Natural selection, on the other hand, by always being strongly opportunistic and by selecting any kind of mechanism that helps to preserve genetic variability, is ready in each generation to jump off in new directions.

Chance and Accident

Selection is the most important of the factors that induce evolutionary changes by affecting the frequency of genes in populations. It is not, however, the only one, as was pointed out at an early stage in the history of evolutionary research. Among the other possible factors is one, namely chance, the significance of which is still greatly disputed. Let me illustrate, on the basis of a single example, how important chance is. The human male produces many billions of gametes during his lifetime, and the human female many hundreds during hers. Still, one human couple can produce, at most, only about a score of children. It is largely a matter of chance which among the countless gametes will form the few successful zygotes. Since virtually all gametes differ from one another genetically, owing to the almost unlimited number of possible combinations of the parental genes, it is obvious that accident plays an important role in determining the genetic constitution of the F_1 of a set of parents.

Chance affects every step in the life cycle of an individual. Mutation is largely governed by chance. So is crossing over and the distribution of the chromosomes during meiosis. Success of the gametes is largely a matter of chance, as is the difference in the genetic constitution of the two gametes that form a new zygote. Sewall Wright (1931, 1960) calculated the effects of accidents of sampling in combination with other factors, such as mutation pressure, size of the population, selective values of the respective genes, and so on. He showed that, regardless of the direction of selection, one of two alleles in a population might be lost entirely under certain calculable circumstances, with the other reaching a frequency of 100 percent. This event he termed *random fixation.*

Genetic Drift

Random events (stochastic processes) occur in evolution at every level, from mutation and recombination upward. The importance of chance phenomena as determinants of the direction of evolution is still controversial. From about 1935 to 1955 it was fashionable to attribute almost any puzzling evolutionary change to *genetic drift,* defined by Dobzhansky (1951) as "random fluctuations in gene frequencies in effectively small populations." A history of the concept of genetic drift and a review of its many untenable applications were given by Mayr (1963:204–214). This survey should be consulted by all those who want to go into the subject more deeply. Ignoring phenomena to which the term drift was improperly applied (they were actually aspects of selection), one can distinguish three types of genetic phenomena where randomness may have significant effects. Unfortunately, these stochastic processes have sometimes been described as alternatives to natural selection. It cannot be emphasized too strongly that this is not the case. Random phenomena affect the samples that are exposed to natural selection but nowhere interfere with the process of selection itself.

(1) *Random fluctuations.* Accidents of sampling occur universally in natural populations. The evolutionary significance of such random fluctuations depends on the contribution to fitness by the genes involved and on the size of the effective population, its isolation, and its permanence. The best available evidence comes from the study of human populations. When neighboring native tribes living essentially in the same environment differ in the frequency of certain genes, it is never possible to separate the effects of selection or migration from those produced by errors of sampling. Studies of religious isolates provide the best evidence because they live in the same environment as their outside neighbors and generally keep excellent records of gene migration.

Among such studies of human isolates the most detailed and interesting one is that made of the Hutterites by Steinberg (MS). They are a religious sect that originally migrated from Switzerland to Russia and from there in the 1880's to North America, where they now form a series of colonies in the Dakotas, Montana, and adjacent parts of Canada. Blood group O occurred in about 29 percent of the individuals investigated, whereas it occurred in well over 40 percent of most European and American populations. Blood group A had a frequency of about 43 percent, well above the normal 30–40-percent level of European and American populations. Blood group B was reduced in frequency and had completely disappeared from two of the colonies studied. Even more significant is the amount of difference between various colonies of Hutterites, even though all of them have descended from closely related founders. For instance, gene A varied from 32 percent (colony 85) to 52 percent (colony 80). Among the Rh genes, R^1 varied from 27 percent to 68 percent, R^2 from 4 percent to 32 percent, r from 27 percent to 64 percent, to cite merely a few examples given by Steinberg (MS). The MN frequencies of all but one of the Hutterite colonies differ from those of nearly all known white populations but agree closely with those of the Dunkers. The data for the Kell blood group differ, so far as I know, from those of any other known human population. The frequency of the K gene is zero in Negroes, Mongoloids, and Indonesians, and varies between 2 and 6 percent in all known white populations. In six colonies of Hutterites it is 13, 20, 21, 22, 23, and 34 (21.2) percent.

The occurrence of accidents of sampling in these human isolates can no longer be doubted, but this does not affect the probability that selection has also contributed to some of these shifts in gene frequencies. This consideration brings us back to the question, what is the relative importance for evolution of these accidents of sampling? Here is another area where the atomistic approach of beanbag genetics has beclouded the issue. Virtually all discussions and calculations of the probability of drift were conducted in terms of the frequency of single genes. Emphasis was placed on the replacement of one gene by its allele, until fixation was reached. When not helped by selection, this process was called *random fixation*. The chance that this would happen was appreciable under only three conditions: a very small size of the population, completeness of isolation, and selective neutrality of the genes involved.

One of the traditional assumptions of those who favor drift is that many genes and genotypes are selectively neutral. As far back as the 1870's, Gulick claimed for the local races and species of *Achatinella* snails in Hawaii that their phenotypes were obviously the result of chance, uninfluenced by any selection. One case after another of reputed selective neutrality of genes has been refuted in recent years (Mayr 1963:207–211; Ford 1965). It must never be forgotten that a gene is not necessarily selectively neutral merely because it does not seem to make an adaptive contribution to the visible phenotype.

Selective neutrality can be excluded almost automatically wherever polymorphism or character clines (gradients) are found in natural populations. This clue was used to predict the adaptive significance (previously denied) of the distribution pattern of the gene arrangements in *Drosophila pseudoobscura* (Mayr 1945) and of the human blood groups (Ford 1945). The haphazard distribution of white and blue flowers of *Linanthus parryae* in the Mojave Desert was originally interpreted as evidence for drift. Wright (1943), however, showed by an analysis of variance that systematic pressures were involved. The stability of the distribution pattern during the past 20 years also indicates the presence of highly localized selective factors, even though it is not yet known whether these consist of the chemical or the physical properties of the soil or are related to ground-water level, exposure, or some unknown biotic factor. Similar "area effects" have since been described for British populations of the snail *Cepaea nemoralis* (Cain and Currey 1963). The occurrence of melanistic races of lizards (*Lacerta*) on small islets in the Mediterranean was first ascribed to drift. Kramer (*in litt.*), however, points out that all melanistic races are restricted to islands of a single type, namely small and rocky ones, and that wherever endemic races had formed on such islands they invariably tended to have more pigment. If random fluctuations had been involved, about half of these races should have been paler and only half of them darker than those of the adjacent large islands. The black color helps the lizards to warm up during the cool seasons and in the morning. Black races do not occur on similar islands in areas where it is continuously hot (as in the Red Sea). Subspecific characters of many island birds, at one time regarded as "neutral," have now been shown to be favored by selection.

One of the classical cases of variation that has often been attributed to drift is the variation in the color patterns of the banded snails *Cepaea nemoralis* and *C. hortensis*. These snails are unusually polymorphic, with the ground color either yellow, pink, or brown and with various banding patterns ranging from unbanded to five-banded with intermediate band numbers in various combinations. It is now known that these phenotypic differences have a measurable selective significance on various substrates (Cain and Sheppard 1950) and that the genotypes producing the different color patterns have various cryptic physiological properties that affect viability differently under different microclimatic conditions. The fecundity of various morphs of *C. nemoralis* is different at different temperatures (Wolda 1967).

All recent researches indicate numerous differences between genotype of *Cepaea* with respect to various components of fitness. Unbanded yellow snails are more resistant to heat and cold than banded ones. There is a positive climatic correlation, in France, between the frequency of unbanded snails and increasing July temperatures, and between the frequency of yellow and decreasing January temperatures (Table 8.2). Different morphs may differ in fecundity, migratory tendency, and habitat selection.

The distribution of the human blood groups has frequently been cited in the past as a classical demonstration of drift, on the basis of the assumption formerly made by leading geneticists that these genes are selectively "neutral." This interpretation has now been largely abandoned because numerous effects on viability have been established since 1953 for blood-

Table 8.2. Temperature and polymorphism of *Cepaea nemoralis* in different districts of France (from Lamotte 1959).

Mean July temperature (°C)	16–18	18–19	19–20	20–21	21
Average frequency of unbanded (percent)	22	24	26	29	30
Mean January temperature (°C)	4	4–2	2–0	0	
Average frequency of yellow (percent)	57.9	61.0	73.4	78.8	

group genes (mainly the ABO system). The selective significance of the blood groups is further substantiated by the fact that only a very few of the possible frequencies actually occur among human races (Brues 1954; Fig. 8.1).

That sampling errors can affect the frequency of genes is unquestionable. Yet the shift in frequency of an "effectively neutral gene" from, let us say, 50 to 60 percent would hardly qualify as an event of evolutionary significance. As long as one is concerned with single genes, complete fixation of one allele (or loss of its alternate) is the only kind of fluctuation that counts. Such fixations are, indeed, frequently found in small colonies. Yet their evolutionary significance is doubtful. Such small isolated colonies are only temporarily withdrawn from the free gene exchange of the species. As soon as contact with the parental species is reestablished, the lost gene will be restored. For all these reasons, it appears probable that random fixation of single genes is of negligible evolutionary importance. The situation is quite different with stochastic processes that affect whole genotypes or major parts of them. Such cases are described under 2 and 3 below.

(2) *Founder principle.* This term (Mayr 1942:237) designates the establishment of a new population by a few original founders (in an extreme case, by a single fertilized female) that carry only a small fraction of the total genetic variation of the parental population. The descendant population contains only the relatively few genes that the founders brought with them, until they are replenished by subsequent mutation or by immigration.

The founder principle is often responsible for the genetic and also the phenotypic uniformity of animal colonies, of peripherally isolated populations, or of colonies in temporary bodies of water, in short, of any population established by one or a few founders. While in the case of random fixation there is a secondary decay of the originally present variation owing to gametic randomness, in the case of the founder principle there is a primary poverty of genetic variability owing to zygotic randomness. The evolutionary role of such founder populations is discussed in Chapter 17.

(3) *Selective equivalence of genotypes.* Perhaps the greatest single source of randomness and indeterminacy in evolution is the selective equality of different genotypes. Since phenotypes are the product of the interaction and collaboration of many genes, it happens not infrequently that different assortments of genes may produce phenotypes that react in an essentially identical manner to a given selection pressure. Various models have been suggested to help us visualize the phenotypic equivalence of different genotypes. Wright (1956) has made a special study of such models and has discussed their evolutionary consequences. When four dominant genes affect a single selectively important character, which genotypes will be favored in a given population will depend upon the interaction of these four genes, their favorable or unfavorable effect on other characters, and finally their interaction with the residual genetic background. Selection

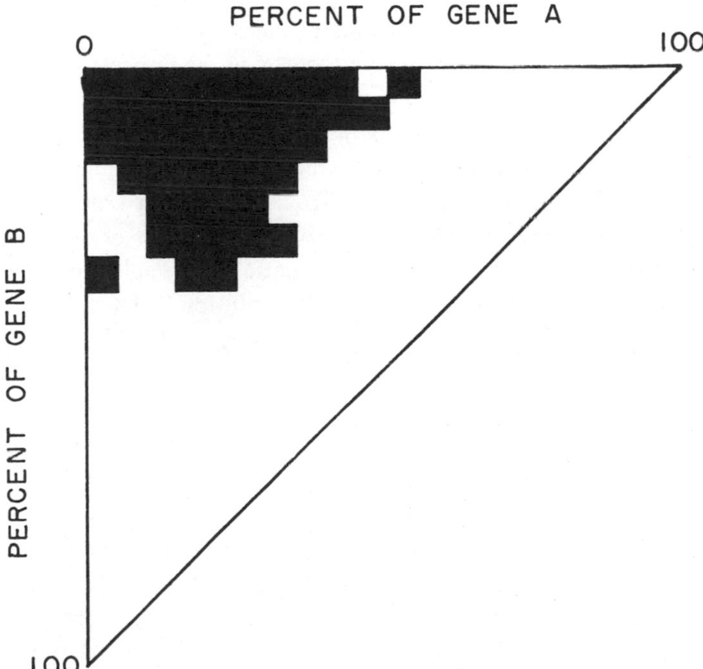

Fig. 8.1. ABO blood-group frequencies actually recorded in various populations throughout the world in relation to the complete possible range. (From Brues 1954.)

may, for instance, favor equally all the genotypes that contain only two dominants, that is, *ABcd, AbCd, AbcD, aBCd, aBcD,* and *abCD.* In another model one peak may be *ABCD* and another of equal selective value *abcd,* with the more heterozygous genotypes inferior. These simple models do not even begin to describe the complexity of the actual relation between genotype and phenotype. It is obvious that chance may play a considerable role in determining which of various equivalent genotypes reaches the greatest frequency in a population and is available to interact with shifts in the residual genetic background and the changing environment. It seems to me—and this has likewise been stressed by Wright—that it is this indeterminacy of the selective aspects of genotypic recombination that introduces the greatest element of chance into evolution. The truth of this assertion has been abundantly established in recent selection experiments. Variation in response was found in almost every case in which several daughter populations of a single parental population were exposed to the identical selection pressure. The uncertainty produced by the indeterminacy of genetic recombination and by the phenotypic equivalence of different genotypes is the most important random factor in evolution.

There has recently been renewed advocacy of the importance of genetic drift (King and Jukes 1969; Fitch and Margoliash 1970). King and Jukes designated evolution, which they believed to be steered entirely by mutation pressure, by the inappropriate name of "non-Darwinian evolution." Considering that Lamarckism, Geoffroyism, orthogenesis, and Goldschmidt's macrogenesis are also non-Darwinian theories of evolution, it is evident that this new designation for genetic drift might be misleading. The evidence cited for such "random walk evolution" (Dobzhansky, orally) is threefold: (1) Owing to the degeneracy of the genetic code certain amino acids (serine, leucine, arginine) are coded for by as many as six different codons, others by four, three, or two codons; mutations producing a new codon that codes for the same amino acid do not necessarily result in a change of fitness (unless differently interacting with the transfer RNA). (2) Not all amino acids in a molecule, it is believed, have a unique function; at certain sites two or three amino acids may be functionally equivalent and mutations resulting in such replacements would be "neutral" mutations. (3) Amino acid replacements in phylogenetic lines appear to occur at a steady rate. Several other molecular phenomena have been cited in support of an assumption of "neutral" mutations and of a great deal of random fixation owing to mutation pressure (for example, Kimura 1969).

At the same time a number of considerations largely deprive the cited evidence of its cogency. (1) Reverse mutations will control the relative frequency of the various codons that code the same amino acid, and there is no evidence that this codon polymorphism exerts any evolutionary pressure. (2) More and more sites even in the largest molecules are found to have specific functions. A "functionless site" is simply one the function of which has not yet been determined. Even when two or three residues may have almost the same function, one or the other may be slightly superior, particularly when subjected to catastrophic selection pressures in times of an environmental crisis. (3) There is no need to postulate a high rate of neutral mutations in order to escape an overwhelming genetic load, provided one rejects previously maintained erroneous concepts of the genetic load (Chapter 9).

It is not valid to evaluate the fitness of an isolated molecule solely on the basis of the function of its active site. In the living organism each molecule interacts with numerous other molecules and membranes, which form its *milieu intérieur*. This *milieu intérieur* continues to change—as the thousands or millions of molecules of which it is composed change in the course of evolution—and thus continuously exerts a changing selection pressure. A given molecule, in addition to being selected for its main function, is thus exposed to a constant lower-level selection pressure for adjustment to the structural changes in other molecules with which it interacts. The result is a fairly constant rate of change, not primarily as a result of random effects of mutation pressure, but rather as the result of steady selection for coadaptation. Similar rates of change in different molecules can very well be due to response to similar selection pressures, rather than to an internal phyletic mutation clock, ticking off new mutations at a specified rate.

Available observations on allele polymorphism are not consistent with the neutrality theory. There should be high levels of polymorphism at all "neutral" loci (but these are *not* found) and where polymorphism is found it should—owing to chance—be highly variable from locality to locality. However, most existing polymorphisms can be demonstrated to be maintained by mechanisms other than mutation pressure (Chapter 9), and these are often remarkably constant over wide areas.

A random replacement of amino acids unquestionably occurs occasionally in evolution, but it appears at present that it does not anywhere near approach selection in importance as an evolutionary factor. Any molecular change that has any effect whatsoever on the development, physiology, or behavior of the organism is bound to be affected by selection.

Random Phenomena and Genetic Drift

Application of the term "drift" to nondirectional random fluctuations is unfortunate, since in the daily language we generally use the term "drift" for a passive movement that is more or less unidirectional. To speak of the drift of icebergs or clouds conforms to this usage. The indeterminacy of the sampling of genes by recombination and by the founder principle is precisely the opposite of what is colloquially understood by "drift."

Random phenomena like recombination (with selective equivalence of different genotypes) and the founder principle introduce a considerable degree of indeterminacy into evolution. Temporarily they may even be stronger than selection in completely isolated and at least initially small populations. How important such populations are for speciation and ultimately for evolution is still rather obscure (Chapter 17).

Interaction of Chance and Selection

The effect of chance on genetic variation is ambivalent. On one hand, it may lead to a depletion of genetic variation through random fixation. On the other hand, by counteracting selection, it may delay the elimination of temporarily disadvantageous genes. It certainly enhances the indeterministic aspect of evolution. Consequently it would be entirely misleading to say that chance directs the course of evolution. Nor does chance cause "drift" in the sense of a steady movement in one direction. On the contrary, if chance has any influence on the direction of evolution, it is that it "jars" it at frequent intervals and may occasionally be responsible for a jump to another track.

Let us remember that evolutionary change is a two-factor process. One stage is the generation of genetic variation. It is on this level that chance reigns supreme. The second stage is the choosing of the genotypes that will produce the next generation. On this level natural selection dominates, while chance plays a far less important (although not altogether negligible) role. Chance causes disorder, selection causes order. Chance is disoriented, selection is directional (including stabilizing selection). Chance is often destructive, selection is frequently creative. Yet both chance and selection are statistical phenomena and thus they not only coexist but also, one might even say, collaborate harmoniously.

9 · Storage and Protection of Genetic Variation

Every natural population that has been carefully analyzed was found to contain a rich store of genetic variation. This high variability surprised the Mendelians. Accepting the gene as the unit of selection ("beanbag genetics") and rating all genes (and new mutations) as clearly either superior or inferior, the classical theory of genetics could find no reason for so much genetic variability: superior mutations would be incorporated into the genotype of the species while inferior ones would be eliminated. As a result, the uniformity of species would be maintained and genetic variation kept at a minimum. But since the high genetic variability of natural populations is an observed fact, it is evident that there must be some basic flaws in the assumptions of the classical theory. What are they?

The classical theory neglected two important factors:

(1) The target of selection is always an entire individual, that is, an entire interacting system of genes. Since the selective value of a gene varies with its genetic milieu, a far greater store of genetic variability will be tolerated in the gene pool than the classical theory allowed for.

(2) The environment is variable in space and time, not uniform as was tacitly assumed in the classical theory. Consequently, there is no "best" genotype. Under certain conditions one genotype is better, under others another. The temporary fitness values of genes are relative and therefore changeable.

The obvious consequences of both 1 and 2 are that, contrary to older beliefs, it is not at all advantageous for the gene pool to get rid of its genetic variability. On the contrary, it is of great selective advantage to preserve as much of it as possible, in order to be prepared for fluctuations of the environment and to maintain genes that can produce viable gene combinations with alien genes of immigrants.

Concealed variation. That a remarkable store of genetic variation is indeed concealed in natural populations has been documented by numerous investigators (Dobzhansky 1951:chap. 3). For instance, Spencer (1947) found in a study of two populations of *Drosophila immigrans* that 110 flies carried 47 visible mutants in 1944 and 51 in 1946, of which 33 (1944) or 24 (1946) were at three loci. Similar high frequencies of visible mutations were found in other species. Special techniques for testing the contribution to fitness of whole chromosomes or chromosome sections have revealed an astonishing richness of variable loci affecting viability (up to and including lethality). (For a summary see Lewontin 1967a.) It is extraordinary how much genetic variability even a single diploid individual may carry in its genotype. For instance, no fewer than 11 of 21 wild pairs of *Drosophila melanogaster*, when inbred, gave rise (among 1000 F_2 progeny) to at least one individual with the rare genetic defect crossveinlessness (Milkman 1960).

The techniques for revealing concealed genetic variation by inbreeding and examination of the F_2 generation are laborious and have severe limitations. It would be far preferable to develop methods that permit direct genic analysis of the phenotype without breeding experiments. The study of the karyotype (the chromosome complement) is one such method, and it revealed a high degree of chromosomal polymorphism in many natural populations. A population study of blood-group genes (ABO, Rh blood groups, and so forth) revealed the same. Direct genic analysis of the phenotypes has, however, received its most powerful boost through the introduction of the technique of enzyme electrophoresis and its application to population analysis by the Texas school of *Drosophila* geneticists (see p. 155).

How important it is to protect this genetic variability from depletion by natural selection is apparent from the number of devices, acquired during the course of evolution, that achieve such protection. This chapter is devoted to a discussion of the known mechanisms for the storage and protection of genetic variation in natural populations.

In a discussion of the factors responsible for the genetic variation of populations, Haldane (1954a) lists ten. Several additional ones have been mentioned by other authors. All, however, fall into three major categories:

(1) *Cytophysiological and developmental devices*, which thwart selection by combining the effects of a gene on the phenotype with that of other genes in such a way that different genotypes in which a given gene may participate have unequal selective values. These devices include recessiveness, modification of penetrance, heterosis, linkage, and epistasis.

(2) *Ecological factors*, consisting of variations of the environment in space and time, which often lead to a neutralization of opposing selection pressures and thereby reduce the effectiveness of selection.

Table 9.1. Protection of genetic variation from elimination by selection.

1. Cytophysiological and developmental devices
(a) Complete recessiveness
(b) Control of penetrance and expressivity
(c) Superiority of heterozygotes
(d) Prevention of free recombination
2. Ecological factors
(a) Inefficiency of natural selection
(b) Change of selection pressure in time
(c) Mosaicism of the local environment (Ludwig effect)
(d) Geographic variation of the environment and gene flow
(e) Heterogamy (inverse assortative mating)
(f) Selective advantage of rare genes

These factors are summarized in Table 9.1.

(3) *Regulation of the size of the gene pool,* that is, an increase or decrease in the amount of outbreeding (see Chapter 14).

Cytophysiological and Developmental Devices

The phenotype is the point of attack of natural selection. Consequently, any developmental process which reduces the phenotypic expression of genes that might otherwise be deleterious shields these genes from selection, at least until changes in the environment improve the selective value of these genes. Diploidy and the organization of the genetic material into chromosomes provide various opportunities to reduce the exposure of genes to selection. Let us consider some of them.

Complete Recessiveness

The less a gene affects the phenotype, the less it is exposed to selection. The completely recessive allele is fully protected while it is in heterozygous condition. According to the classical theory of genetics, the heterozygous genotype *Aa* and the homozygous dominant genotype *AA* have identical phenotypic properties. The presence of the recessive gene *a* in the heterozygote is completely concealed and heterozygous *a* will escape all attacks by natural selection even if it is greatly inferior to *A*. Such an inferior gene is exposed to selection only when it becomes homozygous (*aa*) through recombination. The frequency of a completely recessive allele in a population is thus determined by the balance between its rate of mutation and its elimination in homozygous condition. The early population geneticists ascribed most genetic variation in natural populations to the segregation (becoming homozygous) of previously concealed recessives.

Various findings in recent years have undermined the belief in a high frequency of loci with completely recessive alleles. Every gene produces a gene product and there is no reason why the product of gene *a* should not have an effect on the heterozygote *Aa*. That there is a difference between homozygotes and heterozygotes even when there is no visible difference has been established by paper chromatography and electrophoresis, by immunochemistry, and by comparison of the viabilities of heterozygotes and homozygotes. The difference in fitness between heterozygote and dominant homozygote is not necessarily expressed on all genetic backgrounds or in all environments, and where it occurs it may be in favor of either the homozygote or the heterozygote. Deviations that result in a superiority of heterozygotes are discussed on pages 134–136. Deviations producing less fit heterozygotes are often due to incompletely recessive genes that are lethal or semilethal in homozygous condition. Even such heterozygotes may have normal viability in a fluctuating environment.

Dominance and recessiveness are not inherent qualities of genes, but refer to their effect on characters. Yet new mutations are often from the very beginning strongly dominant or recessive. In other cases the degree of dominance is determined by the remainder of the genotype and then natural selection will favor such modifying genes as will make the heterozygote more similar to the phenotype of the superior homozygote (Sheppard 1958).

Suppression of Phenotypic Variability

The phenotypic manifestation of genes can be regulated by means other than recessiveness. For instance, there are otherwise dominant genes that fail to express themselves phenotypically when placed on certain genetic backgrounds or when their bearer lives in certain environments. Such genes are called *incompletely penetrant*. Incomplete penetrance characterizes many deleterious human genes and this characteristic greatly impedes the analysis of their mode of inheritance. There will be natural selection in favor of background genes ("suppressors") that suppress the phenotypic manifestation of deleterious genes (Chapter 10). Penetrance modifiers are only one set in a large class of genes, all of which contribute to the prevention of deviations from the "normal" phenotype of the species. The greater the number of such homeostatic devices in a gene pool, the greater the amount of genetic variation that can accumulate in the population under standard conditions without being exposed to selection. Such phenotype-stabilizing genes presumably play a dual role. They may suppress environmentally caused variations of the phenotype and may also buffer against the effects of mutation. Their ability to canalize developmental processes presumably gives them high selective value.

The amazing phenotypic uniformity of many species, in spite of the great amount of concealed genetic variation, suggests that there must be a high selective premium on the standardization of the phenotype. The morphology of all individuals is essentially identical in these species even though the morphology in each case is produced by a different combination of genes and each genotype possesses different physiological properties. The high frequency of sibling species (Chapter 3) proves that genotypes can be entirely reconstructed, and that new isolating mechanisms and an entire new ecology can be acquired, without leaving a visible imprint on the morphology. The same is demonstrated by the abundance of isoalleles, which differ in their physiology but do not affect the visible phenotype.

What mechanisms permit preservation of the favored phenotype in spite of the extensive reconstruction of the genotype? Waddington (1957) has discussed the nature of this developmental stability. He postulates that the genotype is buffered in such a way that the development is "canalized." No matter which genes in the gene pool come together, a series of developmental tracks will guarantee a standardized end product. This developmental stability can cope not only with variation of the gene pool but also with that of the environment. As long as the environment varies within "normal" limits, it will not be able to make the phenotype jump out of its normal developmental track. The term *canalization* stresses the inevitability of the developmental pathway, in spite of disturbances and temporary roadblocks caused by the external and genetic environment. The term *developmental homeostasis,* used by other authors for essentially the same phenomenon, stresses the dynamic aspect provided by the homeostatic mechanisms that restore development to its normal course after deviations, no matter how they were caused. In the case of canalization, the end product is stressed; in the case of homeostasis, the process of producing it. Much work has been done in recent years in the attempt to analyze the genetics of canalization. The role of major genes and modifiers and of various feedback mechanisms, as well as the effect of one-sided selection on such genic control systems, has been ably summarized by Rendel (1967). The more precise the analysis of the genetic basis of phenotypic characters, the more evident is the great complexity of this genetic basis. Development of a character is normally controlled by a complex system of genes and supergenes (chromosome segments) whose interaction results in a normalized development of the respective phenotypic character.

It is now evident that each population contains a great deal of concealed variability. Some of it consists of relatively recent mutations that have not yet been eliminated owing to their recessiveness in heterozygous condition. Another part of the variation is directly maintained by natural selection. This portion of the genetic variance may be called "selected heterozygosity." Heterozygote superiority is the most important among the various devices that fall in this category.

Superiority of Heterozygotes

The mechanism by which two alleles can be permanently maintained in a population was first called attention to by R. A. Fisher. He showed that two alleles can be maintained in a population at high frequencies if the fitness of the heterozygotes (*Aa*) is higher than that of either homozygote (*AA* or *aa*). Even if *aa* should be considerably inferior to *AA*, and possibly even lethal, the gene *a* would be retained in the population, provided that heterozygote *Aa* has a higher selective value than has homozygote *AA*. The reservoir of gene *a* will be replenished continuously by segregation of the favored heterozygous genotype *Aa*. Polymorphism (see Chapter 7) maintained by such "overdominance" of the heterozygotes has been designated by Ford (1945, 1965) *balanced polymorphism.*

The potency of overdominance may be illustrated by an example. If a gene *a* is completely lethal as homozygote (*aa*), it will nevertheless be retained in the population with an equilibrium frequency of 0.01 if it raises the fitness of the heterozygote (*Aa*) by 1 percent over that of the homozygote (*AA*). The frequencies of the genes *A* and *a* will reach the equilibrium when the frequency of the lethal allele has reached $s/(s + t)$, where t is the coefficient of selection of the lethal and s that of the nonlethal allele.

Heterosis (heterozygote superiority) is sometimes caused by the interaction of a pair of alleles at a single locus, in other cases by the interaction of whole chromosome sections (gene arrangements) acting as supergenes. Dobzhansky (1951) and his school have made a particularly extensive contribution to the understanding of the working of such balanced chromosomal polymorphism. It occurs not only in *Drosophila* but in many other organisms. Of 27 chromosomally polymorph strains of *Drosophila* maintained in the laboratory at a population size of only 20-40 individuals, 24 were still polymorph after 130–211 generations of random transfers. Indeed, it is becoming more apparent daily that most cases of genuine polymorphism in natural populations are maintained by the superiority of the heterozygotes, regardless of any additional factors (such as mosaicism of the environment) that may contribute to genetic diversity.

By far the most astonishing aspect of balanced polymorphism, to me at least, is the magnitude of the differences in the selective values of the different genotypes. For instance, among genotypes for a melanistic gene in the hamster (*Cricetus cricetus*), calculated coefficients of selection ranged from 0.79, 0.83, 0.89, 0.94, and 0.98 down to −0.82 and −0.86. The melanics were favored (positive values) in some years and districts and discriminated against (negative values) in other years or areas. Dobzhansky (1947) kept a polymorph population of *Drosophila pseudoobscura* from Piñon Flats, San Jacinto Mountains, California, in a population cage under specified conditions, and found that flies homozygous for one of the gene arrangements (Standard) had only 76.2 percent of the fitness of the heterozygotes, those homozygous for the other gene arrangement (Chiricahua) only 37.9 percent. Similar differences have been found for other gene arrangements in the same and other species of *Drosophila*.

In a number of cases clear superiority of the heterozygotes was found, in spite of lethality or near lethality of one of the homozygotes. It is not correct to employ such cases to calculate the mutation rate of the lethal gene in the population, as is possible in the case of strictly recessive genes. Rosin et al. (1958) have shown, for instance, that the widely accepted mutation rate for the hemophilia gene in man must be drastically revised in view of the greatly increased fertility of the female carriers of this gene (about 1.15 compared to 1.00 normally). A fitness for females of 1.22 would be sufficient to maintain the hemophilia gene in the population without additional mutation, even though the observed fitness of the males is only 0.64.

The Effects of Heterozygosity

It is an almost universal observation that severe inbreeding leads to "inbreeding depression," a serious reduction of various components of fitness. Loss of fertility, increased susceptibility to disease, growth anomalies, and metabolic disturbances are among the manifestations of inbreeding depression (see Lerner 1954:22–27 for numerous examples). Countless laboratory stocks have been lost owing to inbreeding. Fitness declines as more and more loci become homozygous, but can be restored dramatically by introducing new genes, which increase the level of heterozygosity.

Of the many recent experiments demonstrating a superiority of heterozygotes, I will cite only one. Carson (1958a) introduced a single wild-type Oregon-R third chromosome into a large *Drosophila* population homozygous for five third-chromosome recessives (*se, ss, k,* e^s, and *ro*). After about 15 generations (1 generation = 14 days) the three recessives that were closely followed had stabilized at the following frequencies: *ro* = 53.4 percent, *se* = 25.3 percent, and *ss* = 12.3 percent. The heterozygosity due to the single introduced chromosome more than tripled the productivity of the parental population (Table 9.2). Indeed, the performance of the new populations (E-1 and E-2) was in all respects superior even to the wild-type Oregon-R laboratory population, which, with its long history of inbreeding, was presumably far more homozygous than the new experimental population.

One of the most important findings of recent researches is that there is not necessarily a very close correlation between the fitness of genes in homozygous and in heterozygous condition. A gene that is lethal or semilethal in homozygous condition also very often reduces the fitness of the heterozygote by 2–5 percent (or more). There is, however, an unexpectedly high proportion of genes lethal in homozygous condition that produce heterozygotes with normal or even superior fitness. The same has been demonstrated for chromosome portions (gene arrangements).

Table 9.2. Size and production of experimental populations of *Drosophila melanogaster* (from Carson 1958a).

Population	Population size		Production	
	Mean number of individuals (weekly count)	Mean wet weight measured weekly (mg)	Mean number of individuals per week	Mean wet weight (mg/week)
Controls:				
C-1 se ss k e^s ro	161.6 ± 6.4	90.3 ± 3.0	100.8 ± 4.4	48.4 ± 1.9
C-3 se ss k e^s ro	154.4 ± 4.4	88.7 ± 2.0	61.0 ± 3.3	27.6 ± 1.5
Experimentals:				
E-1 se ss k e^s ro with *n* Oregon autosomes	457.4 ± 13.7	292.5 ± 9.1	171.0 ± 10.6	91.0 ± 5.9
E-2 se ss k e^s ro with *n* Oregon autosomes	502.6 ± 14.9	318.6 ± 8.9	201.1 ± 14.4	105.2 ± 8.0

The Causes of Heterozygote Superiority

The reasons for the selective superiority of heterozygotes, where it occurs, are manifold, ultimately physiological, and not yet fully understood.

(1) *Heterosis due to dominance.* The traditional interpretation of heterozygote superiority is that increase in viability observed in hybrids between two stocks is the result of a restoration of dominant genes in the hybrids at loci where more or less deleterious recessives had become homozygous in the parental strains. This interpretation is founded on the probability that numerous deleterious recessives are concealed in all lines and that different inbred lines would become homozygous for different, nonallelic recessives. If one line is *AAbb,* the other *aaBB,* the hybrid would be *AaBb,* with the dominant gene suppressing the deleterious recessive in the heterozygote at both loci. This interpretation has been proved valid in many situations observed by animal and plant breeders, but is not sufficient to account for all of the superiority of heterozygotes in polymorph populations.

(2) *Heterosis due to overdominance.* According to this theory a locus is overdominant if the heterozygote has higher fitness than the homozygotes.

There are several possible explanations for cases of heterozygote superiority. One is that heterozygosity gives greater biochemical versatility. A heterozygote, having a combination of different gene products available, is able to cope with a greater diversity of developmental needs than the homozygote, which has only a single gene product. In view of the ability of a single allele at many loci to produce in a "single dosage" all the gene product necessary for the development of a normal phenotype, it would appear that the presence of a second allele at the same locus, with an optimal activity at different temperatures or under other conditions, might result in a combination that would operate as a physiological team. In each set of environmental conditions either one or the other allele will have superior developmental efficiency, resulting in optimal viability. Heterozygotes are specially favored, according to this interpretation, in fluctuating and adverse environments.

Another possibility is that the heterozygote is less exposed to selection pressures than either homozygote. The prototype of such a situation is presented by balanced lethals, a less extreme case by the gene for sickle-cell anemia. This gene produces the abnormal hemoglobin *s,* and the resulting anemia of its carrier causes a reduction in fitness of all least 90 percent among the homozygotes. Indeed, few of them reach even the age of 5 years. On this basis one would expect a rapid elimination of the gene until it reached the level of maintenance by recurrent mutation. Yet there are large areas in tropical Africa where some 20 to 40 percent of the natives are heterozygous for the gene. This is due to a high selective advantage of the heterozygote. The areas of the highest gene frequency coincide, on the whole, with areas with an excessively high rate of morbidity due to subtertian malaria caused by *Plasmodium falciparum*. Children heterozygous for sickle-cell anemia have a greatly lowered rate of infection by *P. falciparum* and heterozygous adults are less susceptible to artificial infection. Although additional factors enter the picture, there can no longer be any doubt that the clinically nonapparent inferiority of the blood cells of heterozygotes is a protection against subtertian malaria and thus gives the heterozygotes superior fitness. Italian doctors had previously suggested the same explanation for the high frequency of thalassemia (another hemoglobin disease) in Italy in districts with a high rate of endemic malaria, and this has since been substantiated.

The continued persistence of polymorphism has often been ascribed to a fluctuating balance of opposing selection pressures against one or the other homozygote. This is possible for rapid fluctuations but unlikely in the long run, since one or the other allele will be lost sooner or later owing to errors of sampling (Kimura 1955). A permanently balanced polymorphism can be maintained only if the opposing selection pressures add up to a superiority of the heterozygote.

The presence of heterozygous loci in a population, guaranteed by heterosis, is of double advantage. First, it produces highly viable individuals that are buffered against environmental fluctuations. More important, it gives the population a great and much-needed diversity, because the number of genotypes in the population is more than simply proportional to the number of alleles: if the number of alleles is n, the number of genotypes is $\frac{1}{2}(n^2 + n)$. A locus with three alleles (A, a, a') produces six genotypes (AA, Aa, Aa', aa, aa', $a'a'$), each optimal at somewhat different environmental conditions. A locus with 10 alleles produces 55 genotypes. Such versatility makes genetic variability available at all times for an immediate evolutionary response to a change in the environment. The population as a whole thus possesses great evolutionary plasticity. Furthermore, being composed of several genotypes, a population that is polymorphic owing to heterosis can better utilize different components of the environment (different subniches). Populations of *Drosophila pseudoobscura* that are polymorphic for two gene arrangements produce more individuals than equivalent monomorphic populations; they produce a greater total biomass and show less phenotypic variation (Beardmore, Dobzhansky, and Pavlovsky 1960). Ayala (1968) summarizes much additional evidence for the superior performance of populations with richer genetic variability.

Heterozygotes may have still another advantage. A study of phenotypic variance showed unexpectedly that it is lower in heterozygotes than in homozygotes (inbred lines) (Falconer 1960). Animal and plant breeders have long been familiar with this phenomenon. It is evident that the increased phenotypic variation in homozygotes or inbreds is largely nongenetic. They are more strongly affected by variation of the environment, they are less able to compensate physiologically for the unbalancing impact of environmental factors, they are less well "buffered." Why? It has been suggested that the developmental superiority of the heterozygotes is due to their greater biochemical versatility. The homozygote, on the other hand, is more one-sided, both alleles coding for the same gene product. It has only a single answer for the vagaries of the environment, and its developmental track is therefore more easily disturbed.

The superior homeostasis of the heterozygotes under varying conditions is one of the reasons for their selective superiority. A tight system of developmental homeostasis helps to shield the organism against environmental fluctuations. However much genetic variation there is in a gene pool, the less of it penetrates into the phenotype, the smaller the point of attack it offers to selection. The developmental mechanisms that guarantee a uniform standard phenotype regardless of the amount of underlying genetic variation are thus another device contributing to the maintenance of genetic variability in populations.

The Origin of Heterosis

The role of natural selection in the production of heterosis is still rather obscure. Two extreme viewpoints are possible, the relative merits of which I have discussed previously (Mayr 1955). According to one view, heterozygous loci are always potentially superior to homozygous loci. The other extreme view is that heterozygotes are superior to homozygotes only in those cases in which there has been selection for such superiority, that is, where the genetic background has been selected to give the heterozygotes maximal fitness for all components of fitness affected by the locus.

The best evidence against the extreme selectionist viewpoint comes from population crosses. Starting from the consideration that the genes of a gene pool are the product of a long history of selection for optimal interaction, one would always expect to find a drastic loss of fitness in interpopulation crosses. Surprisingly, the exact opposite is sometimes the case (see Fig. 10.1). Selection cannot account for interpopulation heterosis (see Chapter 10).

Yet there is also much evidence that selection is important (Dobzhansky 1951; Ford 1964). For instance, the relative superiority of heterozygotes often increases sharply in experimental populations. Also, heterozygotes may lack heterosis in inbreeding species, as has been demonstrated for many plant species. Finally, nothing could be more heterozygous than species hybrids, yet most hybrids are decidedly inferior with respect to most components of fitness. Heterosis therefore cannot be an automatic by-product of heterozygosity.

The understanding of heterosis is greatly facilitated if one remembers that new genes become incorporated into populations in heterozygous condition. Homozygotes become frequent only long after the heterozygotes have already become very frequent. It is during this initial period, as Parsons and Bodmer (1961) correctly point out, that an overdominance of the heterozygotes will be favored by selection. Modifiers that have the greatest fitness-enhancing effect on heterozygotes will be selected during this early phase.

Prevention of Free Recombination

If there were no chromosomes, there would be no limit to the possible assortment of genes. However, with the genes linked together on chromosomes and with the amount of crossing over in every generation restricted, recombination between the parental genomes is severely limited. It is important to find out to what extent and by what mechanisms recombination can be reduced or prevented altogether and how such reduction or prevention affects the storage of genetic variation in populations. The mixing of the gene contents of a species or population is determined at two levels, the gametic and the chromosomal. It is advisable to restrict the term *recombination* rigidly to the factors controlling the degree of mixing of the gene contents of parental chromosomes. The various phenomena, such as hybridization, isolating mechanisms, dispersal, and population size, that determine the degree of genetic difference of the parental gametes are of a very different nature and are discussed in Chapter 14 under outbreeding.

On the chromosomal level, there are mainly two sets of factors that determine the amount of recombination: the number of chromosomes and the frequency of crossing over.

Chromosome number. The various nonhomologous chromosomes assort independently during meiosis. The greater the number of chromosomes on which the genes of a gamete are distributed, the greater the possible number of combinations. The increase is exponential, so that, for instance, for a haploid number of chromosomes $n = 7$ the number of possible gametic chromosome combinations is $2^7 = 128$; for $n = 14$ it is 16,384. Among plants, annuals have on the average lower chromosome numbers, perennials and woody plants higher chromosome numbers. For most groups of animals, as for plants, there exists a "typical" chromosome number, which happens to be high in birds and low in most dipterans. The reasons for these differences are not yet certain and it has not been possible so far to relate them with other components of the breeding system (see Chapter 14).

Frequency of crossing over. The recombination of genes linked on the same chromosome is favored by three factors: a high chiasma frequency, a random distribution of chiasmata, and structural homozygosity of chromosomes. Conversely, the amount of crossing over is reduced by a lowering of the number of chiasmata, by localization of chiasmata, and by structural hybridity of chromosome sections (particularly through inversions), which prevents crossing over in the particular section. A chromosome section that is protected from crossing over acts, for the purposes of recombination, as a single gene and has therefore been called a *supergene.* Such supergenes may have two advantages. They permit indefinite preservation of a particularly valuable assortment of genes; and when several supergenes are simultaneously present in the gene pool, they permit the assemblage of heterotic combinations. The advantages of single-locus heterosis are thus potentially expanded to an entire chromosome section. Furthermore, the genetic loads caused by deleterious homozygotes at the various loci are thereby made to coincide. Finally, supergenes permit the preservation of epistatically balanced gene combinations. See Ford (1964) for a perceptive discussion of the supergene concept.

Linkage. The importance of linkage in evolution is still rather uncertain. Allele a' on locus A and allele b' on locus B, both on the same chromosome, will be linked until they are separated by crossing over. The shorter the distance between the loci, the less likely it would seem that there would be crossing over between them, that is, the more tightly they would be linked. Yet it was generally assumed, on the basis of two-locus analyses, that even the closest linkage would be broken rather rapidly in a large randomly breeding population unless there was complete cross-over inhibition. Linkage was therefore considered unimportant in evolution. Recent analyses, making use of computer simulation, indicate that the classical assumptions are irrelevant, if not misleading, where multi-locus systems of balanced fitness are involved (Franklin and Lewontin unpublished). This suggests that it will be more meaningful in evolutionary analysis to work with the properties of whole chromosomes or chromosome arms than with summed individual gene frequencies. A complete reassessment of the role of linkage in evolution and in the maintenance of the unity of the gene pool is made necessary by these revolutionary findings.

Ecological Factors

The diversity of the environment likewise reduces the impact of natural selection. The number of selective values a phenotype can have is determined by the number of physical and biotic environments in which it occurs. A gene that makes a negative contribution to fitness in certain environments may be retained in a population provided its contribution is positive under other conditions. This multiple capacity increases the chance that the gene will be preserved in the gene pool or at least that its elimination will be greatly retarded.

Opposing Selection Pressures

Since the genotype as a whole is an interacting, integrated system, virtually all aspects of the *phenotype* are *a compromise between opposing selection pressures.* An increase in body size, for instance, might be an overall advantage for an organism, but, since it would necessitate changing the relative size of numerous organs, opposing selection pressures would be set up until the various proportions had been reconstructed. An increase in tooth length (hypsodonty) was of selective advantage to primitive horses shifting from browsing to grazing in an increasingly arid environment. However, such a change in feeding habits required a larger jaw and stronger jaw muscles, hence a bigger and heavier skull supported by heavier neck muscles, as well as shifts in the intestinal tract. Too rapid an increase in tooth length was consequently opposed by selection, and indeed the increase averaged only about 1 mm per million years (Simpson 1944). Genetic variation will not be rapidly depleted when opposing selection pressures are so nearly balanced.

Moreover, many selection pressures are severe only during crises—epidemics, exceptionally severe winters, droughts, and the like (catastrophic selection). Such crises, however, tend to be localized and the genes lost during a crisis may subsequently be restored by gene flow unless a peripherally isolated population is involved (Lewis 1962). All these factors combine to soften the impact of natural selection and to prevent it from depleting genetic variation too rapidly.

Diversity of the Environment in Time

Temporary and secular changes in selection pressure. The storage of genetic variation in populations is greatly affected by shifts in selection pressures. These shifts may involve either the intensity or the direction of the selection pressure. Any sudden change in population size nearly always results in a change in the intensity of selection. A sudden contraction may not only increase homozygosity, but as a consequence may also permit the emergence of deviant phenotypes (Lerner 1954) that have a selective advantage under special conditions. Increased selection pressure and reduced population size unquestionably lead to an increased rate of selective elimination of genes from a population. Conversely, temporary relaxation of selection pressure will permit survival of otherwise inferior genotypes, facilitating the occurrence of rare genetic recombinations that would be impossible during periods of severely adverse selection. If one of these "improbable" genotypes represents a new adaptive peak, it will be retained in the population when a period of higher selection intensity is reestablished. There is observational evidence that cycles in population size affect the genetic variability of populations in this manner (Ford 1964).

Shifts in the direction of selection pressure. All environments are changing; this means, in terms of evolution, that the selection pressures shift continuously. Such a shift might ease the selection pressure on those genes and genotypes that had previously been under the heaviest pressure or even favor them selectively. Detailed samplings of polymorph or otherwise genetically variable populations have indeed established in many cases a parallelism between environmental and genetic fluctuations. Here one can really watch evolution at work. These fluctuations demonstrate that genes may have very different selective values under different environmental conditions (Ford 1964).

Temporal changes in selection pressure can be classified into seasonal changes, secular fluctuations and cycles, and long-term trends, depending on the time interval involved in the change.

Seasonal changes. One of the first seasonal changes to be described in detail concerns a population of ladybird beetles (*Adalia bipunctata*) near Berlin (Timofeeff-Ressovsky 1940). The population occurs in two phenotypes, which may be designated "black" and "red." The frequency of blacks among adults dropped from 55–70 percent (in different years) at the beginning of the winter to 30–45 percent at the end of the winter, owing to the higher mortality of blacks during hibernation. However, the superior viability of blacks during the hot season permitted them to return by the end of the summer to the original frequency of 55–70 percent. (See also Creed 1966.) A fluctuating polymorphism cannot be maintained by a balance of selective advantages of the two phenotypes, because, owing to random fluctuations, one or the other type would sooner or later become fixed (Kimura 1955). It can be maintained only by a superimposed superiority of the heterozygotes.

By far the best evidence for cyclical fluctuations comes from *Drosophila*. Seasonal changes in the relative frequency of gene arrangements were first discovered in Russia by Dubinin and associates and were thoroughly studied by Dobzhansky and his school. For instance, in a population of *D. pseudoobscura* occurring at Piñon Flat, Mount San Jacinto, California, the gene arrangement Standard decreased regularly from March to June while the frequency of another arrangement (Chiricahua) increased correspondingly. The reverse change took place between June and November (Dobzhansky 1951).

Secular changes. In nearly every case in which polymorphic populations were studied carefully over a series of years, shifts in the frequencies of morphs were found. For instance, in a population of *Cepaea hortensis*, Schnetter (1951) found an increase of 16.2 percent in the frequency of yellow unbanded snails during a series of years (1942-1950) that were warmer and drier than average. There was a reversal when the climatic trend changed. In most of these secular changes that is a fairly well-defined correlation between climatic cycles and shifts in morph ratios.

Long-term-trends. Some shifts in gene frequency are long-term trends extending over a considerable number of years. The selective factor leading to such a shift in gene frequency is only rarely understood. The most thoroughly studied case is that of so-called *industrial melanism* (Ford 1964). In a number of species of moths the normal grayish or whitish morph is being replaced in the manufacturing districts of England and western Europe by black or dark forms. The black morphs were rare or absent 100 years ago (before the period of industrialization): now the light forms are rare. In some cases the shift was amazingly rapid, requiring less than 50 years. The swiftness of the shift indicates a selective advantage of the dark gene amounting to somewhere between 10 and 30 percent. The magnitude of its sudden selective advantage was explained by Kettlewell

(1961), who found strong predation by birds on the adult moths resting on the bark of trees. In unpolluted areas with pale, lichen-covered tree trunks, the light-colored moths are cryptically colored; in sooty areas, the black moths (Table 9.3).

Climatic trends seem to be responsible for other shifts of polymorph ratios observed in many species (Mayr 1963:242). The best-studied case of shift in morph ratio occurs in the tiger moth *Panaxia dominula*, of which several small and isolated colonies in the south of England were studied (Ford 1964). A rare gene producing the "*medionigra*" phenotype in heterozygous condition increased in frequency from the 1920's to 1940 at one of these stations until it reached a frequency of 11.1 percent and since then has steadily decreased to a level of about 3 percent. It is not known why the currently disadvantageous gene should have been so highly advantageous before 1940.

The cases of strong shifts in morph ratios cited here are selected from an extensive literature. They are more than compensated for by the evidence of high stability in gene ratios of human blood groups, Neolithic *Cepaea*, *Xiphophorus maculatus*, *Sphaeroma serratum*, and many other cases (Mayr 1963:211).

Diversity of the Local Environment

Ecological mosaicism. In the preceding section I examined evidence for the heterogeneity of the environment in time, and in this section I will consider heterogeneity in space. It would evidently be of great selective advantage for a species to be able to utilize simultaneously a number of different aspects of the environment. Since there are limits to the ecological tolerance and efficiency of a given genotype, a population will be more

Table 9.3. Differential mortality of the Peppered Moth (*Biston betularia*) released in different woodlands (from Kettlewell 1961).

No. released		Type of woodland	No. eaten by birds		Percent recaptured alive	
Melanic	Pale		Melanic	Pale	Melanic	Pale
Equal		Grayish	164	26	—	—
Equal		Sooty	15	43	—	—
473	496	Grayish	—	—	6.3	12.5
447	137	Sooty	—	—	27.5	13.0

successful if it is able to diversify genetically and thus to broaden its utilization of the environment and expand into various subniches. Such genetic diversification does, however, pose the problem of maintaining the harmonious coadaptation of the gene pool while producing a number of different phenotypes, each specializing in a particular aspect of the environment.

Ludwig's theorem. Since the time of Darwin and his predecessors various naturalists have theorized that it would be advantageous for a species to have "varieties" utilizing different subniches. The more widespread and numerous a species, the greater the chance that it would have such ecological variants. This thesis was placed on a firmer basis by Ludwig's (1950) calculations that a genotype utilizing a novel subniche could be added to the population even if it were of inferior viability in the normal niche of the species. A number of other investigators have since calculated under what circumstances such a system of ecological polymorphism can be maintained without superiority of heterozygotes.

A possible case of such ecological polymorphism is that of the gene arrangements in *Drosophila.* Mayr (1945) suggested that new inversion types might be "able to occupy ecological niches that are inaccessible to other members of the ancestral population." Dobzhansky found much evidence (summarized in 1951) to support the thesis that each gene arrangement is adapted to a different subniche within the general habitat of the species. Hence the greatest amount of this polymorphism is found in the most favorable areas of the species range (Chapter 13). However, this is only one of several possible interpretations of chromosomal polymorphism.

Substrate polymorphism. The situation is comparatively simple when phenotypic polymorphism is correlated with a polymorphism of the substrate. The best-analyzed case is that of the banded snails (*Cepaea*) of Europe, formerly often cited as a case of neutral polymorphism. In *Cepaea nemoralis* Cain and Sheppard (1950), studying strikingly different colonies often only a few meters apart, found a close correlation between certain morphs and definite habitats, for instance unbanded yellow and shortgrass downs, unbanded reddish and beech woods, or banded yellow and hedgerows (Fig. 9.1). They postulated that predation by the Song Thrush (*Turdus philomelos*) leads to the elimination of the most contrasting morphs and conversely to the preservation of those that best blend with the substrate.

Substrate polymorphism entirely controlled by predator pressure would seem a rather wasteful method of adaptation. There is evidence, however, that several mechanisms greatly mitigate the effect of the predators, such as differential survival of the different genotypes owing to physiological differences and habitat selection (Mayr 1963; Wolda 1967).

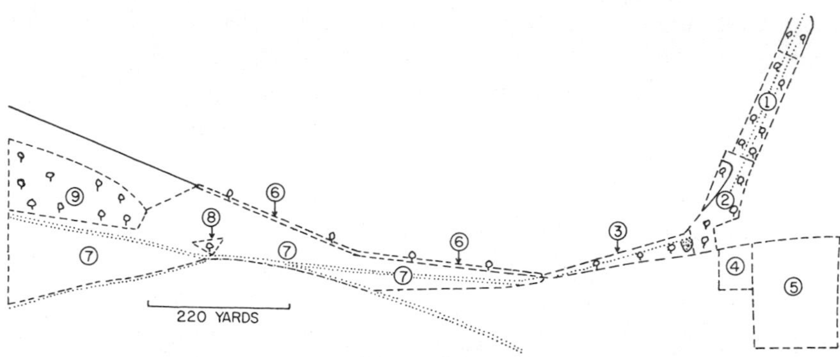

Fig. 9.1. Composition of *Cepaea nemoralis* populations in neighboring habitats at Wiltshire, England. Areas 1, 2, 8, 9 are beechwoods, 5 and 7 shortgrass downs, 3, 4, 6 intermediate. The percentages of yellow snails in the nine areas are: 1: 28.5; 2: 24.5; 3: 35.7; 4: 35.5; 5: 47.7; 6: 15.7; 7: 45.1; 8: 20.0; 9: 16.0. (From Sheppard 1952.)

The more closely a species is studied, the more likely it is that some evidence will be found for ecological polymorphism or gradual ecological variation. The variation is still largely ignored by ecologists, most of whom discuss the ecological requirements of species in a strictly typological manner.

Mimetic polymorphism. In many species of (mostly) tropical butterflies several color types occur within a single population, each mimicking a different species (its "model") distasteful to its predators and thereby providing protection from insect-eating birds. Such mimicry works best when the mimic is sufficiently rarer than the model to prevent the development of counterconditioning. The frequency of butterflies as food of birds and the speed of conditioning of birds against distasteful prey are now sufficiently well established by observation and experiment to dispel all doubts as to the selective advantage of mimetic polymorphism. There have been various illuminating discussions of this phenomenon in the recent literature (Ford 1964, Wickler 1968). In the genus *Papilio* only the females are mimics; in some other genera both sexes may mimic. A good example is the African Nymphalid butterfly *Pseudacraea eurytus,* in which Carpenter (1949) recognized 33 color forms, each mimicking some species of the acraeid genus *Bematistes.* Male and female *Pseudacraea* usually mimic different models. Twelve or more forms of *Pseudacraea* may coexist at a single locality. If under special circumstances the model becomes rarer than the mimic, the latter may "go to pieces." As a result of relaxed selection the precision of the mimicry breaks down and a considerable part of the population consists of intermediates between the elsewhere sharply discontinuous types. This has happened in *Pseudacraea* on some islands in Lake Victoria (Table 9.4). This phenomenon proves that the precision of the mimicry can be maintained only through the continued vigilance of natural selection. It also proves that the now existing broad gaps between the various types of mimics are the result of selection and hence a secondary condition. Even though originally established through small mutational steps, the polymorphism in any local population of *Papilio dardanus* is now controlled by a few major switch genes or supergenes. However, when one crosses different geographic races one gets arrays of intermediates between the morphs. Such an experiment proves that the strict discontinuity of the morphs within a population is the result of selection of the apropriate genetic background. The sharp definition is lost when two such polygenic complexes from different regions are mixed.

Adaptive nature of polymorphism. The great diversity in the forms of polymorphism raises a number of terminological difficulties. Dobzhansky (1951) has remarked correctly that in most cases of observed polymorphism it is quite impossible without experimental evidence to determine what portion is due to superiority of the heterozygotes (balanced polymorphism *sensu stricto*), what portion is due to mutation and immigration pressure, and what portion is due to a selective balance of the phenotypes (Ludwig effect). Yet, regardless of the mechanism, polymorphism "or any other kind of diversity of sympatric forms increases the efficiency of the exploitation of the resources of the environment by the living matter." Dobzhansky therefore refers to polymorphism as "adaptive." Whether or not one considers this designation appropriate, species with ecological polymorphism

Table 9.4. Inverse relation between frequency of mimic and precision of mimicry in *Pseudacraea eurytus* on islands in Lake Victoria (from Carpenter 1949).

Location	Frequency of mimic in model-mimic association (percent)	Frequency of deviant variants among mimics (percent)
Mainland	18	4
Damba Island	62	35
Bugalla Island	73	56
Kome Island, 1914	23	51
1918	68	54
Buvuma Island	43	28

do have two attributes that contribute to their adaptedness. First, as stated by Dobzhansky, their greater genetic diversity permits them to utilize the environment better and more completely. This statement is well supported by observational as well as by experimental evidence. Chromosomally polymorphic populations of *Drosophila pseudoobscura* produced more flies and more biomass under standard conditions than monomorphic populations (Dobzhansky and Pavlovsky 1961). Chromosomal and visible polymorphism are only special forms of genetic variability, and more and more evidence is accumulating to show "that populations with greater genetic variability have larger population sizes" (Ayala 1968). Increased variability usually also enhances the competitive ability of a population in comparison with others. Polymorphism is based on and produced by definite genetic mechanisms, such as genes for differential niche selection and the heterosis of heterozygotes. A population that has not responded to selection for such mechanisms and therefore lacks polymorphic diversity is more narrowly adapted, more specialized, and therefore more vulnerable to extermination. The widespread occurrence of genetic mechanisms that produce and maintain polymorphism is directly due to selection and is in itself a component of adaptiveness. It seems appropriate, therefore, to speak of "adaptive polymorphism" (see also Dobzhansky 1968).

Diversity of the Species Range

For the sake of simplicity I have acted up to this point as if each population were a completely closed system. This is, of course, not the case. Indeed, as will be shown in Chapters 14 and 17, the most important source of genetic variability in any but the most isolated populations is the immigration of genes from other locally adapted populations, each of which has a somewhat different assortment of genes. Such gene flow not only introduces new genes but also restores genes that had been temporarily lost during a catastrophe or owing to accidents of sampling.

Assortative or Random Mating

The existence of definite preferences in the choice of mates may affect gene frequencies. *Homogamy,* the preference for a phenotypically similar mate, will tend to favor inbreeding and the production of homozygotes for the genes controlling that part of the phenotype involved in the homogamy. Other things being equal, this will facilitate the exposure of genes in homozygous condition to natural selection, which in turn may affect gene frequency. On the other hand, heterogamy, the preference for a phenotypically dissimilar mate, favors the increase of rare genes.

Inverse Relation between Fitness and Gene Frequency

The rarer allele is usually at a selective disadvantage not only because its low fitness is the very cause of its rarity but also because the remainder of the genotype (the genetic background) is continuously selected to maximize the fitness of the most common allele. Most deleterious recessives illustrate this rule. There are, however, situations where the rare gene is favored for no other reason than that it is rare. Such a property would automatically help to maintain genetic variation in a population where it occurs (Haldane 1954a). A very simple and previously unexpected mechanism for the maintenance of rare genes in populations was discovered by Petit, who presented her findings from 1951 to 1968 in a series of reports (Petit 1958, 1968). She showed that rare genotypes in *Drosophila melanogaster* cultures have a higher mating success than common ones (Fig. 9.2). Females, which usually exercise the choice in *Drosophila* courtship, mate more readily with males of a rare than of a common genotype. Ehrman (1967) extended these findings to gene arrangements in *pseudoobscura* and other *Drosophila* species. It is evident that the ability to make such a choice is a very efficient mechanism for the maintenance of genetic variability and that it should therefore be greatly favored by natural selection. It is still unknown how widely this mechanism is distributed in the animal kingdom (see Ford 1965 for *Panaxia*) and what the sensory clues are by which females discover which males are rare and which common and to what extent the males may contribute to the unequal mating success.

Additional evidence for the selective advantage of the rarer of two alleles continues to accumulate. For instance, it was found that in a series of isoalleles for an enzyme the selective advantage of the heterozygote increased with the rarity of one of the alleles (Kojima and Yarbrough 1967). The competitive inferiority of larvae of *Drosophila melanogaster* mutants at medium frequencies may disappear or turn into superiority when the mutant becomes rare as compared to the competing genotype (Petit 1968). In the case of polymorph snails and insects, a rare morph is somewhat protected against predators because the predator's "search image" has become conditioned to the more common morph (*apostatic selection*) (Clarke 1962). A rarer gene will also be favored if several genotypes differ somewhat in their ecological requirements and each obeys density-dependent factors independently of the other genotypes. As a model I shall postulate two genotypes in *Drosophila*, one favored in the dry and the other in the moist part of the food medium. Mortality will be increased in the more crowded part of the medium. Whichever gene is rarer will be favored in the part of the food medium for which it is specialized because there

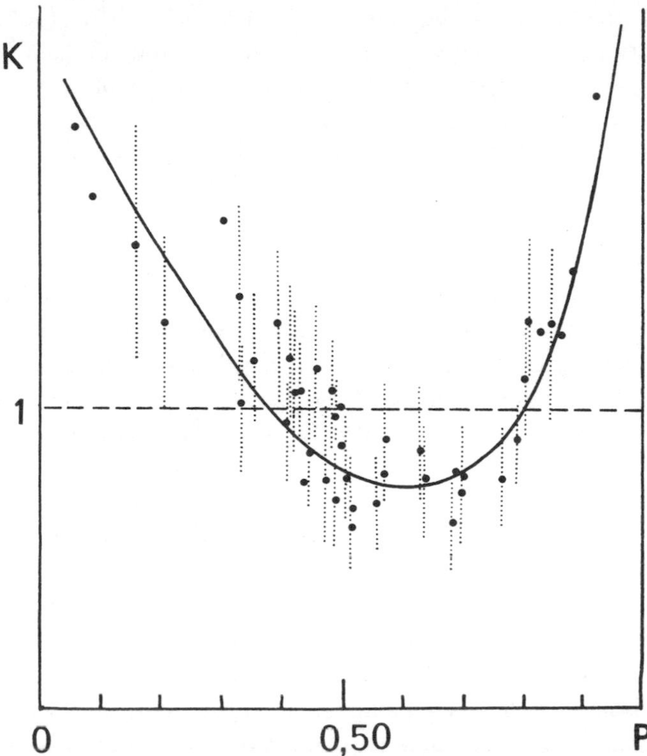

Fig. 9.2. Preference for the white allele as a function of its frequency in the population. Abscissa: frequency (P) of the white allele; ordinate: coefficient of sexual selection (K). $K = A/a \div B/b$. A (females) and a (males) are wild type, B (females) and b (males) are white. (From Petit 1968.)

it will be less subject to crowding. The effect of relative frequency on survival has been stressed by Teissier (1954), who points out that shifts in fitness with changes in frequency permit a polymorphism not maintained by superiority of heterozygotes. It had already been known that a mixing of genotypes leads to increased yield in plants (Gustafsson 1953). The relative importance of this mechanism for the maintenance of genetic diversity of populations is entirely unknown, yet it is evident that wherever it occurs it will contribute to the diversity of the gene pool.

Genetic Load

It is now evident that numerous devices, such as those discussed above, counteract the homogenizing tendency of natural selection to a greater or lesser degree. These mechanisms greatly enhance the evolutionary plasticity of populations, but they are bought at a price. The storage of a great variety of genes results in every generation, through recombination, in the inevitable segregation of an appreciable number of inferior genotypes. Is it not a waste to produce inferior genotypes so prolifically? Should it not be possible to build up a genetic architecture of the population that would avoid such waste in every generation? Those who postulated the possibility of an "ideal genotype," a genotype in which every member of the population has the same maximal fitness, developed the concept of *genetic load.* This was defined as the difference between the mean fitness of the population and that of the optimal genotype. Some six or seven different kinds of genetic load were distinguished (Mayr 1963:253–258) and elaborate calculations were made of the relative contribution to the total load made by mutation, the segregation of superior heterozygotes, and the like.

It is now becoming increasingly evident that this whole approach is misleading. It is based on a set of assumptions that have no real validity, primarily that of the existence of an optimal homozygous genotype. To a considerable extent, this approach presupposes an environment uniform in time and space, and ignores the fact that each genotype is composed of tens of thousands of variable loci, each potentially making a plus or minus contribution to fitness. Finally, it ignores the fact that the production of a certain percentage of deleterious homozygotes or epistatic (synthetic) lethals can be largely neglected, since each successfully reproducing individual produces hundreds, thousands, or even millions of zygotes. (This last statement is not quite valid for species with parental care and a small number of offspring.)

Even if one considers the concept of genetic load as unhelpful or misleading, one nevertheless benefits by dividing the genetic variation of populations into two classes, that caused by the *input* of new genetic factors into a population and that caused by the *segregation* of new genotypes in a population (as for instance homozygous offspring of heterozygous parents). Both types of variability unquestionably occur, and ever since the publication of H. J. Muller's famous essay "Our Load of Mutations" (1950), biologists have argued heatedly about what proportion of variation each source contributes to the observed genotypic variability of populations. See Spiess (1968: 195–199) and Brues (1969) for recent discussions of genetic load.

Input "load." This has been defined as "the presence, owing to mutation and immigration (gene flow), of inferior alleles in a population, and their delayed elimination owing to various protective devices," such as those described above (for example, recessiveness).

Muller and his followers argued that most variation in populations was due to the occurrence of deleterious recessive mutations, and that the preservation of such mutants in a population was severely limited by the "genetic deaths" of the recessive homozygotes. His opponents argued that the genetic variation in natural populations was far too great to be ascribed largely or entirely to genetic input, and that much of it must be the result of segregation of allelically or epistatically balanced genotypes (see p. 155). Schull and Neel (1965) examined the opposing views very carefully but were unable, on the basis of the evidence then available, to decide which was correct.

It became apparent that entirely new techniques had to be developed before the stalemate between the opposing opinions could be broken. Several such methodological advances have been made in recent years and they have indeed helped to settle the issue, but before they are described a word must be said about the second component of variability in natural populations.

Segregation variability. Since the units of natural selection are whole individuals, superior genotypes are often the somewhat improbable result of recombination, which brought together a superior allelic or epistatic balance of genes. In spite of the existence of various protective devices (supergenes, linkage), most of these superior balanced combinations will be broken up by segregation and recombination during the production of the next generation. Every overdominant heterozygote, for instance, will segregate out 50 percent of the inferior homozygotes. In view of this "wastefulness" of balanced variability Muller (1950) expressed the opinion that a mutant gene "advantageous in its heterozygous degree of expression but deleterious homozygously" would usually "become replaced after a while by mutant genes of less deviant expression . . . which give an equivalent advantage when they [are] present homozygously." Since this statement was made, a great deal of evidence has accumulated showing that a homozygote is not necessarily superior in principle to a heterozygote. The latter may not only have greater developmental plasticity and a better capacity for developmental homeostasis but also be favored by selection because it gives the gene pool a greater evolutionary plasticity. By producing three different genotypes in each generation, heterozygotes provide the potential for a rapid shift in case of changing environmental conditions.

That homozygotes are not necessarily the pinnacle of perfection has been shown repeatedly by X-ray treatments. Wallace (1959b) irradiated largely homozygous stocks of *Drosophila melanogaster* and found that the induced heterozygosity of the irradiated stocks resulted in a slight but definite increase in viability. The results of many similar experiments were summarized by Wallace (1968a:253–257). In the meantime the effects of mild irradiation have also been studied in nonhomozygous populations. Irradiated laboratory populations of *D. serrata* decreased in size during the first six weeks, presumably owing to the elimination of carriers of deleterious mutations. Thereafter the irradiated populations became considerably larger than the controls (Ayala 1968:Fig. 2). Even though some of this improvement may be due to the replacement of inferior alleles, much of it is presumably due to genetic enrichment and the establishment of new allelic and epistatic balances.

The main support for the assumption that most genetic variability is input variability came from the assumption that cases of genic overdominance "are so rare and special (although they tend to make themselves conspicuous) as almost to prove that they are contrary to the rule" (Muller 1950). Although there was a great deal of indirect evidence indicating that Muller's assertion was not correct, in 1950 techniques of genetic analysis were not sufficiently refined for a decisive refutation.

What kind of evidence is needed for such refutation? Evidence of simultaneous genetic variation (polyallelism) at very many loci in a population at frequencies that are too high to be explained in terms of not yet eliminated deleterious mutations. The normal genetic analysis concentrates on a single locus at a time, and by inbreeding individuals attempts to determine the frequency of concealed heterozygotes. It determines gene frequencies at single loci through a study of phenotypes, but it cannot undertake a simultaneous analysis of numerous loci of a single individual.

The relatively few exceptions were genes where the genotype can frequently be inferred directly from the characters of the phenotype, like the blood-group genes in man (although it is still impossible to distinguish, for instance, AO and AA or BO and BB); or karyotype variation, like the gene arrangements in *Drosophila* so successfully studied by Dobzhansky and his school, and human chromosome aberrations (for example, Down's syndrome).

The electrophoresis technique has, almost overnight, changed the situation quite dramatically. Proteins produced by different alleles of the same locus may differ in their electric charge and be separable by electrophoresis (Fig. 9.3). This technique permits (with certain reservations) the direct

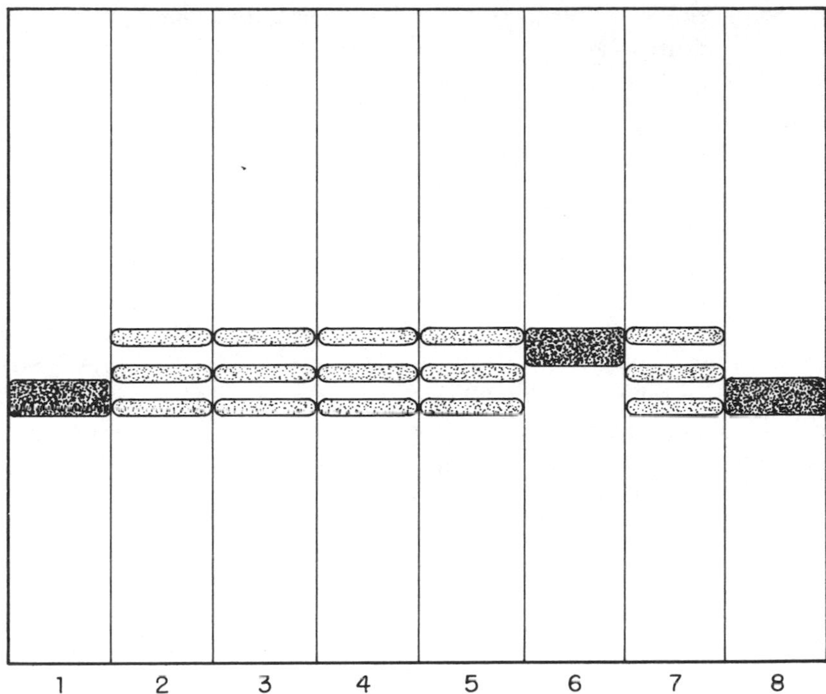

Fig. 9.3. The starch gel pattern of occurrence of different alleles of an enzyme gene in different individuals of *Drosophila pseudoobscura*. Homozygous individuals have only a single band, heterozygous ones have several bands. (After Prakash et al. 1969.)

determination of protein variability of individuals. Hubby and Lewontin (1966) and Lewontin and Hubby (1966) studied electrophoretic variants at 18 loci in 5 populations of *Drosophila pseudoobscura*. Of the 18 loci, 7 (39 percent) were definitely polymorphic in more than one population. Each of the 5 populations was polymorphic for about 30 percent of these loci and about 12 percent of the genome of each individual was estimated to be heterozygous. These are considerable underestimates, for a variety of reasons, "by perhaps as much as 4:1 of the amount of allelic variation actually present" (Lewontin 1967a). If that is the case, then about 50 percent of the loci of every individual are heterozygous, and essentially every locus in a population polymorphic with a mode of three or four alleles segregating. These figures are remarkably close, (certainly of the same order of magnitude) to Wallace's (1958) old estimate that about 50 percent of the loci of a given individual in an open, natural population may be occupied by dissimilar alleles. This work does not stand alone. Analyses of human populations and wild-mouse (*Mus musculus*) populations (Selander 1969) reveal similarly high levels of genetic variability.

Ever since Wallace (1958) published his high estimates of genetic variability in natural populations, some geneticists have questioned the capacity of populations to put up with such high levels of variability. Obviously, it was much too high to be maintained by recurrent mutation, hence it would have to be ascribed to heterozygote superiority at the variable loci. But, making the standard assumption of classical population genetics that each gene has an independent fitness value, one was forced into calculations such as this: "If our estimate is correct that one-third of all loci are polymorphic, then something like 2000 loci (in *Drosophila*) are being maintained polymorphic by heterosis. If the selection at each locus were reducing population fitness to 95 percent of maximum, the population's reproductive potential would be only 0.95^{2000} of its maximum or about 10^{-46}. If each homozygote were 98 percent as fit as the heterozygote, the population's reproductive potential would be cut to 10^{-9}. In either case, the value is unbelievably low" (Lewontin and Hubby 1966:606). Here then was an impasse: theory did not accord with the established levels of genetic variability in natural populations. How can this conflict be resolved?

The solution is actually quite obvious. It consists simply in abandoning the erroneous assumption, made by beanbag genetics, that each gene has an independent fitness. This is, of course, not true, since whole individuals and not single genes are the target of selection. Furthermore, density-dependent factors will tend to keep the population near a standard level, thus always guaranteeing the survival of a certain percentage of the offspring (see below, pp. 159–160).

The fact that whole individuals and not single genes are the target, and hence the unit, of selection forms the basis of Milkman's (1967) arguments. Let us say that a pair of a certain species of animals produces 200 offspring, of which an average of 2 survive, as cofounders of the next generation. If each of the heterozygous loci slightly adds to fitness, the two survivors will—on the average—be the most heterozygous zygotes produced by these parents. A similar favoring of heterozygous offspring will obtain for all the other reproducing members of the deme. The fact that individuals with higher average homozygosity will tend to fall by the wayside will not in any way reduce the reproductive potential of the population, which is presumably determined by the carrying capacity of the habitat and further regulated by density-dependent factors. Considering the high number of variable loci (the estimate of 2000 loci made for *Drosophila* by Lewontin and Hubby is surely a minimum), many of the progenitors of the next generation of every deme will certainly be homozygous at some of the "2000" variable loci. It is unlikely that the selective superiority of the heterozygotes is so great at every locus that only heterozygotes will be selected for. Local demes are far too small to produce a sufficient number of all-heterozygotes. Furthermore, there may well be a gradually diminishing contribution to fitness from adding heterozygous loci. "The upper limit of fitness could well be as low as two or three times the mean fitness" (Sved, Reed, and Bodmer, 1967:473).

If there is complete inbreeding in a small deme, even relatively strong selection in favor of heterozygotes cannot prevent the occasional fixation of certain loci owing to accidents of sampling. It is owing to a steady, rather high gene flow between demes that such homozygous states are of rather short duration, and it is such gene flow that is ultimately responsible for the maintenance of high levels of genetic variability throughout the range of most species.

It is now obvious that the whole concept of genetic load, resting on an assumption of the independent fitness of individual genes, is meaningless. Yet one should not go to the other extreme and think that the ideal level of fitness is provided by heterozygosity at all loci. Indeed, many loci seem to conform to the classical ideal of an invariant "wild type." Cytochrome c, for instance, has not been found to vary within any one species, and is often totally invariant through an entire order of animals. Many other genes controlling biologically important molecules are completely invariant, except for an occasional rare, and obviously deleterious, recessive. There is now much research going on to determine the ratio between invariant loci and such that participate in balanced polymorphism (see, for instance, Perutz and Lehmann 1968 for hemoglobin).

There is still another way to reduce the potential loss due to deleterious homozygotes. If there are only two alleles (of equal frequency) at such a locus, 50 percent of the zygotes are homozygous and will be lost. However, since the number of homozygotes is $1/n$ of the number of alleles (n), any accumulation of (iso)alleles at this locus greatly reduces the number of homozygotes.

The Cost of Evolution

Using the term "genetic load" is only another way of saying that stabilizing selection continuously eliminates genotypes that deviate too far from the optimal genotype. The structure of the gene pool determines the proportion of individuals that will fall outside the limits of "normalcy," that is, outside the group of genotypes that will have a good chance *not* to be selected against. Such selection will go on continuously even in a stable environment, because novel genotypes are continuously segregated owing to recombination and because there is a steady input of genes owing to immigration and mutation.

A new selection pressure is added whenever directional selection favors the replacement of an existing allele by a new allele. How rapidly can such a replacement take place and how much of a "cost" does it involve? Haldane (1957) was the first to recognize this problem and to try to supply a quantitative answer to it. He assumed that gene substitution in a species occurred through the "death" in each generation of a constant fraction of individuals carrying the particular gene. On that basis he calculated that the total number of "selective deaths" required to complete one substitution is of the order of 30 times the population size, independent of the intensity of selection. Haldane concluded from this that the replacement of one gene by another had to be a slow process, requiring an average of 300 generations per substitution. Furthermore, he and Kimura (1960) estimated that the survival of the population would be jeopardized if gene substitution took place at more than about a dozen loci at any one time. The more general conclusion was that evolutionary change is an exceedingly slow process. If two species differ at 1000 loci, Haldane estimated that it may have taken at least 300,000 generations to complete speciation.

Although much evolution is, indeed, slow, naturalists know many cases of extremely rapid speciation, involving no more than several thousand generations (see Chapter 18). Zimmerman (1960), for instance, has pointed out that five endemic Hawaiian "species" of the pyraustid moth genus *Hedylepta* are restricted to the banana plant, which the Polynesians introduced to Hawaii only 1000 years ago. These five allopatric species seem to have evolved in this extraordinarily short time from a parental *Hedylepta*, host-specific on a palm.

It is evident that some of the basic assumptions made by Haldane and Kimura cannot be correct. These false premises were first pointed out by Mayr (1963:260–261), and the objections were subsequently stated more concisely by Brues (1964), van Valen (1965), Sved (1968), and Maynard Smith (1968). Some of Haldane's questionable assumptions may be paraphrased as follows:

(1) *Selection for gene substitution at different loci is entirely independent.*

The contributions to genetic death made by each gene substitution are treated as if they were independent of each other. It is quite likely, however,

> that they are partially synergistic, when coming together in a genotype. Intuitively I would assume that an individual with three deleterious genes, each with a 1-percent loss of fitness as compared to the normal allele, would have a loss of fitness far in excess of 3 percent, perhaps amounting to as much as 10 percent. But even if fitness was decreased only by 3 percent, the probability of the elimination of this genotype is greatly increased. To be sure, if each of these genes has a frequency of 1 in 1000, the probability that all three genes would come together in a single individual would be only 1 in 1 billion. This, evidently, would be negligible. However, since populations contain numerous genes with minor deviations (+ or −) from mean fitness, there always will be some genotypes with several poor genes. A single death of one such genotype will remove several deleterious genes simultaneously (Mayr 1963:260).

Maynard Smith (1968) likewise supports the idea of a synergistic effect of the simultaneous replacement of genes at several loci and likewise points out that the number of "genetic deaths" is drastically reduced by selection against individuals with several deleterious genes. He terms this "threshold selection." It is essentially the same principle that was applied above (p. 157) to losses supposedly caused by the genetic load.

(2) *Each gene replacement improves the fitness of the population as a whole.* Haldane makes the tacit assumption that gene substitution leads to a steady increase in the fitness of the population. Actually, the competition among genotypes within a deme may have little bearing on the fitness of the given deme.

> It is probable that the overall "fitness" of a population over long periods is essentially constant (with the limits set by the capacity of the niche) and that either the elimination of deleterious genes or the addition of superior new genes simultaneously changes the selective values of the remaining genes. This effect will mitigate the impact of gene substitution and permit an increase in the rate of substitution. If the total fitness remains constant and likewise the number of zygotes eliminated (not reproducing) in each generation, then the concept of "load" becomes rather nebulous. Under these premises gene substitution could be accelerated or retarded, regardless of "load," as long as much of the mortality is density dependent and does not result in lowering the ability of the species to compete with other species (Mayr 1963:260).

When the various selection pressures have been analyzed far more completely than heretofore, it may be possible to partition the "cost of selection" into various components. If the carrying capacity of 1 square kilometer of a certain habitat is 1000 individuals of species A, the intrademe selection determining which particular 1000 individuals will "represent" the species in the next generation may have to be charged to a cost component that may not be of great evolutionary consequence.

> In a species where one pair normally produces 100 juveniles, 98 die on the average prior to reproduction. Mean population size will remain fairly constant; consequently it is irrelevant what factors are responsible for the mortality of the 98. This part of the offspring is expendable and will be eliminated regardless. If an allele is converted from advantageous to deleterious, it may simply be charged to this expendable surplus. The earlier in the life cycle such mortality occurs, the less effect it will have on the fitness of the population as a whole, by reducing intrapopulation competition at later stages in the life cycle. The higher a percentage is killed by gene substitution, the lighter will be the "ecological load" on the remainder of the population (Mayr 1963:261).

Gene substitution, under these circumstances, will not create a formidable "load." Sved (1968:292) comes to very similar conclusions: "The selective value consistent with a given death rate or set of fitness differentials is not independent of the rate of substitution but is approximately inversely proportional to it. The result is little affected by the degree of dominance or by the initial gene frequency."

Wallace adds another argument. He believes that selection operates to stabilize the average fitness (measured in surviving daughters per mother) of a population at 1.00. On this assumption the importance of density-dependent factors for the Darwinian fitness of populations is clear: "novel sources of premature death need not be met by a corresponding increase in the number of progeny produced. They are met rather by a reduction in population size . . . Genetic load is unrelated to population size and, consequently, to the fitness of a population" (1968b:106–107).

Brues goes one step further. She attacks with convincing arguments the propriety of referring to the evolutionary substitution of genes as "cost of natural selection" (Haldane) or "cost of evolution" (Mayr). Incorporating a new favorable mutation into a genotype improves its viability and "costs" the population less than not to replace the inferior gene. Survival will actually be enhanced, unless there is a very tight density-dependent control of population size. This consideration also makes implausible Haldane's estimate that gene substitution can take place only at a few loci at a time. "More likely limiting factors on multiple gene replacement would be paucity of useful mutations, and the fact that mutations which were all individually advantageous in a particular genetic background might conflict with one another when they became common enough to interact" (Brues 1964).

It is now obvious that Haldane's conclusion of extremely slow evolutionary rates is not correct. Nevertheless his 1957 essay called attention to a previously neglected area of evolutionary theory and has had an enormously stimulating impact. When some of Haldane's erroneous assumptions are replaced, it becomes possible to make numerical estimates of the rate of gene substitution that seem consistent with observed evolutionary rates.

Conclusions

Genetic variation in a population is controlled by three sets of factors: (1) the input of new genetic material through mutation and immigration, (2) the erosion of this variation by selection and errors of sampling, and (3) the protection of the stored variability by cytophysiological devices and ecological factors. The diversity of these protective devices indicates that they are strongly favored by natural selection. Too uniform an "optimal" genotype would evidently favor too uniform a survival and would result in the double disadvantage of high density-controlled mortality in favorable periods and of extinction under adverse conditions.

The genetic variation of a population is not merely a "mutational load," because heterozygote superiority and selection for ecological diversity greatly add to it. Indeed, heterozygote superiority may well be responsible for a greater portion of the genetic variation of wild populations than all other factors combined.

All natural populations contain abundant genetic variation that serves as potential raw material for evolutionary change. The problems that this variation poses for adaptation and speciation are considered in the ensuing chapters.

10 • The Unity of the Genotype

The procedure of classical Mendelian genetics, of studying each gene locus separately and independently, was a simplification necessary to permit the determination of the laws of inheritance and to obtain basic information on the physiology of the gene. When dealing with several genes, the geneticist was inclined to think in terms of their relative frequencies in the population. The Mendelian was apt to compare the genetic contents of a population to a bag full of colored beans. Mutation was the exchange of one kind of bean for another. This conceptualization has been referred to as "beanbag genetics."

It has long been apparent that the thinking of beanbag genetics is in many ways quite misleading. Recent work in population genetics and developmental genetics has strongly confirmed this impression. An individual, the target of selection, is not a mosaic of characters each of which is the product of a given gene. Rather, genes are merely the units of the genetic program that governs the complicated process of development, ultimately resulting in the phenotypic character. To consider genes independent units is also meaningless from the evolutionary viewpoint because the individual as a whole (representing the total genotype), not individual genes, is the target of selection. To regard genes as independent units is meaningless from the physiological viewpoint because genes interact with each other in producing the various components of the phenotype.

That the different parts of an organism are inseparable in development as well as evolution was well understood by the fathers of modern biology. Darwin spoke of the mysterious laws of correlation. When referring to correlation of growth, he said, "I mean by this expression that the whole organization is so tied together during its growth and development, that when slight variations in any one part occur, and are accumulated through natural selection, other parts become modified" (1859:143). He further stated that "if man goes on selecting, and thus augmenting any peculiarity, he will almost certainly modify unconsciously other parts of the structure, owing to the mysterious laws of the correlation of growth." It had long been known to animal and plant breeders that various "correlated responses" may occur as a consequence of selection for a particular character and result in changes in seemingly independent aspects of the phenotype.

In virtually every recent selection experiment such correlated responses produced unexpected difficulties. The occurrence of sterility during selection for changed bristle number in *Drosophila* is a typical example and numerous others are listed in the literature (Lerner 1954). The evolutionist is especially interested in cases where selection for a physiological trait has produced a correlated response in morphological characteristics. This occurs frequently and perhaps invariably when strains of the house fly (*Musca domestica*) and of *Drosophila* are selected for DDT resistance. These strains differ in various morphological characteristics not only from each other but also from nonresistant strains. The diversity of the responses indicates that it is not the genes which give DDT resistance that are responsible for the morphological changes, but the reconstruction of the entire genotype. Important in all of these selection experiments is not only the occurrence of morphological changes (without ad hoc selection) but also the rapidity of the response. The elapsed time is negligible in terms of geological age.

The most frequent "correlated response" of one-sided selection is a drop in general fitness. This plagues virtually every breeding experiment. An exceptionally well-analyzed case is that of hybrid turkeys (*Meleagris gallopavo*), in part derived from domestic stock, that were released in Missouri to "strengthen" the depleted native stock. It was found that the feral birds were inferior to the wild birds in every aspect of viability studied (Leopold 1944). They also had lower relative brain, pituitary, and adrenal weights. Native turkeys raised larger broods than hybrids and a larger proportion of wild hens was successful in raising broods. The commercial qualities of the domestic birds had been bought at the price of abandoning qualities favoring survival in the wild.

The working of this interaction of genes had to remain "mysterious," as Darwin called it, until the mechanisms of inheritance had been elucidated. A full understanding of the complexity of interactions among genes during development still eludes us. Yet much of it has become clear and it has become possible to ask meaningful questions concerning unsolved problems. Perhaps the most important point to remember is that in all higher organisms the gene is removed by a long developmental pathway from the components of the phenotype that it affects. At each step along the pathway, the proteins produced by the gene interact with proteins produced by other genes, and it is this interaction of proteins that determines growth and differentiation of the cells involved. What particular effect a particular gene will have depends to a very large extent on its genetic milieu, that is, on the specificity of the other genes with which it is associated in the genotype. This universal interaction has been described, in deliberately exaggerated form, in this statement: *Every character of an organism is affected by all genes and every gene affects all characters.* It is this interaction that accounts for the closely knit functional integration of the genotype as a whole.

The Mechanism of Gene Interaction

The gene itself, a unit of DNA, programs the particular protein the manufacturing of which it controls through the agency of RNA, which governs the assembling of the amino acids. The proteins thus produced will either induce or inhibit activity at other gene loci and will also diffuse into other cells, affecting their growth and differentiation. The cells in different tissues become increasingly different from each other in their potencies during the process of differentiation even though all of them normally have identical chromosome sets. Obviously, some genes are active in a given cell while others are not. Precisely what the mechanisms are by which genes are induced to become active or are repressed into inactivity is still poorly understood (Jacob and Monod 1961). It is becoming increasingly evident that a gene is characterized not only by the chemical nature of its product but also by the period in the developmental process when it is most active. Not only may a gene product serve as substrate for the activity of other genes but it may also serve as an inducer or repressor for other gene loci. It seems probable that normal development depends on a very precise balance of the proper amounts of specific gene products. This is the reason that the interaction of genes is of such vital importance for the viability of the phenotype, which is the end product of the developmental process.

The old preformistic concept of the relation between gene and character is, thus, completely wrong. No white eye is encapsuled in a white-eye gene nor is a clutch of four little eggs boxed into the chromosome of a bird laying a four-egg clutch. Development is strictly epigenetic: genes merely give the potentiality to produce or to contribute to the production of a given phenotype; they supply a program for its production, "all other things being equal." These "other things" include the external as well as the genic environment.

One of the aspects of the multiple interactions of genes during development is discussed in Chapter 7 (p. 93). This is *pleiotropy*, the capacity of a gene to affect several different aspects of the phenotype. It is doubtful whether any genes that are not pleiotropic exist in higher organisms. Since the primary gene action in multicellular organisms is usually several steps removed from the peripheral phenotypic character, it is obvious that non-pleiotropic genes must be rare if they exist at all. This realization "leads us to the idea of the genotypic milieu which acts from the inside on the manifestation of every gene in its character. An individual is indivisible not only in its soma but also in the manifestation of every gene it has" (Chetverikov 1926).

Polygenic Characters

In the early days of genetics emphasis was placed on genetic factors that caused discontinuous phenotypes such as were discussed in Chapter 7 under polymorphism. The alleles responsible for such conspicuous morphs as eye-color genes in *Drosophila*, banding patterns in *Cepaea* snails, or spotting in beetles and fishes illustrate Mendelian inheritance almost diagramatically. On the genetic background of a single population they seem to confirm the one gene–one character hypothesis. But such genes are very much in the minority in a population. Most genes affect characters that vary continuously, such as size, proportions, or general tone of coloration. At first it was thought that such continuous variation did not obey the principle of Mendelian inheritance. It is now clear that continuous variation is due to the fact that several, usually very many, genes contribute to a single character. Continuous variation, thus, is not an exception to particulate inheritance; it differs from discontinuous variation only in the number of genetic factors controlling the particular character. Control of the variation of a character by several or many genes is called *polygeny*.

Let us study this with the help of a very simple model. Assume that in an organism large body size is determined by contributions from two loci *A* and *B*, each with two alleles (*A*, *a*, *B*, *b*). Let the heterozygotes (single dose) be intermediate in size between the homozygotes and let either locus make a roughly equal contribution to increased size. The nine genotypes fall into five classes of phenotypes: four size genes (*AABB*), three size genes

(*AABb, AaBB*), two size genes (*AAbb, AaBb, aaBB*), one size gene (*Aabb, aaBb*), and no size genes (*aabb*). With a third locus (with two alleles) affecting size, the number of classes of phenotypes rises to seven, with a fourth, to nine, and so on. Since it is unavoidable that each member of each class is subject to a certain amount of nonheritable environmental modification, it becomes obvious that, as the number of genes rises, sharp classes are increasingly difficult to distinguish because the genetic classes will overlap phenotypically with each other. The final result will be a more or less smooth curve of continuous variation even though the genetic basis consists of discrete loci and alleles exactly as in the case of completely discontinuous polymorphism. Since well-defined classes of phenotypes cannot be distinguished in the progeny of crosses, special biometric techniques (Falconer 1960) are required to separate the heritable from the nonheritable portion of variation and to determine the various genetic components (multiple factors).

It is sometimes suggested that there are two kinds of genes, major genes and modifiers or polygenes. It is now doubtful whether such a distinction is either correct or helpful. When a specific character is studied, it is often found that some genes make a smaller contribution to its phenotypic expression than others. But it is quite possible that these "minor genes" simply reflect pleiotropic effects of genes which themselves make major contributions to other characters. Even where the phenotype is affected by a large number of genes, it is possible that one gene will serve as a *switch gene* that shifts the phenotype into an entirely new direction, such as a sex-determining gene in certain organisms or a gene involved in mimetic polymorphism. The switch gene on one hand and the large array of small polygenic factors on the other are, however, only two extremes of a broad spectrum of possible genetic determination, not two separate classes. The disintegration of mimetic morphs when individuals of different populations are crossed demonstrates the strict dependence of switch genes on their genetic environment. Some geneticists, however, believe that polygenes, major structural genes, and so forth, are clearly separable categories of genes (Rendel 1968).

Interactions of the Genotype

Pleiotropy and polygeny are terms that stress the two endpoints in the developmental pathway, the gene and the character. In reality, genes interact with each other all along the epigenetic pathway in numerous direct and indirect ways. This abundant and continuous interaction greatly strengthens the internal cohesion of the gene pool.

The only gene interactions that are well studied are *allelic interactions,* the interactions among alleles at the same locus. Such interactions, indicated by the terms dominance, recessiveness, and overdominance (superiority of the heterozygote), were discussed in Chapter 9. There is, however, a second class of interactions, that between genes at different loci.

Epistatic Interactions

It has been known since the earliest days of Mendelism that different loci may interact with each other in their effect on the phenotype. Two loci, each with a dominant and a recessive allele, segregating independently, will normally produce phenotypes in a ratio of 9:3:3:1. Aberrant ratios such as 15:1 or 9:7 or 9:3:4 indicate the existence of other modes of interactions. A rather elaborate terminology used by the early Mendelians for various types of interactions was found not to be helpful, and the term *epistatic interactions* is now used for all interactions among genes situated at different loci. An extreme case of epistasis exists, for instance, when an albino gene prevents the manifestation of any other pigmentation gene. More often the effect is not absolute and consists only in a modification of the degree to which a character is shown, such as the interruption of a wing vein in *Drosophila.* Incomplete penetrance (Chapter 9) is a frequent manifestation of epistatic interactions.

Suppressor genes are one of several special types of interacting genes. Glass (1957) has analyzed such a gene, situated on the second chromosome of *Drosophila melanogaster,* which suppresses the effects of the *erupt* gene on the third chromosome. The erupt gene, when present without the suppressor, produces a wartlike eruption on the eye. Both the erupt and the suppressor genes were found to be widespread in natural populations. The evidence indicates that a whole set of isoalleles of different strengths exists on each locus. The erupt gene was not found in any population without the suppressor gene, but in a number of populations suppressor genes were present without the erupt genes.

What may be the evolutionary interpretation of this curious system of interactions? Glass quite rightly suggests that the erupt gene is maintained in the populations because it adds to fitness through some other unknown pleiotropic manifestations. This maintenance sets up a selection pressure in favor of suppressor genes that remove from the phenotype the undesirable effect of the erupt gene on the morphology of the eye. On the other hand, the frequency, if not the universality, of the suppressor gene indicates that it also has pleiotropic manifestations for which it is maintained in the gene pool even in the absence of the erupt gene. Numerous other occurrences of suppressor genes have been reported in the genetic literature.

Suppressor genes are only one illustration of the numerous types of interactions that exist among different loci of the same genotype. Such interacting systems are highly sensitive to selection and they permit numerous feedbacks and systems of regulation during ontogeny. The students of development have various terms for these regulatory powers, such as *buffering, canalization,* and *developmental homeostasis.* These terms apply to models that help us to visualize the action of genes in the developmental process, but they should not blind us to our basic ignorance of the exact mechanisms by which the universally observed regulation during development is achieved. (For further details on the physiology of differentiation of tissues and organs in relation to gene action, refer to books on epigenetics.)

In order to make an optimal contribution to fitness, a gene must elaborate its chemical gene product in the needed quantity and at the time when it is required for normal development. The total genotype can be considered a "physiological team," an analogy that has considerable illustrative value. Some of the best-known athletes are poor team players, or might star as members of one team but not of another. Some musical virtuosos unexcelled as soloists are only mediocre in an ensemble. Genes are never soloists, they always play in an ensemble; and their usefulness, their "selective value," depends on their contribution to the fitness of the product of this ensemble, the phenotype.

Each gene has numerous pleiotropic manifestations that add to or detract from the fitness of the phenotype. Other loci are able to enhance those manifestations that make the greatest positive contribution to fitness and to suppress other manifestations having a deleterious effect, such as the erupt phenotype of the erupt gene. The selective value of a gene is thus only partially determined by what one might call its own direct effect on the phenotype. An equally or even more important component of the selective value of a gene is its effect on or contribution to the fitness-enhancing qualities of other genes. Plant geneticists were perhaps the first to fully appreciate this multiple function of genes, a point particularly stressed by Harland (1936) and his school. The analysis of hybrids and of mutator genes, the work on polygenic characters, on dominance, on sex determination by genic balance, and on dosage compensation, all indicate the universal occurrence of genic interactions regulated by their contribution to the total fitness of the phenotype.

Gene Interactions in the Gene Pool of a Population

The contribution that a gene makes to the fitness of an individual is tested against the unvarying genetic milieu of a single individual. During evolution, however, every gene is placed in every generation on a new genetic background owing to sexual reproduction (new sets of parental chromosomes) and recombination (owing to crossing over), and this genetic background will be different in every individual that carries the gene. The scope of genetic interactions is thus greatly expanded: The gene is now tested for its contribution to fitness in combination with all the other genes of the gene pool. Since the contribution to fitness that a gene makes depends on the particular team of other genes with which it is associated in the production of a particular phenotype, it is evident that the fitness of a gene is "statistical." It is averaged from the fitness values of the gene on all the different genetic backgrounds in the many genotypes in which the gene is exposed to selection. The selective value of a gene is the mean of the selective values of all the combinations in which it appears in a given population (gene pool). As long as this mean value is higher than that of other genes competing with it, the gene will be favored by selection even if it is highly deleterious or lethal in certain combinations.

These two new insights into the role of the whole individual as the target of selection and the contribution of a gene to fitness have resulted in a new interpretation of the evolutionary meaning of genes and genotypes. From the evolutionary point of view, genes can no longer be considered disconnected entities to be studied in isolation. Their selective value is no longer considered absolute. On one genetic background a given gene may add to the fitness of the genotype; on another genetic background the same gene may create an unbalance and produce a severely deleterious effect. To the two well-known classes of environment, the physical and the biotic, must be added a third, the genetic environment. The genetic environment of a given gene includes not only the genetic background (Chetverikov 1926) of the given zygote on which it is temporarily placed, but also the entire gene pool of the local population in which it occurs. The evolutionary fate of this gene will depend, in the long run, on how well it cooperates with the other genes of this gene pool, how well it is coadapted to them.

Epistatic Interactions and Fitness

It is evident that natural selection will favor those genes which harmoniously interact with each other in the gene pool of a population. Such genes are *coadapted*, to use Dobzhansky's felicitous term. This process of coadaptation goes on steadily, from generation to generation, since each natural population is exposed to a steady influx of alien genes and to all sorts of environmental disturbances. Dobzhansky and his school have made this area their particular field of research and have greatly advanced our understanding through a series of brilliant experiments.

There is abundant evidence that a given gene has much higher fitness in certain gene combinations than in others. When recombination dissolves such a superior genotype, a gene may lose its selective advantage. This is why any mechanism will be favored that preserves a constellation of genes which enhances their contribution to fitness. The so-called gene arrangements (paracentric inversions) in *Drosophila* are examples of such supergenes, because crossing over between different gene arrangements is strongly inhibited, if not totally prevented. As a result each gene arrangement can acquire a different set of coadapted genes. For instance, Prakash and Lewontin (1968) showed that at Mather, California, the *Pt-10* gene was represented in the "Pikes Peak" arrangement by the 1.04 allele (100 percent) and the "Tree Line" arrangement by the 1.06 allele (100 percent). A similar specificity was found for other genes and alleles. Furthermore, a given gene arrangement tends to have the same gene content throughout its range. At Bogotá, Colombia, for instance, "Tree Line" likewise has allele 1.06 (100 percent) of gene *Pt-10*. This strongly suggests that the genic differences date from the origin of these inversions and thus conceivably date back millions of years. Since there was abundant opportunity during this long period for genic equalization through mutation and occasional cross over in heterozygotes, the strict maintenance of the difference must be ascribed to selection for the particular genic combination found in the arrangement, that is, to coadaptation.

Coadaptation is likewise revealed by the response of populations to chronic irradiation (Wallace 1956). Two experimental populations of *Drosophila melanogaster* that received about 5 r per *hour* rapidly accumulated a great deal of genetic radiation damage. About 80 percent of all the second chromosomes tested after 70 generations were lethal when homozygous. Considering the number of accumulated lethals, one would expect heterozygous combinations of chromosomes from these populations to display evidence of greatly lowered viability. Yet when such tests were made it was found that flies carrying two randomly chosen second chromosomes from one of the irradiated populations showed on the average only a slight (1–8 percent) reduction of viability compared to the standard population. Moreover in a certain percentage of such crosses viability was above the average of wild flies. It is evident from these results that there must be a high selection pressure in these closed experimental populations for all genes and chromosome sections that combine into harmonious, efficient genotypes. Whenever a chromosome is damaged by radiation, this damage is compensated by the selection of a mutant at another locus that restores the balance. Some bypass is found for every interrupted metabolic pathway. The average viability of recombinants remains nearly normal in spite of the steady increase in the frequency of "deleterious" genes.

It was discovered in recent decades that such deleterious chromosomes are quite frequent also in natural populations. As long as there is an abundant supply of compensating genes on other chromosomes, the frequency of the deleterious chromosomes will be reduced only very slowly. The probability of two such chromosomes meeting in a single zygote, resulting in a homozygote for the deleterious condition, is extremely low. Since particular lethals reach comparatively high frequencies, it is probable that they are actually heterotic in heterozygous condition. Such heterosis would explain why such deleterious chromosomes accumulate in most widespread, numerous species. With the help of ingenious techniques, it is possible to study in *Drosophila* the relative viability of individual chromosomes in homozygous and heterozygous condition. When testing wild-type flies from certain localities in Russia, Dubinin et al. (1934) discovered that between 10 and 20 percent of the second chromosomes in wild flies are lethal when homozygous. This figure has since been confirmed by Dobzhansky and his associates for natural populations of five or six species of *Drosophila;* indeed, the figure is around 30 percent in the majority of the populations studied. Nearly all the remaining chromosomes, although not lethal, are at least somewhat deleterious when homozygous. It appears probable that the lethality of some of these chromosomes is not the result of lethal genes, but of lethal gene combinations. Evidence for this conclusion comes from experiments in which a recombination of viable chromosomes resulted in lethals.

Synthetic Lethals. It has been found that chromosomes which are of approximately normal viability both as heterozygotes and homozygotes may become lethal after crossing over. An experiment by Dobzhansky (1955) may be used as illustration: Ten second chromosomes were taken from a population of *Drosophila pseudoobscura* collected near Austin, Texas. These chromosomes produced normally viable to subvital homozygotes. Females heterozygous for all possible (45) combinations of these ten chromosomes were obtained, and ten chromosomes from the progeny of each heterozygote were tested for viability effects in homozygous conditions (450 chromosomes in all). Among these 450 chromosomes 19 were lethal and 57 were semilethal in double dose; 30 of the 45 combinations produced at least one lethal or semilethal among the 10 chromosomes tested. Most remarkable of all, the combinations including the chromosomes that were known to be subvital when homozygous gave actually fewer lethals and semilethals than the normally viable to supervital chromosomes. It must be remembered that crossing over does not take place in *Drosophila* males and that there is no question that the lethal chromosomes were derived from the viable chromosomes of the females.

What is the explanation of the atonishing change in fitness of these chromosomes? Normal recessive lethal loci can be excluded because two generations earlier the same chromosomes had been of normal viability when homozygous. Their frequency is far too high to be attributed to newly arisen lethal mutations. Therefore one must conclude that these lethal chromosomes are the result not of lethal alleles but of epistatic interactions due to lethal gene combinations, newly created by recombination. This explanation is confirmed by the fact that such lethal chromosomes can be restored to normal viability through crossing over. The more genic diversity there is in a species (owing to gene flow among many widely distributed local populations), the greater the risk of disharmonious combinations, that is, the greater the probability of the occurrence of "synthetic" lethals or detrimentals.

Allelic versus epistatic interactions. Both kinds of interactions are universal, but it is still unknown what relative contribution each makes to fitness. Vetukhiv (1956), Brncic (1954), and Wallace (1955) investigated this question by crossing *Drosophila* flies from different localities. They then compared the viability of intrapopulation F_1 and F_2 hybrids with that of F_1 and F_2 interpopulation hybrids. Viability was measured for several components of fitness, such as fecundity of females and survival of larvae in severe competition with larvae of a standard strain of another species of *Drosophila*. The results of these crosses are entirely consistent; F_1 hybrids between different populations are significantly more viable than intrapopulation F_1 hybrids. The situation is reversed when the F_2 cross is made (Fig. 10.1).

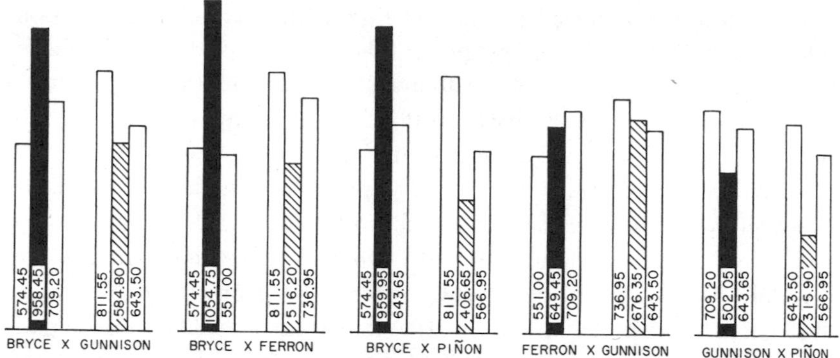

Fig. 10.1. Fecundity of geographic populations of *Drosophila pseudoobscura* and of hybrids between them. The height of each column gives the mean number of eggs deposited per life per female. The white columns represent the parental populations, the black columns the F_1 hybrids, and the crosshatched columns the F_2 hybrids of these populations. (From Vetukhiv 1956.)

Two conclusions are obvious. The first is that the superiority of the F_1 interpopulation hybrids cannot be due to a previous selection of a suitable genetic background because the populations tested came from localities far distant from each other. The increase in the fecundity of the females shows that more than mere "luxuriance" (somatic vigor) of the larvae is involved. Thus we have a case of genuine heterosis that is not the product of selection. The second and even more important conclusion is that this superiority is not due to mere high heterozygosity, because the F_2 lose this superiority in spite of being far more heterozygous than the parental populations. The conclusion is inevitable that the loss of viability in the F_2 is due to a loss of epistatic balance among interacting loci and that this overrides the beneficial effects of high heterozygosity. A number of other tests confirm this conclusion, particularly the results of three-way and of double crosses. The integration of the gene complex, its coadaptation, therefore depends on the presence of two kinds of harmonious balance. One of them, a balance among alleles leading to overdominance and through it to balanced polymorphism, Mather (1953) has called *relational* (= allelic) *balance*. The other balance, that among different loci, is best designated *epistatic balance*. The actual architecture of the epistatic balance is not yet understood. In *Drosophila subobscura* neither F_1 heterosis nor F_2 breakdown is observed and in *D. paulistorum* these phenomena are only weakly developed. Each species seems to have its own balancing devices.

The constancy of the phenotype. The genetic variation of populations is far greater than one would judge from the variation of the visible phenotype. Only a small fraction of the abundant genetic variation that is stored in the gene pool expresses itself visibly. The extraordinary frequency of sibling species (Chapter 3) demonstrates that even gene substitution so extensive as to lead to the origin of new species is possible without visible effect on the phenotype. Each gene substitution during the origin of these species was somehow compensated for and thus prevented from affecting the end product of development, the phenotype. How great the genetic differences between sibling species are is sometimes revealed by incompatibilities in their hybrids. In F_1 hybrids between *Drosophila melanogaster* and *D. simulans*, Sturtevant (1929) found many disturbances in bristle development, even though the bristle patterns of the two parental species are identical. In 167 hybrid females (with 334 bristle sites on both sides of the dorsum) there were 142 anterior dorsocentrals (192 missing), 181 posterior dorsocentrals (153 missing), 192 anterior scutellars (142 missing), and 198 posterior scutellars (136 missing). The developmental pathways leading to the phenotypically identical bristle patterns of the parental species are evidently very different.

It would lead too far afield to discuss here the genetic and developmental system that maintains the constancy of the phenotype in spite of gene substitution. The best analysis is that of Rendel (1967) for the scutellar bristles in *Drosophila.* The phenotype "normal number of scutellar bristles" is so tightly canalized that even a considerable amount of genetic substitution will not result in a visible change of the phenotype. Only after the character has been moved outside the zone of canalization will it begin to respond again to selection. The normal bristle number is four, but it takes about eight times as much genetic change to move from three bristles to five as it does to move from one bristle to three.

Gene substitution without effect on the morphological phenotype is analogous to the restoration of a Gothic cathedral. In the course of the never-ending repair work, many of the stones have been replaced, sometimes repeatedly, since the original construction some seven centuries ago. Yet the "phenotype" of the medieval edifice has remained unchanged. The tendency of the morphological phenotype not to respond to a far-reaching reconstruction of the genotype is manifested even above the level of sibling species. The same "character" may go through an entire genus, family, or even higher category. In many instances, there is no obvious reason why a particular aspect of the phenotype should be favored by selection since a different phenotype seems to serve equally well in a related taxon that coexists in the same environment. Interpreting these characters as the by-product of the entire well-integrated gene complex held together by genetic homeostasis, not as products of specific genes, greatly facilitates an understanding of such *conservative characters.*

A well-knit system of canalization tends to restrict the evolutionary potential quite severely. It accounts for parallel evolution which, for instance, induced several separate lines of mammallike reptiles to cross the borderline to the mammals independently (see Fig. 19.2). Which gene will mutate and at what time is "random," but the subsequent fate of such a mutation is strongly controlled by the gene complex in which it occurs. The direction of evolution, although not predictable, is by no means random.

Selection in a Stable Environment

The phenotype is produced by genes that are intimately welded together into a single closely-knit ensemble through pleiotropy, dominance (and overdominance), epistasis, and polygeny. Since every individual is genetically unique, selection is offered a vast array of choices. The nature of the selection pressure determines which genotypes will be favored.

Stabilizing selection. Let us begin with the simplified assumption of a constant environment. This does not at all mean a cessation of selection. Even in a stable environment there will always be *stabilizing selection* (Chapter 8). In a population with high genetic variation, the processes of recombination and of segregation will inevitably produce numerous individuals that deviate from the optimal (average) phenotype in one respect or another and that are therefore of lower fitness. The term "stabilizing selection" (Schmalhausen 1949) refers to the fact that natural selection will tend to eliminate such deviant individuals from the population, the selection pressure normally being more severe the greater the deviation.

The literature records numerous instances (Mayr 1963:282) of such discrimination against *phenodeviants* (Lerner 1954). Snakes and lizards, for instance, do not change the number of scales or scutes between birth and adulthood. Yet herpetologists have recorded many cases in which the variability in scale number is much smaller in adult samples of reptiles than in juvenile samples of the same species. This is evidently due to the higher mortality in juveniles of the extreme types.

Waddington (1957) has pointed out correctly that stabilizing selection has two different aspects. The negative one, the elimination of all individuals that are phenodeviants, he calls *normalizing selection.* The positive one, a selection in favor of all sorts of feedback mechanisms that would produce the standard phenotype in spite of considerable substitution of genes in the genotype and of environmental fluctuation, he calls *canalizing selection.* These two processes are, in a way, merely two aspects of a single process, since canalizing selection by necessity operates with the help of normalizing selection.

One of the main objects of stabilizing is to minimize the effects of disharmonious epistatic interactions. One way to achieve this is to favor genes that I have dubbed "good mixers" (Mayr 1954), that is, genes that make a positive contribution to fitness on the greatest number of possible genetic backgrounds. Second, supergenes will be favored, that is, well-balanced combinations of genes that are protected against recombination (through suppression of crossing over).

Effects of a One-directional Selection Pressure

The maintenance of a genic balance composed of allelic and epistatic interactions appears comparatively simple as long as the environment remains constant. Indeed canalizing selection will tend to build up internal balances that will maintain a rich reservoir of genetic variation under the cover of a uniform phenotype. An entirely new situation arises, however, when a new environmental factor exercises strong directional selection and attempts to shift the phenotype away from its present norm. The occurrence of such a new selection pressure poses some interesting questions. To what extent can the gene pool respond to such a new pressure? How strong must the pressure be to imperil the maintenance of a harmonious cohesion? All animal and plant breeders employ one-directional selection in order to increase yield per acre in a crop plant or the annual production of milk in cows. The most refined analysis of the effects of one-directional selection has, however, been made with certain laboratory organisms (Falconer 1960).

In most breeding work, selection is simultaneous for a great many factors, particularly for most of the components of fitness. In such cases it is usually quite impossible to determine what happens during the process of selection. Therefore I have chosen for analysis a case which, though complex, is easier to analyze because it deals with severe selection for a single character in a laboratory animal. I refer to the pioneering and now classical studies of Mather and his associates on bristle number in *Drosophila.* The object of the selection was an increase in the number of bristles (chaetae) on the ventral surface of the fourth and fifth abdominal segments in *Drosophila melanogaster.* Two selection experiments were run, one for increase and one for decrease in bristle number (Fig. 10.2). In the starting stock, the combined average bristle number of males and females on these segments was about 36. Selection for low bristle number was able to lower this average after 30 generations to 25 chaetae, after which the line soon died out owing to sterility. A mass "low" line (maintained without selection) was started with 32 chaetae and remained nearly stable for 95 generations. However, all attempts to derive from this line others with lower bristle numbers proved failures. The lines invariably died out owing to sterility before selection had made much progress. The evidence indicates that the resistance of the low line to further selection was not due to an exhaustion of genetic variability but was in part due to a balanced sterility system. Yet the possibility could not be excluded that the genes of this stock favoring low bristle numbers had a depressing effect on viability when in homozygous condition.

In the "high" line (selection for high bristle number), progress was at first rapid and steady. Within 20 generations bristle number had risen from

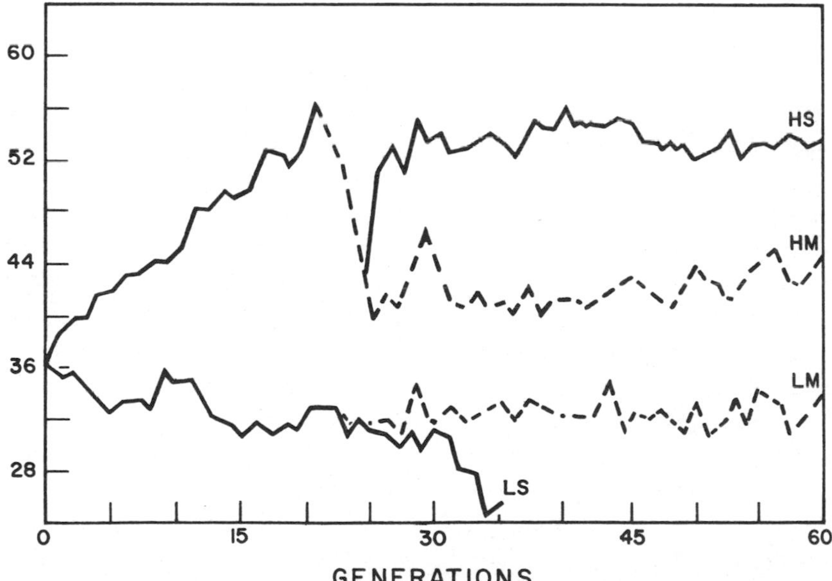

Fig. 10.2. Bristle number in four lines of *Drosophila melanogaster* derived from a single parental stock: *HS*, high selection line; *LS*, low selection line; *HM*, unselected mass culture derived from the high line; *LM*, unselected mass culture derived from the low line. (After Mather and Harrison 1949.)

36 to an average of 56, without marked spurts or plateaus. At this stage sterility became severe and a mass culture (without selection) was started. Average chaetae number fell sharply and was down to 39 in 5 generations, a loss of over 80 percent of the gain of the preceding 20 generations. Without selection, this line fluctuated somewhat up and down (going at the 29th generation as high as 46), but finally settled approximately at an average of 40. New high selection lines were repeatedly taken from this "high" mass line. The first two (started at the 24th and 27th generations) regained the previous high bristle number as quickly as the line had lost it when selection was stopped. In these lines viability was much improved and a viable line could be maintained under constant selection pressure for high bristle number (without, however, much further phenotypic response). A mass line taken off this new high line some five or six generations later maintained its high bristle number rather than dropping precipitously as had the earlier mass line. Other reselection lines taken off the original high mass line at later periods were much less successful, a fact indicating that the high mass line had restabilized itself in a new way, with a loss of the combinations leading to higher bristle numbers.

A number of generalizations can be derived from this and similar studies. The first point is the immense capacity of the genotype for storing variability. Selection was able to double bristle numbers even though the population was closed and no new genetic material was added after selection had started. The second point is that such a well-defined character as the number of abdominal bristles depends on a considerable number of genes. The genetic analysis showed that genes affecting bristle number could be found on every chromosome and indeed at several loci on some chromosomes. However, the brilliant analyses of polygenes by Spickett and Thoday (1966) demonstrate that about 80 percent of the heritability for sternopleural chaetae in *Drosophila melanogaster* is contributed by rather few loci (about 5–10). Robertson (1968), Rendel (1967), and others have likewise come to the conclusion that, even in the case of polygenic characters, the major contribution is made by a limited number of principal genes. Many other genes have some effect on these characters, but in a circumscribed gene pool their contribution is minor. The geographic variation in the genetic basis of the genes responsible for the mimetic polymorphism of butterflies suggests that the "switch action" of major bristle genes may be a canalized system that might disintegrate if placed in a different genetic milieu. The most interesting result, however, is the third finding, that any intense selection results in various correlated effects. Some of these, like changes in pigmentation and in the number of spermathecae (which increased in some of the stocks), had no particular effect on viability. Others affected fertility, fecundity, and larval survival. The counterpressure of natural selection finally became so strong that the stocks reached a plateau where they would not respond further to the artificial selection. This limit (and the same observation has been made by animal breeders) was not due to the exhaustion of genetic variability, since genetic variance remained considerable. A relaxation of selection (it would be more precise to say a shift to selection for high fitness) sometimes resulted in a rapid loss of the "artificially" acquired character. An extreme phenotype, then, (high bristle number) can be produced only by very specialized genotypes. Such genotypes are bound to be unbalanced in more ways than one, as, for instance, with respect to various components of fitness.

All the results of strong one-directional selection experiments confirm that the response of the genotype is not that of a set of independent genes but that of a single interdependent genetic system. The nature of this system results not only in severe limits in the response to the selection pressure but also in the breakdown of epistatic balances. Such selection experiments provide graphic illustrations of the cohesive complexity of any genotype or gene pool.

Unrepeatability. Since the response to selection depends on a highly intricate interaction of a very large number of genes, it is not surprising that no two genetic stocks will give exactly the same response when exposed to the same selection pressure. The rate of response to the selection pressure, the final level of response, and the occurrence of correlated responses will depend not only on the original constitution of the gene pool but also on the largely accidental choice of one of the many alternate pathways available during each generation. Dobzhansky (1951) records numerous instances of such unrepeatability, particularly with relatively small founder populations. For instance, in six parallel experiments with *Drosophila pseudoobscura* in which chromosomes with the *ST* gene arrangement from California were mixed with chromosomes with the *CH* gene arrangement from Mexico, there was a rapid elimination of the *CH* chromosomes during the first 100–150 days of the experiment (Fig. 10.3). After that period, each population behaved differently. In four populations, *CH* reached the vanishing point sooner or later, while in two populations heterosis between the *CH* and *ST* evolved at the 69-percent and 80-percent level of *ST* frequency. The foundation stock of these experiments was a mixture of twelve strains from Mexico and fifteen from California. In two out of six

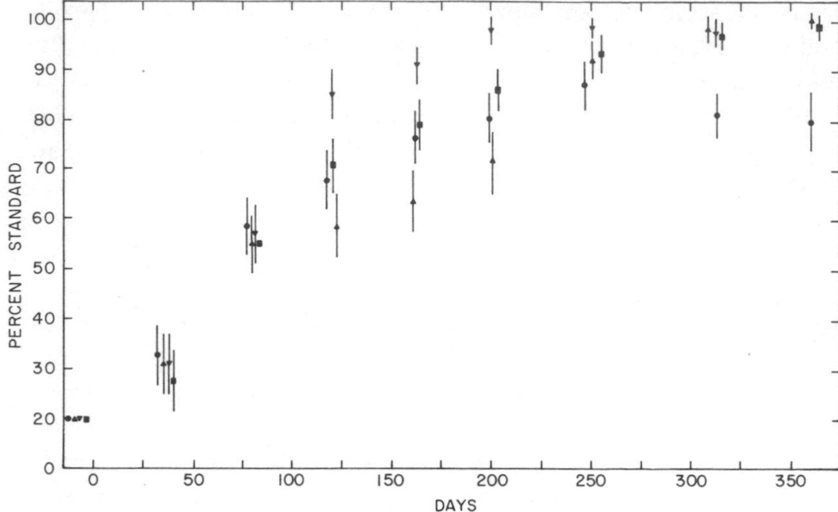

Fig. 10.3. Divergent behavior of four experimental populations of *Drosophila pseudoobscura,* with the *ST* gene arrangement from California and the *CH* arrangement from Mexico. Although all four populations were started with a 20-percent frequency of *ST* chromosomes, each population responded differently to natural selection, as shown by rate of change and final level. (From Dobzhansky 1960.)

cases, recombination between these numerous chromosomes permitted the piecing together of new balanced chromosomes that displayed heterosis in the inversion heterozygotes. In one of the populations, the effects of the coadaptation became apparent after only three generations and reached equilibrium after ten, in the other, after four and nine generations. Lewontin (1967b) has shown, in the case of a fluctuating environment, that the exact sequence of the fluctuations is important. Because no sequence of climatic fluctuations in nature is ever quite the same as any other sequence, it is clear that evolutionary events are very often unique.

Fitness itself is not apt to show much response to artificial selection because it is merely a continuation of a constant selection pressure and it is to be expected that a plateau has long since been reached. However, individual viability factors may respond to artificial selection, particularly in a newly established closed population placed in a new environment. In such cases one must assume that part of the improvement in the stated characters has been made at the cost of a loss of general viability or of specific resistance to various factors of the physical environment. Obviously any drastic improvement under selection must seriously deplete the store of genetic variability. An open, natural population has to cope with numerous conditions to which the closed and sheltered laboratory population is not exposed. Protected against the effects of immigration and an adverse and variable environment, such a population can afford to develop genotypes that would be of inferior viability in nature.

Limitations on the storage capacity of gene pools. In the early days of genetics Johannsen came to the conclusion that the genetic variability of populations was quickly exhausted by selection. This finding was vigorously contested by many animal and plant breeders who, through selection, achieved steady progress for scores of generations. The reason for the apparent discrepancy is now evident. Johannsen based his conclusions on his own work with beans (*Vicia*), which are exceptional in that they are self-fertilizing and virtually homozygous stocks; the breeders were dealing with highly variable stocks of cross-breeding organisms. All selection experiments confirm that there is a rich reservoir of genetic variability in all populations. Yet much of this variability is not available to selection because it is needed to maintain allelic or epistatic balances. Only those genes can be accommodated in a gene pool that add to fitness and help to produce a harmonious and well-adapted phenotype. Epistatic interactions narrow the capacity of the genetic reservoir more severely than any other factor. The best evidence for the limitations of genic diversity comes from crosses between populations of the same species, as described above. Each local population is a separate integrated system. If the gene contents of two such systems are mixed, many incompatible combinations occur and these are eliminated by normalizing selection. The narrowness of secondary hybrid belts (Chapter 13) is further evidence of this incompatibility of genes. Where an isolated hybrid population has been formed between two dissimilar but fertile parental populations, a period of "coadaptation" may lead to the elimination of the most disharmonious elements. In artificial hybrid populations between the "species" *Drosophila arizonensis* and *mojavensis*, the *mojavensis* chromosomes generally proved superior. The *arizonensis* *X*-chromosome was usually lost altogether and the second and third chromosome retained only because of heterosis with the *mojavensis* chromosomes (Mettler 1957). The final result was a great reduction of the total genetic variability.

Genetic Homeostasis

One of the most interesting findings of recent selection experiments is the tendency of phenotypes exposed to a severe selection pressure for a specific phenotypic character to return to their original condition when that pressure is discontinued. Examples are experiments involving drastic selection pressures for increased bristle number or body size in *Drosophila* and for increased egg number or egg size in the domestic fowl. Lerner (1954) designated this phenomenon *genetic homeostasis*, defining it as "the property of the population to equilibrate its genetic composition and to resist sudden changes."* It has also been referred to as "genetic inertia."

The reason for genetic homeostasis should be evident from the preceding discussions. A naturally existing phenotype is the product of a genotype that has been built up during a long history of selection for maximum fitness. Any selection for a new phenotype will force the abandonment of the previously integrated genotype and will thus lead to lowered fitness, due either to an accumulation of homozygous recessives or to a disharmony between the newly favored genes and the remainder of the genotype. Relaxation of the selection for the new phenotype permits at least a partial return by natural selection to the historical combination that had given maximum fitness, particularly heterozygous combinations. As a by-product there will be a partial restoration of the original phenotype. Genetic homeostasis may well provide the solution for many previously unsolved phenomena. If two related but long-isolated human races retain the same frequency of fingerprint patterns or blood-group genes, in spite of numerous selection pressures favoring a shift in frequency, the maintenance of the original frequency may well be due to the superiority of this particular frequency on the common genetic background. Indeed genetic homeostasis may be the reason for all instances of evolutionary "stagnation."

Genetic homeostasis determines to what extent a gene pool can respond to selection. The less associated with general fitness a particular aspect of the phenotype is, the greater the probability that the phenotype will respond to ad hoc selection pressure on that aspect. If the character does not contribute to fitness in nature, like high bristle number in *Drosophila* or fancy color in pigeons and parrots, natural selection would not previously have taken advantage of the possibilities revealed by artificial selection. The more specific a character and the more monogenic its basis, for example resistance to a specific toxic substance, the more rapid will be the response to selection. Resistance to insecticides is proof of this.

The Structure of the Epigenotype

Traditionally all genes are treated as of one kind—whether dominant or recessive, whether pleiotropic or not, whether selectively advantageous

*The term *homeostasis* (Cannon 1932) was originally applied to the capacity of an organism to hold certain physiological steady states at definite optimum levels.

or deleterious. Only for polygenes was it once suggested that they might reside in the heterochromatin and be qualitatively different from the regular genes. All discussions in this chapter (as well as in Chapters 8, 9, 15, 17, and 19) adopt this assumption for the sake of simplicity even though it is now obvious that it is not correct.

Mammals have enough DNA in their nuclei for more than 5 million cistrons (functional genes). Yet protein studies indicate that there are only 10,000 or at most 50,000 structural genes. What is the function of the other 4,950,000 functional genes? The conviction is spreading that there are different kinds of genes and that the largest class consists of regulator genes, which control the activity of structural genes. McClintock (1961) found the first evidence for the existence of such regulator genes in maize in the early 1940's, and her findings largely anticipate later similar studies in bacteria. Wallace (1963) was the first to propose a consistent theory, based on regulator genes, to explain genetic homeostasis, phenodeviants, synthetic lethals, and incompatibilities in hybrid populations. The rapid accumulation of new information on "kinds of DNA" has now permitted a far more elaborate model (Britten and Davidson 1969).

The development of the new techniques, permitting (at least the beginning of) a distinction between structural and "other" (largely regulatory) genes, is a breakthrough in evolutionary biology of major magnitude. I fully agree with Britten and Davidson's statement (1969:356): "At higher grades of organization, evolution might indeed be considered in terms of changes in the regulatory systems." Much that is now explained as "epistatic interactions between different loci" might well be due to the activities of regulatory genes. The fact that the macromolecules of most important structural genes have remained so similar, from bacteria to the highest organisms, can be much better understood if we ascribe to the regulatory genes a major role in evolution. Since they strongly affect the viability of the individual, they will be major targets of natural selection. Regulatory genes, far more so than structural genes, are part of a delicately balanced system, and therefore they are presumably a major mechanism for coadaptation. The rate of evolutionary change in the macromolecules of important structural genes is presumably largely controlled by the system of regulatory genes. The number of new questions this opens up is legion.

The day will come when much of population genetics will have to be rewritten in terms of the interaction between regulator and structural genes. This will be one more nail in the coffin of beanbag genetics. It will lead to a strong reinforcement of the concept that the genotype of the individual is a whole and that the genes of a gene pool form a unit.

Summary

Our findings on the genetic cohesion of gene pools can be summarized in the following statements:

(1) A considerable number, perhaps the majority, of loci of a species is represented in each population by several alleles. Many of these are so-called isoalleles and produce indistinguishable phenotypes. Each individual is therefore normally highly heterozygous. Since many genes occur in natural populations primarily in heterozygous condition, that kind of genetic background will be favored by selection which enhances the selective value of these genes in heterozygous condition.

(2) The phenotype is the product of the harmonious interaction of all genes. There is extensive interaction not only among the alleles of a locus, but also among loci. The main locale of these epistatic interactions is the developmental pathway. Natural selection will tend to bring together those genes that constitute a balanced system. The process by which genes that collaborate harmoniously are accumulated in the gene pool is called "integration" or "coadaptation." The result of this selection can be called epistatic balance. Each gene will favor the selection of that genetic background on which it can make its maximum contribution to fitness. The fitness of a gene thus depends on and is controlled by the totality of its genetic background.

(3) The result of the coadapting selection is a harmoniously integrated gene complex. The coaction of the genes may occur at many levels, that of the chromosome, nucleus, cell, tissue, organ, and whole organism. The nature of the functional mechanisms of physiological interaction are of only minor interest to the evolutionist, whose main concern is the viability of the ultimate product, the phenotype.

(4) There is a definite upper limit to the amount of genetic diversity that can be accommodated in a single gene pool.

(5) Many devices tend to maintain the status quo of gene pools, quantitatively and qualitatively. The lower limit of genetic diversity is determined by the fact that heterozygosity is often advantageous and that under variable or severe conditions homozygotes are often very inferior. The upper limit is determined by the fact that only those genes can be incorporated that are able to "coadapt" harmoniously. No gene has a fixed selective value; the same gene may confer high fitness on one genetic background and be virtually lethal on another (genetic theory of relativity).

(6) The phenotype is the by-product of a long history of selection and

is therefore well adapted. The effect of selection will normally be to stabilize or to normalize this phenotype. Since this well-integrated phenotype is adapted to make a maximal contribution to fitness, it will resist change (inertia, genetic homeostasis) in the face of new selection pressures.

(7) The result of the close interdependence of all genes in a gene pool is tight cohesion. No gene frequency can be changed, nor can any gene be added to the gene pool without causing an effect on the genotype as a whole, and thus indirectly on the selective value of other genes.

(8) A sudden rise in the input of genes into a gene pool, for instance by hybridization, inevitably results in a disturbance of the internal balance and in the production of many genotypes of lowered viability. Disharmonious combinations will be eliminated by natural selection until a new balance is reached.

(9) The cohesion of the gene pool results in various characteristic responses to new selection pressures. The amount of response to selection is unpredictable because different genotypes have different correlated phenotypic responses, particularly with respect to fitness. The uniqueness of every chromosome, of every individual (in sexual species), and of every population results in an immense and unpredictable diversity of responses to selection.

11 · Geographic Variation

The study of individual variation (Chapter 7) has revealed how erroneous it is to regard all individuals of a species as replicas of the type. This typological concept of the species is further undermined by the fact that variation occurs not only within populations but also between populations. The occurrence of differences among spatially segregated populations of a species is called *geographic variation.* The emphasis on "geographic" stems from the fact that the phenomenon was first discovered when specimens from geographically far-distant populations were compared. We know now that even neighboring populations differ from each other, indeed that, in sexually reproducing organisms, no two demes can ever be identical. Some authors prefer a separate term for local variation among neighboring demes, calling it "microgeographic variation" or "spatial variation." In this volume the term geographic variation is used in the broadest possible sense, to include all population differences in the space dimension.

The number of species of animals in which even a rather elementary analysis has failed to establish the occurrence of at least some geographic variation is very small. Geographic variation can therefore be considered a nearly universal phenomenon in the animal kingdom.

The genetic basis of geographic variation. The existence of geographic variation was a source of considerable annoyance to De Vries, Bateson, and other early Mendelians, who attributed evolution to spectacular mutations. That populations from different portions of the range of a species should differ from each other by "gradual" characters, and that these differences should be greater the greater the distance, was so completely in conflict with the mutationist interpretation of speciation that it forced the mutationists to deny the genetic nature of this variation. The present generation of evolutionists can hardly appreciate the enormous impact of the demonstration by Schmidt (1918) for the fish *Zoarces*, by Goldschmidt (1912–1932) for the gypsy moth (*Lymantria dispar*), and by Sumner (1915–1930) for the deermouse *Peromyscus* that the slight differences between geographic races have a genetic basis. Some of the vast literature on the genetics of geographic variation is summarized in Mayr (1963). The genetic uniqueness of each local population is an inevitable consequence of sexual reproduction. Because no two individuals are genetically identical, no two groups of individuals will be identical. If they live in slightly different environments, as most geographically distributed populations do, one would expect such differences to be accentuated.

Descriptive Aspects of Geographic Variation

The study of geographic variation began as a by-product of taxonomic research. The working systematist, when comparing different specimens or population samples, wants to determine whether or not they belong to the same species. In the course of his studies he is often forced to compare specimens from different parts of the range of a species and he will record the differences he finds. In view of the simplicity of the method, it is not surprising that we find records of geographic variation as far back as the Linnaean period and earlier. Interest in the phenomenon increased rapidly during the period 1830–1870 as the size of the study collections grew and different parts in the ranges of species were increasingly well represented. The study of geographic variation became during the ensuing century one of the primary concerns of the students of the better known taxa, particularly birds, mammals, fishes, butterflies, and snails (Mayr 1942, 1963). Today no substantial taxonomic work dealing with the better-known groups of animals fails to include information on the geographic variation of the species treated.

The study of geographic variation in fresh-water organisms is made particularly difficult by the great capacity of many species for nongenetic modification of the phenotype. Recently many workers have tried to separate the nongenetic from the genetic contribution, but in only a few cases have they been able to do so with complete certainty. Except for important food fishes, notably herring and sardine, the study of geographic variation of marine animals has been rather neglected. The pioneer work was done by Heincke and Schmidt. The widespread occurrence of geographic variation among plants, expressing itself both in local (often ecotypic) and in broadly regional variation, is abundantly substantiated (Stebbins 1950; Grant 1963).

Microgeographic variation. One of the rather unexpected results of recent studies is the extreme localization of phenotypically distinct populations in some species. Gulick (1905) pointed out long ago that every valley or ridge on Oahu (Hawaii) had its own characteristic *Achatinella* snails. It is perhaps not altogether surprising to find high localization in animals as sedentary as these snails. In extreme cases it is possible to demonstrate differences between populations living only a few meters apart. In the banded snail *Cepaea nemoralis,* for instance, Sheppard (1952a) found that a population in a hedgerow was quite different from the population in an adjacent meadow (see Fig. 9.1), and Lamotte (1951) calculated that the diameter of a deme of this species, even in relatively homogeneous terrain, is only about 50 meters. Habitat selection by genotypes (Chapter 9) may accentuate local differences. High localization of populations has also been demonstrated for flightless insects (grasshoppers, beetles) and for other animals with low dispersal facilities. Unexpectedly high localization has, however, also been found among many highly mobile animals (Chapter 18).

Comparative amount of geographic variation. No two species or characters agree in their pattern of geographic variation. Generalizations on the factors controlling geographic variation can be derived from comparative studies of taxonomically well-known groups of animals. When an entire family or the members of a local fauna are quantitatively analyzed, it is usually found that from one half to two thirds of the species show geographic variation in one or several characters (Chapter 13).

Absence of geographic variation. The population structure of species and the need for local adaptation would lead one to suppose that geographic variation is a universal phenomenon. Yet in all groups of animals there are some species that display no geographic variation. In most of these cases, variation undoubtedly exists, but has not been revealed owing to insufficient study. Adequate biometric analysis has revealed geographic variation even in the genus *Drosophila,* noted for its morphological uniformity. In some species where morphological variation has not yet been definitely established, pronounced geographic variation in many cryptic characters has been demonstrated by genetic and cytological analysis.

The comparatively few cases of an apparent total absence of geographic variation (also most cases of very slight variation) can be explained by one of the following four possibilities, or a combination of them:

(1) The total range of the species is so small that there is no opportunity for geographic variation. The environment in the species area is essentially uniform. Rothschild's Starling (*Leucopsar rothschildi*), for instance, is restricted to a limited area in northwestern Bali (itself a small island). Kirt-

land's Warbler (*Dendroica kirtlandi*) occurs only in a few counties in Michigan (Mayfield 1960) (Fig. 11.1). Several species of fresh-water fish are found only in a single spring.

(2) The means of dispersal are so great that the species is almost panmictic, regardless of the geographic extent of its range. This may well be the case for many of the very small "cosmopolitan" fresh-water organisms, such as protozoans, rotifers, tardigrades, cladocerans, and the like (see Fig. 18.2). The same seems to be true for certain ducks and for most migratory species of the North American warbler genus *Dendroica.*

(3) The phenotype is stable. The case of sibling species proves that little of the genetic variation penetrates the phenotype when strong homeostatic devices are present. Presumably the greatest part of the apparent absence (or slightness) of geographic variation is due to such developmental homeostasis.

(4) The genotype is stable. It is sometimes postulated that ancient forms, like the horseshoe crab (*Limulus*), have lost all mutability. There is no evidence available to support this frequently proposed "explanation." It would seem more reasonable to assume that the uniformity in space and time is due to a highly perfected "buffering system," in other words, to genetic homeostasis.

Lack of geographic variation, then, can be attributed either to an essentially panmictic condition of the species population or to low penetrance of genetic variation, combined with stabilizing selection. Why some species should have pronounced phenotypic variation while others with similar population structure lack it (owing to factors 3 and 4) is still a complete mystery.

What Characters Are Geographically Variable?

Any character, be it external or internal, morphological or physiological, may vary geographically. Many instances of geographic variation are cited in the literature (Mayr 1942, 1963; Huxley 1942).

Size is the character most easily shown to be subject to geographic variation. It varies in virtually every species of animal that has an extensive geographic range. Proportions also are highly variable in the great majority of animal species, particularly the relative length of extremities and appendages.

Color pattern is particularly important in those animals in which vision plays a significant role, such as birds and certain reptiles, fishes, and insects,

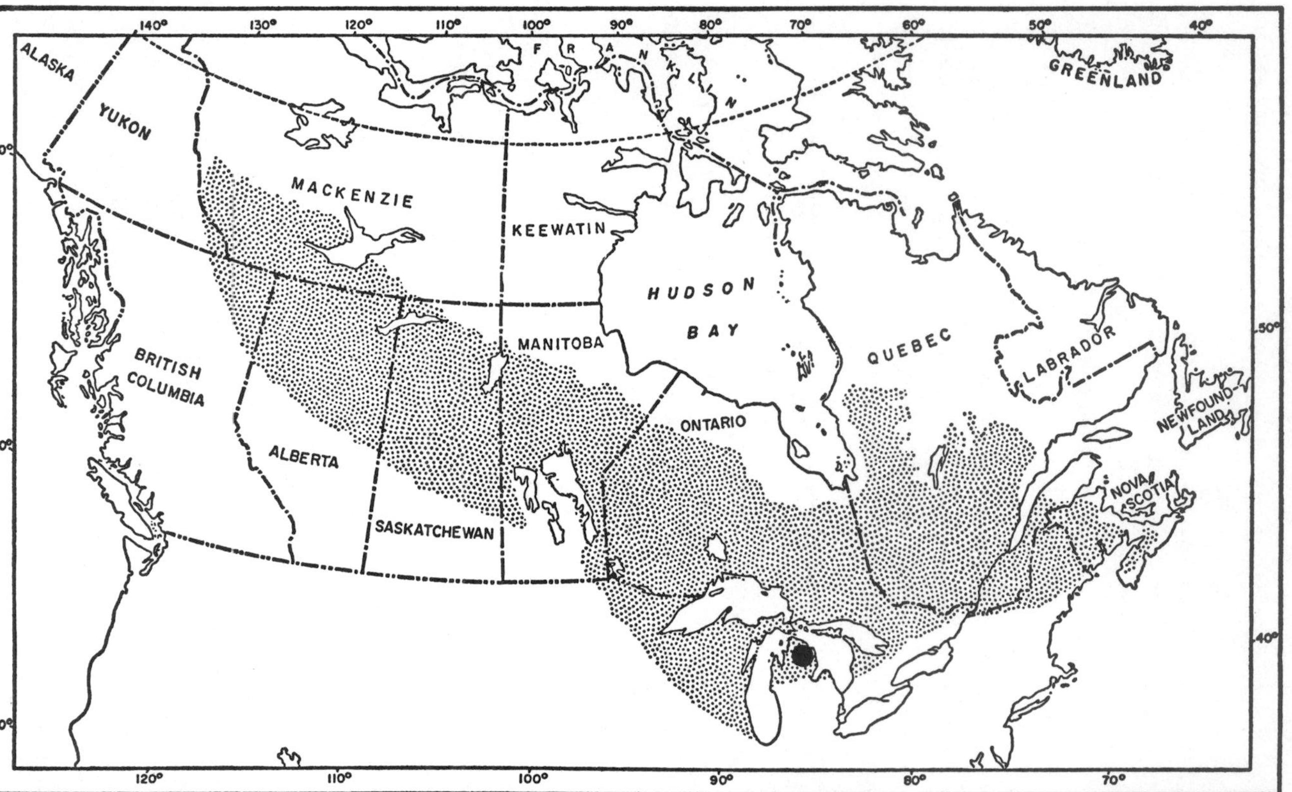

Fig. 11.1. The large dot indicates the nesting area of Kirtland's Warbler (*Dendroica kirtlandi*), comprising only a minute fraction of the range of its habitat, jack-pine forests (dotted). (From Mayfield 1960.)

but a certain amount of geographic variation in tone of color is found in almost every group of animals. That striking differences may occur within a very limited area is illustrated by the variation within *Zosterops rendovae* in the central Solomon Islands. (Table 11.1; see Fig. 18.1). The color or color pattern of a given animal consists of many unit elements. These often vary independently in the species range (Fig. 11.2).

There is much intraspecific variation of chromosome structure. The presence and frequency of inversions, the fusion of chromosomes, and the existence of supernumeraries are all subject to geographic variation. These phenomena are discussed in Chapters 9, 10, and 17.

More and more cases are found of geographic variation in behavioral and ecological characters. Much of this variation is potential raw material for the formation of isolating mechanisms and is discussed in Chapter 16.

The Significance of Geographic Variation

When taxonomists first became aware of geographic variation, they studied it as a phenomenon of purely practical taxonomic interest. Only rarely did they inquire why not all populations of a species are identical. The majority opinion was, unquestionably, that most of this variation is irrelevant and biologically insignificant. It has been asserted quite recently that variation below the generic level is nonadaptive, yet it had been realized surprisingly early that there is close correlation between the environment and much geographic variability. In 1833, Gloger devoted an entire book to climatic effects on birds, *The Variation of Birds under the Influence of Climate.* Bergmann (1847), J. A. Allen (1877), and others, in

Table 11.1 Conspicuous geographic variation on a small archipelago (*Zosterops rendovae*).[a]

Subspecies	Eye-ring	Bill	Black on forehead	Belly	Green on breast
(*a*) *vellalavellae*	Large	Yellow	None	White	Little
(*b*) *splendida*	Large	Black	Much	Yellow	None
(*c*) *luteirostris*	Large	Yellow	Little	Yellow	None
(*d*) *rendovae*	Absent	Black	None	Yellow	Much
(*e*) *tetiparia*	Absent	Black	None	White	Much
(*f*) *kulambangrae*	Small	Black	Little	Yellow	Little

[a] This polytypic species is sometimes treated as a superspecies with four species: *a*, *b*, *c*, and *d–f*.

See also Table 18.1.

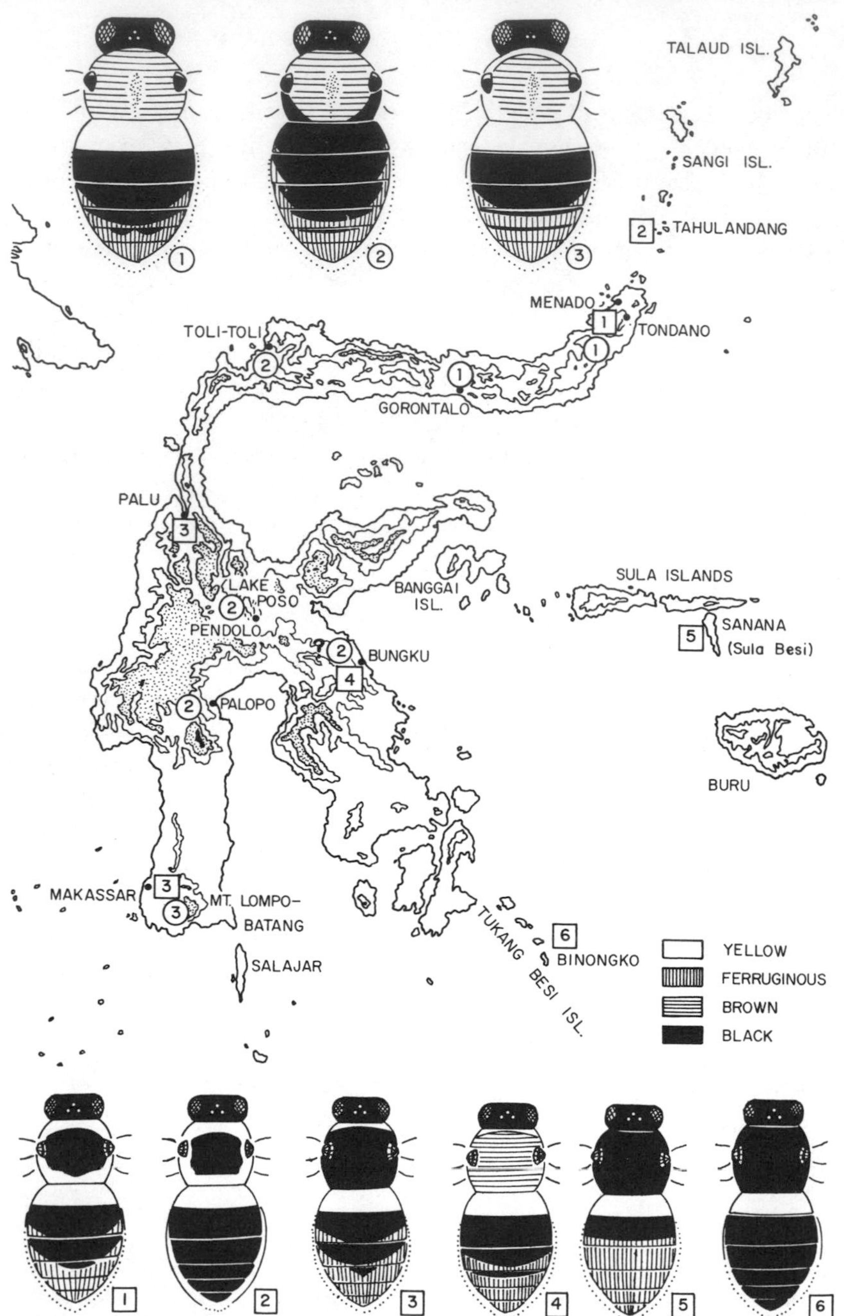

Fig. 11.2. Geographic variation of color in two species of carpenter bees on Celebes and adjacent islands. *Xylocopa diversipes* (above) with 3 subspecies (circles), and *X. nobilis* (below) with 6 subspecies (squares). (From van der Vecht 1953.)

classical contributions to the subject, came to similar conclusions long before 1900.

That specific as well as subspecific characters are, on the whole, adaptive and acquired through natural selection was maintained by many leaders of evolutionary research at an early date. Wallace (1889:142) stated: "It has not even been proved that any truly 'specific' characters—those which either singly or in combination distinguish each species from its nearest allies—are entirely unadaptive, useless, and meaningless; while a great body of facts on the one hand, and some weighty arguments on the other, alike prove that specific characters have been, and could only have been, developed and fixed by natural selection because of their utility." And Darwin (*Life and Letters*, 3:161) wrote to Semper, November 30, 1878: "As our knowledge advances very slight differences considered by systematists as of no importance in structure, are continually found to be functionally important . . . Therefore it seems to me rather rash to consider slight differences between representative species, for instance, those inhabiting the different islands of the same archipelago, as of no functional importance, and as not in any way due to natural selection."

A strong reaction to these adaptationist views developed in the 1880's and dominated evolutionary thought during the ensuing three or four decades. Gulick (1873) concluded that there was no correlation between the various phenotypes of the *Achatinella* snails found in Hawaii and the respective environmental conditions in which the various local varieties occurred. Zoologists in the last quarter of the nineteenth century and the Mendelians during the first two decades of the twentieth denied almost unanimously the adaptive significance and even the genetic nature of geographic variation. Leading systematists continued to maintain the opposite viewpoint throughout this period, vigorously defending the contention that much of variation found in nature was adaptive and directly correlated with the local conditions of the environment. In retrospect it is evident that much of the heated discussion was based on semantic misunderstandings and insufficient analysis.

The sources of misunderstanding were manifold. The major ones may be listed as follow:

(1) Use of the expression "effect of the environment" does not necessarily connote a Lamarckian interpretation. The selective "effect of the environment" is now recognized as an evolutionary force of primary importance.

(2) There has been considerable confusion over the significance of the terms "character" and "phenotype." Some visible components of the

phenotype may in fact not be adaptive, but may be the by-product of a genotype selected for its invisible, cryptic contributions to fitness.

(3) Different assortments of genes may produce populations with similar phenotypes or viabilities.

Each local population is the product of a continuing selection process. By definition, then, the genotype of each local population has been selected for the production of a well-adapted phenotype. It does not follow from this conclusion, however, that every detail of the phenotype is maximally adaptive. If a given subspecies of ladybird beetles has more spots on the elytra than another subspecies, it does not necessarily mean that the extra spots are essential for survival in the range of that subspecies. It merely means that the genotype that has evolved in this area as the result of selection develops additional spots on the elytra.

Evidence for the Adaptive Nature of Geographic Variation

The geographic variation of species is the inevitable consequence of the geographic variation of the environment. A species must be adapted in each part of its range to the demands of the local environment. Every local population is under continuous selection pressure for maximal fitness in the particular area where it occurs. In the external environment there are principally two groups of factors that may exert a selection pressure on the phenotype: (a) climatic factors and (b) habitat and biotic factors. These two sets of factors usually manifest themselves differently in their effects on geographic variation. Climatic factors generally change rather slowly over wide areas (except where altitude is involved) and this results in a variability expressed in regular gradients. Biotic factors, and even more so habitat factors, are often very local and irregular. The extreme in this respect is adaptation to the color of the substrate, which often shows a veritable checkerboard of distribution, utterly different from the clinal variation of characters under climatic control. The evidence for the adaptive nature of geographic variation can be summarized under four headings: (1) geographic variation of physiological characters, (2) ecogeographical rules, (3) substrate adaptation, and (4) geographic variation of balanced polymorphism.

(1) *Geographic Variation of Physiological Characters*

Each local environment exerts a continuous selection pressure on the localized demes of every species and molds them thereby into adaptedness.

As a consequence, local populations differ not only in morphological characters, but also in numerous genetically controlled adaptive features of habit, ecology, and physiology. The universal occurrence of adaptive geographic variation of physiological traits is indicated by much of the literature on comparative and environmental physiology (see Mayr 1963).

In many butterflies and moths there is geographic variability in the number of broods per year and this entails many physiological adjustments. Dawson (1931) compared two populations of the Polyphemus moth, a southern one reaching north to Nebraska and a northern one characteristic of Minnesota. He found that in the southern race the moths are larger, chestnut brown; the cocoons are of coarser silk; the egg diameter is larger (2.8 mm); the number of eggs per female is smaller (average 236); and there are two broods per year. In the northern race the moths are smaller, yellowish brown; the cocoons are of finer silk; the egg diameter is smaller (2.2–2.4 mm); the number of eggs per female is larger (average 291); and there is only one brood per year. In the southern population, the larvae of the second brood complete their growth in the middle of September and spin their cocoons by the end of the month, when the seasonal isotherm of 60°F reaches that locality. Northern larvae of the single annual brood complete growth and pupate in August, when the seasonal isotherms of 69–63°F pass through the region. Exposing the northern larvae in their last stages to lower temperatures produces dormant pupae. The same temperature drop fails to induce dormancy in the southern larvae. The northern population is prevented by its physiological mechanisms from starting a second generation in the fall, a generation destined not to mature successfully. Climatic races have since been described for many species of insects.

Fresh-water animals appear to be even more sensitive to changes in temperature than land animals. As a consequence, they show much geographic variation in temperature tolerance, growth rate, and other physiological constants that adapt them for living in waters of particular temperatures.

That temperature races occur in marine animals has been known for a long time, particularly as a result of a comparison of Mediterranean populations (Naples) with northern European populations of the same species. The widespread occurrence of local races is particularly well established for oysters, in America *Crassostrea virginica,* and in Europe *Ostrea edulis* (Korringa 1958). On the whole, the genetic component of phenotypic variation induced by changing water conditions in marine animals seems smaller than the genetic contribution to phenotypic differences owing to climatic change in land animals. The demand for phenotypic plasticity is particularly great in sedentary intertidal species and in all species whose pelagic larvae are at the mercy of currents. Successive spawnings of the same local populations may be forced to colonize rather different areas. The very free gene flow will result in a highly panmictic condition. All this counteracts local genetic differentiation and favors developmental flexibility.

The adaptive significance in the geographic variation of many other physiological characters is not nearly as evident as in the case of temperature tolerance. This is particularly true for many of the changes in reproductive pattern through the range of species.

Sex races and developmental stages. Bacci (1950) has shown that in many marine invertebrates certain geographic races are gonochoristic (composed of male and female individuals), while others are hermaphroditic. For a review of this subject see Montalenti (1958).

Many species of marine snails can produce two kinds of eggs, small ones with little yolk that produce pelagic larvae, and large, yolk-rich eggs that produce young snails (omitting the larval stages). In species that are thus dimorphic the percentage of eggs producing pelagic larvae decreases from the warm to the cold parts of the range, and only yolk-rich eggs are produced in arctic populations. The elimination of the pelagic larval stage is an obvious adaptation to arctic water conditions (Thorson 1950).

Seasonal adjustments. The seasons, particularly winter and summer, differ in severity in different latitudes, and seasonal adjustments may vary geographically through the range of a species. In many species of birds the northern races are migratory, the southern races more or less sedentary. Blanchard (1941) made a particularly detailed analysis of the physiological differences between migratory and sedentary races in the White-crowned Sparrow *Zonotrichia leucophrys*. Miller (1960) has summarized much of the information on the geographic variation of breeding seasons. Some species of birds (ptarmigans) and mammals (weasels, hares) molt each autumn into a white winter dress, except the populations in the southern portions of their geographic range where no extensive snow cover prevails during the winter.

Some practical consequences. One conclusion emerges from these observations more strongly than any other: the phenotype of every local population is very precisely adjusted to the exacting requirements of the local environment. This adjustment is the result of a selection of genes producing an optimal phenotype. The discovery of this physiological adaptation of local populations is of considerable practical importance, for instance, in wildlife management. Populations that are well adapted in their native environments are often very vulnerable when transplanted into different environments. The literature on game animals records many instances in which stocks died out rapidly after introduction into a different region. If they survive long enough to breed, introgression of their inferior genes will contribute to the deterioration of the native stock. It is for this reason that some countries now prohibit the import of game birds and mammals. Millions of dollars of taxpayers' money spent on raising and releasing ill-adapted game stocks could have been saved if those in charge had been aware of the physiological differences among local populations.

(2) *Climatic, or Ecogeographical, Rules*

One should find much parallel variation in different species if the assumption is valid that every species adjusts to local conditions and if there is a gradual change ("climatic gradient") of these conditions with latitude and longitude. Such parallelism in geographic variation is indeed widespread and has led to the establishment of a series of generalizations, the so-called climatic, or ecogeographical, rules. Gloger (1833) and Bergmann (1847) were the first to describe this phenomenon. The historical significance of the climatic rules is that they focused attention on the importance of the environment in a period during which many biologists denied that the environment played any evolutionary role.

The climatic rules are purely empirical generalizations describing the parallelism between morphological variation and features of the physical environment. For instance, *Bergmann's rule* states that *races from cooler climates in species of warm-blooded vertebrates tend to be larger than races of the same species living in warmer climates.* The validity of these rules depends, of course, on the statistical validity of the data on which they are based; the rules are not unalterable "laws," even though they may be true for "most" species or races. Bergmann's rule, for instance, is valid only if it is really true that in more than 50 percent of species of warm-blooded vertebrates there is an average increase of size in the cooler portions of the range of the species. Moreover, as some recent critics have failed to recognize, the validity of these empirical findings is independent of the physiological interpretation given to the observed regularities.

The degree of validity of ecogeographical rules varies from one group of animals to another and from one region to another. Rensch, who devoted a long series of investigations (1938, 1960) to a precise analysis of the percentual validity of these rules, found that their application depends on "other things being equal." Hamilton (1961) has made a particularly careful analysis of the interacting and often conflicting environmental factors that affect the direction of these trends.

In view of occasional misunderstandings in the literature, it is important to emphasize that the validity of the ecological rules is restricted to intraspecific variation. A more northerly species is by no means always larger than its nearest more southerly relative. As we shall presently see, separate species have different means of adaptation available than have open populations within species. The climatic rules are ecotypic, not phylogenetic, phenomena.

Variation in size in warm-blooded vertebrates. No ecogeographical rule is more widely known than Bergmann's rule. The amount of difference in size between races from cooler climates and those living in warmer climates can be very impressive, the largest races sometimes being more than twice as big (heavy) as the smallest. For birds, Bergmann's rule shows no more than 10–25 percent exceptions, mostly concerning migratory species or those that display no geographic variation. Among mammals there are some 20–40 percent exceptions. Burrowing mammals, for instance, are well protected against the cold, particularly in areas with snow cover, and for them the amount of food available in winter seems to be the decisive factor determining body size. Bergmann's rule tends to be valid also for changes of size with altitude, particularly in the tropics.

Changes of size with time have been recorded by paleontologists. The races of several species of mammals that occurred in North America, in Europe, and in the East Indies during cold periods at the height of the Pleistocene glaciations averaged larger in body size than the races of the same species now occurring in these regions.

The usual physiological explanation of Bergmann's rule is based on the fact that the volume of the body increases as the cube and the surface as the square of a linear dimension. The larger a body, the relatively smaller its surface. In a cool climate there should be a selective advantage in the relative reduction of surface resulting from increased size, since the metabolic rate is more nearly proportional to body surface than to body weight. In hot climates the premium should be on small body size and relatively large surface. This interpretation has been attacked on the basis that other devices in warm-blooded animals (feathers, fur, circulatory mechanisms, and so forth) prevent heat loss far more efficiently than slight shifts in surface/volume ratios. This argument overlooks the fact that multiple solutions for biological needs are the general rule in evolution. Selective advantages are independent and strictly additive. That a thicker fur or denser plumage reduces heat loss does not eliminate completely the selective advantage of an improved body surface/volume ratio.

A study of exceptions shows that the phenotype is a compromise between conflicting selection pressures, and that in many species of warm-blooded vertebrates the northernmost populations are exposed to conditions tending to neutralize the advantage of increased body size. In many Eurasian and North American species of birds the largest body size is found not in the coldest part of the range (near its northern periphery), but unexpectedly in the highlands of the semiarid subtropics (Iran, Atlas Mountains, Mexican highlands). It is not yet certain whether this "latitude effect" is due to the shortness of the arctic winter day, which depresses size by reducing daily food intake, or due to a need for water conservation in an arid area. In all these cases of exceptions to Bergmann's rule some environmental selective factor other than temperature has come into conflict with the size trend.

It has long been known that in warm-blooded animals protruding body parts like bill, tail, and ears are shorter in cooler than in warmer climates (Allen 1877). *Allen's rule* is actually an extension of Bergmann's rule, dealing likewise with the surface-to-volume relation, and has a similar percentage of exceptions. Because wings and tails of birds do not contribute to heat loss, being composed merely of feathers, they naturally do not obey Allen's rule. On the contrary, birds from northern populations of migratory species normally have wings that are longer in relation to body size than birds from more southerly populations of the same species. The wings of northern birds are often not only longer, but also more "pointed," that is, with the outer wing feathers more elongated. This change in the shape of the wing is correlated with an increase in the efficiency of the wing stroke ("wing rule").

Exceptions to Allen's rule involving the length of the bill in birds are usually related to food requirements. In the crossbills (*Loxia*), bill size is determined by the genus of conifers on which a given population feeds most frequently. If it is *Pinus* the bills are large and robust, if it is *Abies* or *Picea* the bills are intermediate, if it is *Larix* or *Tsuga* the bills are small and slender. In some species of titmice (*Parus*) relative bill length decreases with decreasing environmental temperature; however, no further decrease takes place once a certain bill length is reached, as if this represents minimal functional bill length.

Island birds often show considerable changes in body proportions (relative length of bill, tarsus, and so forth). These, though sometimes influenced by climatic changes, are in most cases caused by a shift in niche occupation. Owing to the considerably reduced faunal diversity on islands, shifts into vacant niches are not infrequent (Grant 1965; Keast 1969).

Pigmentation in warm-blooded vertebrates. Races in warm and humid areas are more heavily pigmented than those in cool and dry areas. Black pigments are reduced in warm dry areas, and brown pigments in cold humid areas. This rule, called *Gloger's rule,* seems to have comparatively few exceptions, but its physiological basis is not at all clear. It cannot, for instance, be ascribed to substrate adapted cryptic coloration, because arboreal and even nocturnal animals obey the rule well. The selective advantage of the genes responsible for these pigmentation differences is not evident.

Ecological rules for other animals. Various regularities, summarized by Mayr (1963), have been observed in geographic variation within widespread, polytypic species of invertebrates and cold-blooded vertebrates. However, the interactions between environment and physiology are rarely as straightforward as with warm-blooded vertebrates. The color of the substrate, the structural diversity of the habitat, and the presence of predators and competitors may affect trends of variation as much or more so than the climate. Insects, curiously, often seem to obey Gloger's rule by being most heavily pigmented in the warmest, most humid part of their range. In this case, as in others in which it has been possible to correlate character gradients with certain environmental gradients, it has not been possible to determine the adaptive significance of the climatic variation. The problem is particularly acute for size variation in invertebrates and cold-blooded (ectothermal) vertebrates. For these, of course, Bergmann's rule, with its implication of conservation of body heat, has no validity. It seems that three factors, to some extent antagonistic, determine the size trend in these animals. In species with a single generation every year, the length of the available growing period ("degree days") determines maximum larval size and hence adult size. Maximum size is usually reached in such species in the warmest, most humid portion of the range. In species in which sexual maturity is delayed for several years, as in certain marine animals, largest size may be reached in the coolest portion of the species range, by individuals having had the greatest number of (shortened!) growing seasons. If individuals of a single population are raised at different temperatures, those exposed to the lower temperature grow more slowly but usually reach ultimately larger size (Ray 1960). Finally, where factors other than temperature (such as humidity, food supply, absence of disease or competitors) affect body size, largest size will be reached in the optimal portion of the species range, independently of temperature. The relative importance of these three factors has not yet been determined for any group of invertebrates or cold-blooded vertebrates.

Conclusions. The climatic regularity in much of geographic variation proves that different species may react to the same factor of the environment in a similar manner. Since it can usually be shown that these differences have a genetic basis, the ecogeographical rules constitute evidence for the selective role of the environment. The multiplicity of largely independent regularities indicates the multiplicity of selective components of the environment. The regularity, the "smoothness," of most character gradients resulting from the climatic rules indicates that geographical changes of phenotype result from the interaction of numerous genetic factors, each with only a small phenotypic effect.

(3) *Geographic Variation in Substrate Adaptation*

The adaptive nature of geographic variation is also demonstrated by cryptically colored animals. An amazing agreement between the coloration of animals and the color of the substrate on which they normally live has been recorded not only for species (Cott 1940), but also for local races in many animals of open lands, lava flows, and deserts. To cite one specific example, most of the animals inhabiting the Namib Desert in southwest Africa display the reddish yellow coloration characteristic of that desert's soils, gravels, and rocks:

> Small succulent rock plants of the genus *Lithops* show the same color adaptation to the substrate as the elephant shrews *Elephantulus intufi namibensis,* the rodents *Gerbillus g. leucanthus* and *G. vallinus,* the larks *Certhilauda curvirostris damarensis, Tephrocorys cinerea spleniata,* and *Ammomanes g. grayi,* the bustard *Heterotetrax rueppelli,* the viper *Bitis peringueyi,* the lizards *Eremias undata gaerdesi* and *Meroles suborbitalis,* and finally a diversified group of wingless grasshoppers (Batrachotettiginae). The phenomenon of a "local coloration" which is shared by a large part of a local fauna is found also in other districts of Southwest Africa, e.g. the Usakos district, the Waterberg area, the Etosha Pan, and the Kaoko Veld (Hoesch 1956).

The more open the country and the more contrastingly colored the substrate, the more striking these substrate races; some of the most characteristic are found on lava flows and white gypsum sands. As to be expected, races evolved on lava flows are darkened, those on gypsum sands pale. However, a pocket gopher living in the sands is dark, presumably because this strictly subterranean mammal is very little exposed to selection by predators. Substrate races are also found in lizards and even in certain kinds of dersert insects.

The older literature abounds in conflicting "explanations" of such cryptic coloration, many of them openly vitalistic. Attempts to explain "desert races" with the help of climatic factors (heat, dryness, solar radiation) are not very convincing in view of the amazing agreement with the color of the substrate and the close proximity, under virtually identical climatic conditions, of black lava and white limestone or gypsum-sand races. The majority opinion now is that substrate races are the result of predator selection. That predators tend to take first the most discordant individuals has been proved by numerous observations and experiments. An analysis of snails (*Cepaea nemoralis*) taken by thrushes clearly shows the selective advantage of cryptic coloration, as does Kettlewell's work (1961) on melanistic moths.

(4) *Geographic Variation of Balanced Polymorphism*

Most polymorphism found in nature, as shown in Chapter 9, is due to a delicate balance between three (or more) genotypes, the heterozygote *Aa* being on the average selectively superior to the two classes of homozygotes, *AA* and *aa*. This delicate balance tends to change with the seasons and is also subject to geographic variation. This is one further demonstration of causation of geographic variation by natural selection. To list the examples of geographically variable polymorphism would be to list all known cases of polymorphism (Mayr 1963:330).

A number of generalizations emerge from these studies. There are some species that show a rather even degree of polymorphism throughout their ranges, others that are highly polymorphic in some areas and rather uniform in others. If one singles out an individual morph gene and follows it through the range of a species, one can make the following generalizations: (1) rare genes often have very restricted, local distributions; the same gene may reappear in far distant parts of the species, usually with equal rarity; (2) where the species range is essentially continuous, gene frequencies usually change clinally and the morph clines generally run parallel to climatic gradients; (3) at the periphery or in isolated parts of the species range some genes are generally lost while others may reach "fixation" (a frequency of 100 percent).

The adaptive nature of this variation is in most cases merely a hypothesis, being based on the known selective differences between genotypes in the cases of balanced polymorphism. There are some cases, however, in which the geographic variation of the visible phenotype is clearly adaptive. This is true of some of the microgeographic variation of such substrate-adapted polymorphic species as the snails *Cepaea nemoralis* and *Littorina obtusata*. It is even more true of mimetic polymorphism, where the frequency of the mimetic types in different regions is determined by the local frequencies of the model species. The parallelism in the geographic variation of models and mimics is particularly compelling evidence of the sensitivity of the selection process.

Conclusions

The findings emerging from the study of geographic variation have had a decisive impact on the development of evolutionary thought. The assertion of the early Mendelians that mutations are drastic and disruptive and that selection is therefore irrelevant forced students of geographic variation to adopt a Lamarckian interpretation. The smallness of the geographical differences, the gradualness of the changes from population to population, the obvious correlation with factors of the environment, in short all the findings of the students of geographic variation, negated De Vries's mutation theory. When the geneticists themselves proved the error of the mutation theory, and not only adopted the idea of small mutations but also began to admit the selective importance of the environment, there was no longer any obstacle to a reconciliation between geneticists and students of geographic variation.

The study of geographic variation has resulted in a number of well-established conclusions.

(1) Every population of a species differs from all others genetically and, if sufficiently sensitive tests are employed, also biometrically and in other ways.

(2) The degree of difference between different populations of a species ranges from almost complete identity to distinctness of almost species level.

(3) The area occupied by superficially identical populations may be extremely small, as is the case with some land snails, or may cover the entire species range.

(4) The various characters of a species may and usually do vary independently. Neighboring populations agree, therefore, in some characters and differ in others.

(5) All characters employed to distinguish species from each other are also known (at least in the better-known groups of animals) to be subject to geographic variation.

(6) The characters of a given population have at least in part a genetic basis and tend in most cases to remain rather constant through the years.

(7) Geographic variation as a whole is adaptive. It adapts each population to the locality it occupies. However, not all the phenotypic manifestations of this genotypic adaptation are necessarily adaptive.

(8) The ecotypic adaptation of local populations is a centrifugal evolutionary force. It leads to an increased genetic diversity of the species and results, as a by-product of gene flow, in a continuous readjustment of the local gene complexes.

Geographic variation is a population phenomenon that has greatly contributed to our understanding of the nature of species. It demonstrates the invalidity of the typological concept of species. It permits the conclusion that much, if not all, of variation is adaptive in response to the varying environmental demands. Some variable components of the phenotype, such as general size, proportions, and general coloration, are usually clearly adaptive. The geographic variation of other components, such as certain color patterns, does not seem directly adaptive, but it can often be shown by an appropriate analysis that these "neutral" phenotypes are the outward manifestations of genotypes that simultaneously control cryptic physiological characters established and maintained by natural selection.

The adaptive response of the different populations of a species affects in many ways the structure of a species as a whole. These various aspects of the population structure of species are discussed in Chapter 13.

12 · The Polytypic Species of the Taxonomist

The biological species concept determines which taxa deserve to be ranked as species. Its application is unambiguous in the local situation where a species is represented by a single local population (deme) that is reproductively isolated from the coexisting local populations of other species. Since the dimensions of space and time are not involved, such a local situation is sometimes referred to as a nondimensional species. The nondimensional species taxon has the virtue of being sharply delimited from other sympatric species, but it has the great weakness of being found only in a local situation. This limitation raises a serious difficulty since most species taxa are *polytypic,* that is, they consist of many local populations which are more or less different from each other. How to apply the biological species concept to such polytypic taxa is the subject of this chapter.

That the nondimensional species was an oversimplification became apparent very soon after its introduction into the biological literature, and was indeed already apparent in the days of Linnaeus. Two developments in particular contributed to the gradual undermining of the assumption that species taxa are necessarily the typologically unvarying and monotypic species. Both were consequences of the extensive explorations in the eighteenth and nineteenth centuries that converted the local naturalist into a widely traveled explorer. This explorer discovered new local populations that deviated somewhat from the population occurring at the locality where he had previously known the species. These new populations were at first called "varieties" and later "subspecies" (see below). A species within which such subspecies were recognized was designated a polytypic species (see Mayr 1969). The number of such polytypic species increased steadily after the middle of the nineteenth century as more and more populations of widespread species were sampled.

The second factor responsible for a reduction in the number of monotypic species was likewise a result of "geographical collecting." As the fauna of the world became better known, it happened more and more often that two allopatric species, originally thought to be completely distinct, were found to be connected by intermediate, intergrading populations. The honest systematist had no choice but to reduce the two "species" to the rank of subspecies and to combine them, together with the intermediate populations, into a single, widespread polytypic species.

Let me illustrate this with the concrete example of the birds of the *Passerella* (+ *Melospiza*) group. The early explorers and pioneer ornithologists discovered four similar species in eastern North America: the Fox Sparrow (*Passerella iliaca* Merrem 1786), the Swamp Sparrow (*P. georgiana* Latham 1790), the Song Sparrow (*P. melodia* Wilson 1810), and the Lincoln's Sparrow (*P. lincolni* Audubon 1834). During the exploration of the West in the middle of the nineteenth century, several additional forms of *Passerella* were discovered, for instance on Kodiak Island (*insignis*), in Alaska (*rufina*), in California (*gouldi*), and in Arizona (*fallax*). These forms were described as "species" because to their describers they seemed as different from each other as the four original species of eastern North America. However, as the ornithological exploration of North America continued, additional populations were found that were intermediate between these four western "species," and between them and the Song Sparrow (*melodia*) of eastern North America. As a result, all five "species" were finally reduced to the rank of subspecies and combined into a single polytypic species, the Song Sparrow (*Passerella melodia*), now comprising more than 30 subspecies (Fig. 12.1).

The change from the typological-morphological species to the biologically defined polytypic species occurred gradually and at different rates in different higher taxa. As far as birds are concerned, it has been essentially completed for several decades; in the more poorly known groups of insects and lower invertebrates it has hardly begun. For further information on the history of this development see Mayr 1942, 1963, and 1969.

The shift from the nondimensional to the multidimensional species poses practical and conceptual problems. The two concepts are radically different (Mayr 1957): the nondimensional species is objectively (nonarbitrarily) defined by the gap separating it from other (sympatric) species, while the polytypic species is characterized by an actual or potential genetic continuity of allopatric populations. In many cases this genetic continuity can be determined only by inference. For a discussion of the practical problems of the working taxonomist see Mayr 1969.

Fig. 12.1. Distribution pattern of a polytypic species of bird. Numbers refer to the breeding ranges of the 34 subspecies of the Song Sparrow, *Passerella melodia*. (From Miller 1956.)

The Value of the Concept of the Polytypic Species

The occurrence of polytypic species taxa has been established for most groups of animals. They are known even in such morphologically uniform animals as *Drosophila.* Claims of some authors that polytypic species are absent in the families in which they specialize are based either on the backward state of the taxonomy of these families (not enough samples of peripheral or peripherally isolated populations available) or on the tendency of some authors to call every morphologically distinct geographical isolate a species.

Combining numerous more or less isolated and morphologically distinct allopatric "species" into polytypic species not only had the practical result of greatly improving the classification of animals, but also paved the way for a better understanding of the process of speciation (Chapter 16). The most immediate and conspicuous effect of the method is a simplification of the system. This is best illustrated by a few figures. The last complete listing of the birds of the world (Sharpe 1909) gave about 19,000 full species. The arrangement of these species (and many hundreds discovered since 1910) into polytypic species has reduced the total number to about 8,600. Several thousand "species" of Eurasian mammals were combined by Ellerman and Morrison-Scott (1951) into about 700. Of much greater significance is the restoration of biological meaning and homogeneity to the species category. There are some 28,000 subspecies in the 8,600 species of birds, and it is evident that it would represent a misleading systematic evaluation if the same rank were given to these subspecies as to full species. Yet such unequal treatment is the rule in many of those groups in which the polytypic species concept has not yet been adopted, and where every distinct population is ranked as a full species.

The application of the concept of polytypic species has been particularly fruitful in the elucidation of complex taxonomic situations. In each case it forces an unequivocal decision as to whether or not two forms should be considered conspecific. The sorting of large numbers of "nominal species" and "varieties" into polytypic, biologically defined species has not only resulted in a great refinement of the taxonomic technique, but has also often revealed the pathway by which populations reach species level.

The method of classifying populations into polytypic species is most apt to run into difficulties where a species or species group is actively evolving. In island regions, the decision as to which isolates are to be combined into polytypic species is often quite arbitrary. The difficulties are particularly apparent where subgroups of a species have changed their ecological requirements and have, by renewed range expansion, come into secondary contact with other subgroups of the species. Behavior of the meeting populations is frequently unpredictable. Furthermore, forms that behave toward each other like good, reproductively isolated species are sometimes more similar phenotypically than are intergrading or interbreeding forms. Such a situation exists in many species groups of South Sea Island birds, as for instance *Rhipidura rufifrons, Edolisoma tenuirostris-morio, Pachycephala pectoralis,* and *Ptilinopus*. The cases of circular overlap (Chapter 16) are further instances of the difficulties that evolution occasionally poses to those attempting to apply the polytypic species concept.

The Terminology of the Subdivisions of the Species

As implied in its name, the polytypic species has subdivisions. The kinds of subdivisions distinguished by various authors and the terms proposed by them are influenced by the philosophies of these authors. Some have as their ultimate objective the facilitation of the purely pragmatic, formalistic task of classifying specimens; others attempt to find units with specific biological or evolutionary significance. Although we are mainly concerned with the evolutionary aspects of species structure, we cannot entirely avoid a discussion of terms with a purely practical taxonomic significance because these terms (for instance, the term "subspecies") have been employed a great deal in the evolutionary literature. For a more detailed discussion see Mayr (1969).

The Variety

The variety (*varietas*) was the only subdivision of the species recognized by Linnaeus and the early taxonomists. A variety was anything that deviated from the ideal type of the species. An analysis of varieties actually recognized by Linnaeus in his taxonomic writings indicates that they were a highly heterogeneous lot of deviations from the species type. Only some were genuine geographic races or subspecies. Much of the argument in the nineteenth century over the biological and evolutionary meaning of the variety was due to the fact that the term concealed two entirely different phenomena: (a) individual variants within a polymorphic population; (b) distinguishable populations in a polytypic species. It became increasingly evident that this confusion could be resolved only by restricting the term variety to individual variants or by abandoning it altogether. The abandonment of the term variety for geographic races is nearly complete in zoology.

The Subspecies

The term *subspecies*, when it came into general usage in taxonomy during the nineteenth century, was a replacement for "variety" in its meaning of geographic race. As a consequence the term "subspecies" was endowed from the very beginning with all the typological shortcomings of the term "variety." It was considered a taxonomic unit like the morphological species, with the same objectives but on a lower taxonomic level. Corresponding to the species defined as a unit comprising those individuals that conform to the type of the species, the subspecies was defined as a unit consisting of those individuals that conform to the type of the subspecies.

This concept of the subspecies is fallacious. Species are not composites of uniform subtypes—subspecies—but consist of an almost infinite number of local populations, each in turn (in sexual species) consisting of genetically different individuals. The difficulties of the subspecies concept are intensified if one considers the subspecies not merely as a practical device of the taxonomist, but also as a "unit of evolution." The better the geographic variation of a species is known, the more difficult it becomes to delimit subspecies and the more obvious it becomes that many such delimitations are quite arbitrary.

It is now apparent that the subspecies is not a unit of evolution except where it coincides with a geographical isolate. In all other cases it is merely a convenient pigeonholing device of the practicing taxonomist, who must be aware at all times of the shortcomings of this category, as discussed by Mayr (1969).

The modern definition of the subspecies is exceedingly different from that of the Linnaean geographic variety. It attempts to meet the various objections listed above and may be worded as follows: A *subspecies is an aggregate of phenotypically similar populations of a species inhabiting a geographic subdivision of the range of the species and differing taxonomically from other populations of the species.* It is important to emphasize certain aspects of this definition:

(1) A subspecies is a collective category because every subspecies consists of many local populations, all of which are slightly different from each other genetically and phenotypically.

(2) Every subspecies has a formal name (trinominal nomenclature) and it would obviously lead to nomenclatural chaos if every slightly differing local population were to be dignified by a trinomen. Therefore subspecies are to be named only if they differ "taxonomically," that is, by diagnostic morphological characters. How great this taxonomic difference ought to be can be determined only through agreement among working taxonomists.

(3) Although it is usually possible to assign populations to subspecies, it is often impossible to do this for individuals because of the individual variability of each population and the overlap of the curves of variation of adjacent populations.

(4) A subspecies inhabits a definite geographic subdivision of the range of the species, a necessary consequence of the fact that subspecies are composed of populations and each population occupies part of the range. The distribution of the subspecies will be determined largely by the correlation between the diagnostic characters and the environment; consequently the range of a subspecies may sometimes be discontinuous (*polytopic* subspecies).

All attempts either to replace the subspecies with a different term or to abandon the category altogether have been found unacceptable by most taxonomists. The category subspecies continues to be a convenient means of classifying population samples in geographically variable species, in particular in those with phenotypically distinct geographical isolates. It must be realized, however, that in many cases the subspecies is an artifact rather than a unit of evolution.

The Temporal Subspecies

Subspecies have been treated in the preceding discussion as subdivisions of the species in the dimensions of longitude and latitude, that is, as spatial units. However, species are as polytypic in time as they are in space, a fact requiring the recognition of temporal subspecies, that is, subspecies in the time dimension of the multidimensional species. A biological difference between the geographical and the temporal subspecies does not exist. Therefore it does not seem advisable to make a terminological distinction in paleontology between geographical and temporal subspecies because it is usually quite impossible, when different subspecies of a fossil species are found at different localities, to determine whether or not they are precisely contemporary. Even when there is a sequence of subspecies at a single locality, it need not necessarily be purely temporal. Subspecies found in succeeding strata may actually be geographical races that replaced each other owing to climatic or tectonic changes.

The Ecotype

Botanists use the term *ecotype* to designate local populations whose characters are determined by habitat, for example, sea-cliff ecotype, sand-dune ecotype, or salt-marsh ecotype. The term has not been used much by the zoologists and has now also fallen out of favor in much of plant ecology owing to its typological connotation (Mayr 1963). Actually most so-called ecotypes are part of a continuum and show furthermore the characteristic variability of any local population.

Those plant taxonomists who divided species into ecotypes believed that doing so would permit a more precise and more biological description of geographic variation than the division of species into subspecies. Unfortunately, ecotypes are rarely discontinuous, rarely well delimited, often polyphyletic, and always full of intra-ecotype variability. The ecotype concept, thus, suffers from precisely the same weaknesses as the subspecies concept. However, the situation is even more serious in the case of the ecotype concept because the subspecies is admittedly an arbitrary instrument of the taxonomist, created purely for taxonomic convenience, while the ecotype was established for the very purpose of getting away from the artificiality of taxonomic categories and of replacing them by something more meaningful biologically. The ecotype has not been successful in this endeavor. For a further discussion of the ecotype concept and a review of some of the ecotypic research, one should consult the botanical literature (Grant 1963).

The Ecological Race

A local population that is particularly conspicuously adapted to a local habitat is often referred to as an *ecological race.* Not all populations thus designated in the literature are truly ecological races; some have recently been unmasked as sibling species, and others are nongenetic modifications of the phenotype (ecophenotypes, Chapter 7). There remains, however, a considerable residue of ecological races characterized by differences in habitat preference. Every field ornithologist is familiar with such cases. The Savanna Sparrow (*Passerculus sandwichensis*) of the eastern United States is found in coastal salt marshes and also on dry uplands in the interior. The Swainson's Warbler (*Helmitheros swainsoni*) lives in the cane brakes of the coastal marshes in the southern United States but also in the southern Appalachian highlands above 3000 feet in thickets of rhododendron, mountain laurel, hemlock, and American holly.

The question most often debated concerning the ecological race is whether or not it is a category distinct from the geographical race. This can now be answered in the negative. The two kinds of phenomena cannot be kept apart. For instance, the ecological races of mammals adapted to live on lava flows have well defined geographic ranges and are thus also geographic races. Subspecies, in many cases, differ by such well-defined habitat preferences that they could be called with the same justification ecological races or geographic races (for examples, see Chapter 11).

There is a special type of ecotypic variation, the *polytopic race,* that does not fit too well into the previous characterization. Many species of plants and animals have the ability to occupy several specialized habitats and to become adapted to them. The banded snail *Cepaea nemoralis* may develop a prevailing color type in moist beech woods (reddish, unbanded). Yet there is presumably a greater total genetic difference between the reddish unbanded types of England, of southern France, and of eastern Germany than between some of the different color types in southern England (red or yellow, banded or unbanded). No terminology is suitable to express simultaneously the independent variation caused by such different factors as substrate and climate.

SUMMARY

A species consists neither of an aggregate of strictly morphologically definable subspecies and varieties nor of purely ecologically definable ecotypes or ecological races. A species is actually composed of populations distributed in space and time that possess similar morphological as well as similar physiological and ecological characteristics. The recent work on geographic variation has led to the reinterpretation of the geographic race as a genetic-physiological response to a local environment. There is no antithesis between geographic race and ecological race (or ecotype) because not a single geographic race is known that is not also an ecological race; nor is there an ecological race that is not at the same time at least a microgeographic race.

In animals as well as in plants, local populations are selected for adaptation to the specific environment in which they live. The principal difference between plants and higher animals is the amount of their direct dependence on the environment and the phenotypic expression of the local physiological adaptation. At one extreme are the warm-blooded birds, highly mobile and highly independent of the direct effect of the environment. At the other extreme are certain species of plants and sedentary invertebrates that are highly dependent on their substrate and fully manifest this dependence in their phenotype. The two extremes are differences of degree and are connected by a complete spectrum of intermediate conditions.

13 · The Population Structure of Species

As a means of simplification, the practicing taxonomist divides species taxa in a typological manner. He implies in his species catalogues that the subspecies and ecotypes into which he divides his species are well defined, more or less uniform over extensive areas, and separated from other similar units by gaps or steep and narrow zones of intergradation. It is now increasingly apparent that this simplified typological picture of the species structure is the exception rather than the rule.

A very different approach, based on the population structure of species, is necessary in a study of the internal variation of species from the ecological and evolutionary point of view. This new approach investigates the degree of difference between neighboring populations, the presence or absence of discontinuities between populations, and the characteristics of those populations that are intermediate between phenotypically distinct populations. It is an objective approach because it does not try to force natural populations into a preconceived framework of artificial taxonomic or ecological units and terms. A new picture of the population structure of species emerges from this new approach. It shows that all populations of a species can be classified under one (or more) of the following three structural components of species: (1) series of gradually changing contiguous populations (*clinal variation*); (2) populations that are geographically separated from the main body of the species range (*geographical isolates*); (3) rather narrow belts, often with sharply increased variability (*hybrid belts*), bordered on either side by stable and rather uniform groups of populations or subspecies.

Nearly every well-studied species manifests more than one of these elements. The best analyses of the species of an entire fauna for these structural components are those of Keast (1961) for the Australian avifauna and Hall and Moreau (1970) for the African avifauna. The rich findings of these studies demonstrate the extraordinary value of this type of analysis.

Clinal Variation

When neighboring populations of a species are compared, one finds that they usually differ from each other, slightly or appreciably, in a number of characteristics. Furthermore, when one traces a character through a series of contiguous populations, the changes usually show a regular progression. Such regular progressions of characters were discussed in Chapter 11 (under the ecogeographical rules) as an indication of the adjustment of populations to local conditions. Huxley (1942) coined the term *cline* for such a character gradient. The study of geographic variation has revealed that much of it is clinal.

There are several reasons for the clinal mode of geographic variation. The first is that the environmental selective factors themselves (such as climate) vary along gradients, and, as a consequence, so do those phenotypic characters that respond to this selection. There are very few features of the environment, such as soil color and other properties of the substrate, that may change abruptly. Another reason is that gene flow between adjacent populations tends to smooth out all sharp differences. The potency of such gene flow is particularly apparent where it spills across natural barriers or where it leads to a discrepancy between coloration of a population and the color of its substrate.

Clines are, ultimately, the product of two conflicting forces: selection, which would make every population uniquely adapted to its local environment, and gene flow, which would tend to make all populations of a species identical. The cohesive force of gene flow gives physiological unity to the species as a whole but increases the necessity of adjustments to local conditions. Reaching a compromise between the "typical" physiology of the given species and the demands of the local environments becomes increasingly difficult toward the periphery of the species range, and the ultimate inability to make this compromise is responsible for the phenomenon of the species border, which the species cannot transgress (Chapter 17).

Clines are widespread and occur in the majority of, if not in all, continental species. This is not surprising since climatic factors such as temperature, rainfall, evaporation, number of days with frost or snow, and so forth, show regular gradients (Huxley 1942). One of the finest analyses of clines in a group of animals is that by Petersen (1947) of Fennoscandian butterflies (Fig. 13.1). He analyzed in 16 species the geographic variation of 59 characters and found that 29 (about 50 percent) varied clinally. The

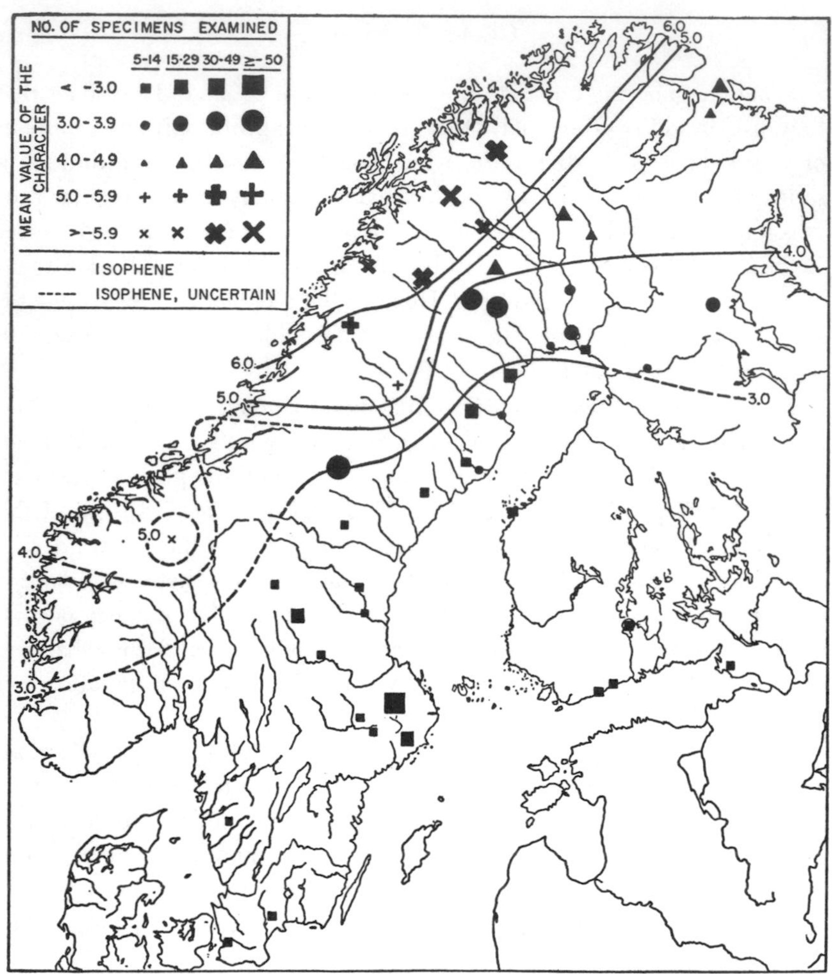

Fig. 13.1. Character gradient concerning pigmentation of the upper side of the wing in females of *Pieris napi* from Fennoscandia. The darkest values occur in the northwest. Size of symbol indicates size of sample. (From Petersen 1947.)

six species that showed no clines were partly migratory. In the 10 sedentary species, 70 percent of the examined characters varied clinally.

Clines and isophenes. A clear distinction must be made between clines and isophenes. A cline is the total slope from one extreme of the character to the other. Clines, when plotted on a map, are crossed at right angles by *isophenes,* the lines of equal expression of a character. For instance, in many species of Indian birds, a size cline runs from the largest populations in the Himalayas south to the smallest populations in Ceylon and Malay Peninsula. This north-south cline is crossed by the isophenes, lines connecting all populations with the same phenotype, for instance, all those with a mean wing length of 180mm.

Size, color, or any other kind of morphological or physiological character may vary clinally. Huxley (1942:206–207) has listed a number of such characters. The cline for each character is theoretically independent of the others. Nearly all Australian birds with size variation, for instance, decrease in size from Tasmania northward to Torres Straits along a regular cline, following Bergmann's rule. Intensity of color, however, changes along a very different cline, leading from the most humid periphery of Australia to the most arid interior.

The term "cline" refers to a specific character, such as size or color, not to a population. A population may belong to as many different clines as it has variable characters. The potential independence of different character gradients makes the cline unsuitable as a taxonomic category. A concordance of the clines for different characters is normally found only where ranges are essentially longitudinal and where the various environmental gradients (temperature and humidity, for example) run by chance more or less parallel.

In many cases gene flow seems more responsible for the maintenance of a cline than an environmental gradient. The importance of gene flow can be demonstrated best by studying isolated populations adjacent to continuous populations and situated on the same environmental gradients. Such studies show that the variation of truly isolated populations is unpredictable and often remarkably independent of the clines found in the adjacent continuous populations.

To what extent a given character gradient is the product of gene flow or of an environmental gradient is usually difficult to say. The clines in the frequencies of the human blood groups, formerly ascribed entirely to gene flow, may well have a substantial selective component. The working hypothesis that the presence of a cline indicates a geographically variable selective factor of the environment has proved very productive. It induced Mayr (1945) to postulate that each gene arrangement of *Drosophila pseudoobscura* had a definite selective value at each locality rather than that this distribution pattern was due to a historical accident.

Geographical Isolates

Not all populations are members of a set of contiguous populations with clinal variation owing to gene flow. Some populations are genetically independent of other populations of the species because they are geographical isolates. The term "isolate" has been used in the biological literature with different meanings. In anthropology and human genetics it is usually applied to the inhabitants of a partially isolated area and more broadly to what the naturalists would call any local and usually only very incompletely isolated population. In the present discussion I define *geographical isolate* as *a population or group of populations prevented by an extrinsic barrier from free gene exchange with other populations of the species.* The essential characteristic of the geographical isolate is that it is separated from the rest of the species by a discontinuity. The degree of discontinuity depends on the efficiency of the extrinsic barrier. The isolation is never complete, since a certain amount of gene flow reaches even an isolated oceanic island (or else it could not have been colonized originally). Virtually every species contains some isolates, particularly near the periphery of the species range. The frequency of isolates increases sharply wherever geographical or ecological conditions produce an insular distribution pattern. This is true not only for oceanic islands, but also for all kinds of ecological islands, be they mountains, forest patches in grasslands, or lakes and streams. The frequency of isolates within a species depends on the structure of the environment and the dispersal facilities of a species. It is well known among taxonomists that species may show great phenotypic uniformity over wide areas where the species range is continuous or else an astonishing production of isolates where barriers break up the ranges. Continental lizards usually have few geographic races and these are not strikingly different; the same species on archipelagos may break up into dozens or scores of highly distinctive isolates. A comparison of the continuous ranges of temperate-zone Eurasian species with the isolated Mediterranean populations of the same species reveals the same phenomenon.

We know, as yet, little about the frequency of genuine isolates in various groups of animals. This is regrettable, considering the great potential importance of isolates for speciation. The minimum number of isolates in several well-analyzed groups of birds is listed in Table 16.1. Keast (1961) found that 425 species of Australian birds had developed 211–226 morphologically differentiated isolates. Even in a continental area there is, then, ample incipient speciation. The number of isolates in an equivalent island area is about five times as great.

Geographical isolates may occur throughout the range of a species, wherever barriers occur, but they are most frequent at the periphery. Taxonomists have long been aware of the importance of these peripheral isolates and have pointed out, again and again, that major deviations from the "type" of a species will most likely occur in such populations. For a further discussion of peripheral isolates see Chapter 17.

Geographical isolates have three possible fates (discussed in more detail in Chapter 16). They may become separate species, die out altogether, or reestablish contact with the main body of the species, forming a secondary zone of contact. For a discussion of the criteria used to determine taxonomic status of isolates, see Mayr 1969.

Zones of Intergradation

The third phenomenon, in addition to clines and isolates, that is characteristic of the population structure of most species is the existence of contact zones between phenotypically different populations. The terminology, classification, and interpretation of such belts has long been a source of disagreement. Taxonomists have sometimes referred to these belts as "subspecies borders" because widespread and comparatively uniform subspecies often meet in such belts. Yet there is no congruence between the two phenomena. Some subspecies intergrade imperceptibly along a cline; on the other hand, distinct steps in clines sometimes separate populations that are not sufficiently distinct to deserve subspecific recognition. It seems probable that these contact zones reflect two phenomena that from the evolutionary viewpoint are rather different (Mayr 1942:99).

Primary intergradation exists if the steepening of the slope developed gradually and took place while all the populations involved were in continuous contact. *Secondary intergradation* refers to cases in which the two units now connected by a steeply sloping character gradient were separated completely at one time and came into contact only secondarily, after a number of differences had evolved. Cases of primary intergradation are believed to be caused by a corresponding change in environmental conditions; zones of secondary intergradation are hybrid belts between populations that had become differentiated during a preceding period of isolation.

Allopatric Hybridization

The interbreeding of two previously isolated populations in a zone of contact has been designated *allopatric hybridization* (Mayr 1942). This term conforms to the definition of hybridization (Chapter 6) as the crossing of individuals belonging to two unlike natural populations that have secondarily come into contact. Two recent surveys of such cases are those of Remington (1968) and Short (1969).

A well-analyzed and carefully described case of allopatric hybridization is that of the Hooded Crow (*Corvus c. cornix*) and the Carrion Crow (*C. corone corone*) (Meise 1928). The all-black Carrion Crow inhabits western Europe; the Hooded Crow, gray with a black head, wings, and tail, inhabits eastern Europe and most of the Mediterranean region. The two forms come into contact in a narrow zone starting in Scotland, extending through Denmark, central Germany, and Austria to the southern slopes of the Alps, and reaching the Mediterranean somewhere near Genoa (Fig. 13.2). Pairing within the hybrid belt seems to be random, and there is every conceivable combination of the parental characters as well as all degrees of intermediacy. Beyond the hybrid belt an occasional bird is encountered that does not appear to be quite "pure," such as a Hooded Crow with some Carrion Crow characters, or vice versa, but on the whole the visible effects of the hybridization are rather localized.

Populations meeting in zones of secondary intergradation may show any degree of difference. They range from those as different morphologically as good species (such as the crows) to populations that can be separated only by biometric or genetic tests. Only the more conspicuous cases are usually recorded in the taxonomic or evolutionary literature, and it is therefore difficult to determine the relative frequency of zones of secondary intergradation in different species. Owing to the numerous discontinuities in the ranges of most species and the never-ending changes of the environment, fusions between previously isolated populations, resulting in zones of secondary intergradation, occur very commonly. Even rather spectacular cases of allopatric hybridization between very different forms (including semispecies) are remarkably frequent.

Cases of Allopatric Hybridization. Allopatric hybridization usually results from the expansion of isolates because of changed environmental conditions. It occurs with particular frequency after periods of climatic change, such as the end of the Pleistocene. During the height of the glaciation, the ranges of many temperate-zone species contracted into small pockets, so-called glacial refuges, which persisted south of the area of glaciation. In Europe,

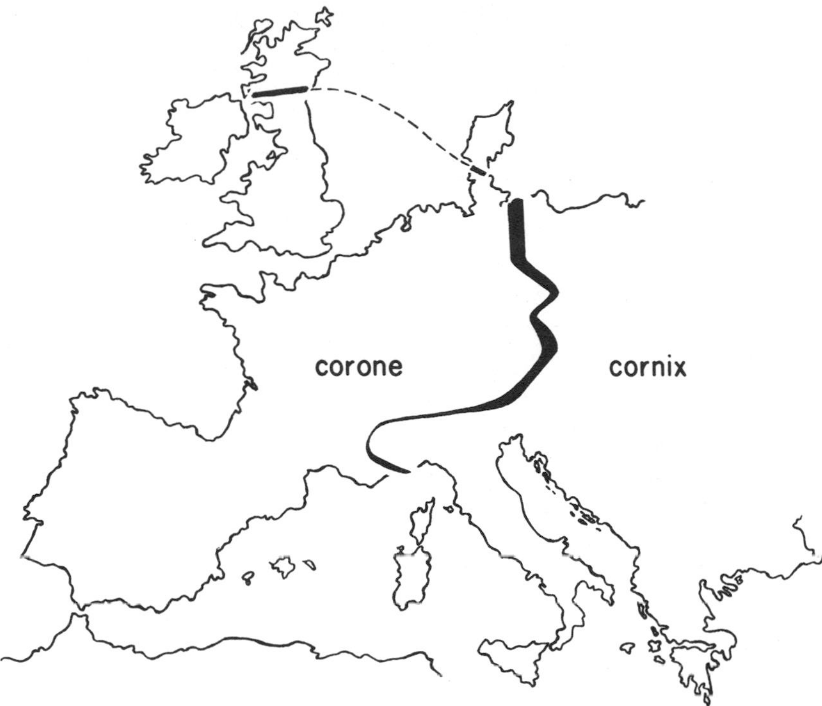

Fig. 13.2. Course of the hybrid zone between the Carrion Crow (*Corvus c. corone*) and the Hooded Crow (*Corvus corone cornix*) in western Europe. Note the relative narrowness and unequal width of the zone. (After Meise 1928.)

for instance, the Alpine and northern ice caps came within 300 miles of each other, separated by icy wind-swept steppes. The forest animals retreated into southwestern or southeastern Europe. When conditions improved at the end of glaciation and the populations in the refuges expanded northward, the isolates in southwestern and southeastern Europe had, in many cases, become sufficiently distinct from each other to form hybrid zones along the line of contact. Such hybrid belts in central Europe have been described for mammals, birds, amphibians, and invertebrates (Mayr 1963).

A corresponding great hybrid belt runs through Asia from Iran and Turkestan north to the Siberian tundra (Johansen 1955). Numerous hybrid zones in North American birds likewise owe their origins to the post-Pleistocene range expansions (Rand 1948; Short 1969; Selander 1965).

Most hybrid zones in the temperate region are the result of the fusion of populations expanding into the areas vacated by the retreating ice. In other cases shifts in vegetation zones, indirectly caused by climatic changes, are responsible for creating hybrid belts. The recent hybridization in the northern American plains in the avian genera *Colaptes, Icterus, Passerina, Pipilo,* and *Pheucticus* (Sibley 1961) is partly due to the planting of trees, which provide avenues of contact across the previously largely treeless prairie. As far as the subtropical and tropical regions are concerned, alternation between arid and humid periods is certainly the main cause of the separation and eventual rejoining of isolates. In Australia, for instance, most birds of the forested areas were squeezed into a number of coastal refuges during a Quaternary drought period. The present distribution and variation of the tree runners (*Neositta*) indicates that this species group had one such refuge in southwestern Australia, one in northwestern Australia, one in northern Queensland, and two or three in eastern Australia between southern Queensland and Victoria. As the rainfall increased at the end of the dry period, trees and tree runners began to spread, and the former isolates came into secondary contact. There are now five or six zones of hybridization (Fig. 13.3). Keast (1961) treated in detail the hybrid zones in the Australian birds.

Allopatric hybridization has been particularly well studied by ornithologists, because of the advanced state of avian systematics. Moreau (1966),

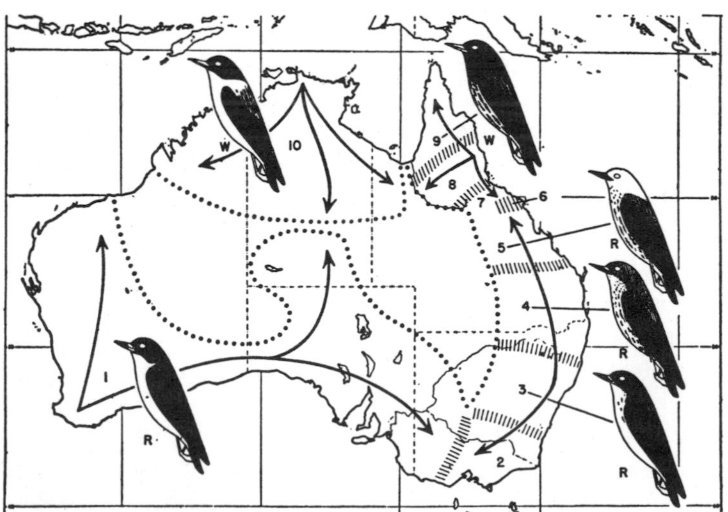

Fig. 13.3. Tree runners (*Neositta*) from Australia. The arrows indicate expansion from post-Pleistocene aridity refuges. Wherever two former isolates have met, they have formed hybrid belts (indicated by hatching). *R*, red wing bar; *W*, white wing bar.

for example, published a masterly analysis of recent population expansions (and contractions) in Africa, many of which resulted in the production of hybrid belts. However, cases of secondary intergradation have also been described for many other groups of animals.

Various Aspects of Hybrid Belts

When two populations are isolated from each other, their gene pools become independent and they diverge steadily in their genetic composition (Chapter 17). When the geographic isolation breaks down and the two populations reestablish contact, we should find evidence in the zone of contact for the degree of genetic differentiation achieved during the preceding isolation. An analysis of the characteristics of populations in such hybrid belts does indeed reveal much evidence for such prior genetic differentiation.

Random or selective mating. If incipient isolating mechanisms had developed prior to fusion, one should expect definite deviations from random mating. This is what Howell (1952) found in the sapsucker genus *Sphyrapicus.* Where the ranges of *nuchalis* and *ruber* meet at Kersley, British Columbia, one form is replaced by the other within the short stretch of 1.5 miles. Among pairs observed in the area, five appeared to be *nuchalis,* three *ruber,* and three pairs either mixed *ruber* × *nuchalis* or *nuchalis* × hybrid, or both hybrids. In such cases of partial breakdown of reproductive isolation a decision on the taxonomic treatment (species or subspecies) is often difficult, but these *Sphyrapicus* forms are best considered species since the majority of pairs are conspecific.

In most cases of secondary intergradation genetic divergence in the former isolates has not yet proceeded to the development of preferential mating. Meise (1928) found no evidence for anything but random mating in *Corvus,* nor does preferential mating occur where the Tufted and the Black-crested Titmouse (*Parus*) meet in Texas. Usually morphological differentiation seems to take place more rapidly than the acquisition of isolating mechanisms. This sequence is shown to a particularly striking degree in the snail genus *Cerion,* where exceedingly different populations interbreed freely and at random in their contact zones (see Fig. 2.1).

Ecological divergence preceding hybridization. Isolates often differ in their habitat requirements or in other ecological characteristics. When they come into contact secondarily, without having acquired reproductive isolation, they form a zone of intergradation along a steep ecological gradient or along an ecological discontinuity. Numerous such cases are described by Mayr (1963:375).

The tenacity with which certain ecological preferences remain tied up with gene complexes has been described in a number of cases. For instance, the pocket gopher *Thomomys bottae pascalis* of the irrigated lands of the San Joaquin Valley in California meets in a very narrow zone (rarely more than ½ mile wide) the subspecies *T. b. mewa* of the wild uncultivated grasslands. The amount of interbreeding in the zone of contact without visible introgression indicates strong selection against introgressing genes.

In the marine snail *Thais lapillus* two chromosomal forms meet in Brittany (Staiger 1954), one with 8 acrocentric and 5 metacentric (= 13) chromosomes, the other with 18 acrocentric chromosomes in haploid condition. The two interfertile forms differ in their ecological requirements; the 13-chromosome type occurs on the rocky coast exposed to surf, while the 18-chromosome type is found in sheltered bays in the shallow-water zone. Heterogeneous colonies with intermediate chromosome frequencies (diploid means of 27–35) are found in intermediate localities. These mixed colonies show "hybrid vigor" in increased shell thickness (but smaller shell size) and high population density, yet they also exhibit a certain amount of hybrid sterility and inviability. The different chromosome numbers and habitat preferences had apparently developed during a previous isolation of the 13- and 18-chromosome types without, however, resulting in reproductive isolation. Only the 13-chromosome type occurs on the Atlantic coast of North America. The ecological adaptation of the gene complexes meeting in these hybrid zones prevents in all these cases a widening of the zone of intergradation through gene flow. This situation is paralleled by most cases of ecotypic variation in plants where secondary contacts are involved.

Incompatibility of gene complexes. The amount of nonrandom mating in hybrid belts and, even more, the maintenance of ecological differences on either side of the line of contact indicate the amount of genetic difference that must have existed prior to the secondary contact. Analysis of the incompatibilities of entire gene complexes is as yet impossible owing to technical difficulties, but much can be inferred from analyzing the phenotypes of hybrid populations.

The narrower the hybrid zone, the greater is the probability of unbalance in the recombined genotypes and the greater the probability that the unbalance will express itself as highly increased variability. Many such cases are reported by Mayr (1963:377) and Short (1969). This variability is due partly to the various grades of back crosses of individuals in the zones of intergradation and partly to a breakdown of developmental homeostasis resulting from mixing of somewhat incompatible genes. Both parental phenotypes may be found in the same area, together with all sorts of intermediates.

In a hybrid belt, some characters may be more variable than others. In some hybrid colonies of the snail *Cerion*, for instance, size and proportions do not show increased variability, while sculpture and pigmentation do. This difference may be due partly to the much greater number of genes controlling size and partly to the stabilizing effect of selection, to which size may be more subject than the ornamental shell characters. When two (sub)species hybridize in several separate areas, such as the Grey-billed and Black-billed Honeyeaters, *Melidectes*, in New Guinea (Mayr and Gilliard 1952), a different character may become stabilized in each area.

Because natural selection is unceasingly at work in these hybrid zones, weeding out the most unbalanced combinations, high variability is maintained only by the continued reintroduction of new parental genotypes. A hybrid population may achieve considerable phenotypic stability when subsequent isolation deprives it of gene flow from the two parental populations. Low phenotypic variability in a population is thus no proof of the absence of former hybridization.

Width of hybrid belts. One of the least-understood aspects of hybrid belts is their width: some are very wide, others amazingly narrow. This narrowness is a great puzzle in hybrid belts that must have existed for thousands of years. One would expect either that reproductive isolation would be acquired as a result of an inferiority of the hybrids, or that gradual infiltration of the hybridizing genes would steadily widen the hybrid belt until it occupied the greater part of the ranges of the hybridizing populations. Apparently there is a third alternative: a vigorous selection against the infiltration of genes from one balanced gene complex into the other, but without the development of any isolating mechanisms as a by-product of this selection. As an example of a narrow hybrid belt I would like to cite the flea *Ctenophthalmus agyrtes*. Where the *eurous* subspecies group meets the *agyrtes* subspecies group in western Germany the total belt of hybridization is only 6.5 kilometers wide. In Normandy, where *eurous* meets the western subspecies *celticus*, Jordan (1938) found that they came within 100 meters of each other, *eurous* being restricted to a wooded hill, *celticus* to open fields (both occurring on several host species).

Narrow but virtually permanent hybrid belts must be interpreted as zones of contact between balanced gene complexes stabilized through selection during a previous isolation. All disharmonious combinations in the hybrid zone will be selected against. Similarly, their penetration into the adjacent populations will be continuously counteracted by selection. This selection will not entirely eliminate gene flow but will greatly reduce its phenotypic effects. It is to be expected that some genes will be less strongly selected against than others, and that these genes may penetrate beyond the hybrid belt. The more closely related the populations coming into contact and the less disharmonious their gene complements, the more likely it is that such penetration will occur.

Geography, Ecology, and Species Structure

We conclude from the findings presented in the preceding section that species are not the uniform typological entities envisioned by classical taxonomy. Species actually have a complex population structure, characterized by series of clinal populations, isolates, and zones of intergradation. The relative frequency, importance, and location of these three components of species structure differ from species to species. It is one to the tasks of comparative systematics to determine what kinds of organisms have what kind of species structure. Various geographical and ecological factors, the genetic and developmental potentialities of a given species, and its past history determine species structure. Our knowledge of these factors is still very elementary, but in the better-known groups of animals it is possible to arrive at certain generalizations.

Geography

Different species structures prevail in different geographic regions. Most populations differ only clinally from each other in continental areas; in island areas most species consist of strong isolates. For instance, among

Table 13.1. Species structure in birds of continental and island regions (after Mayr 1942).

<table>
<tr><th rowspan="2">Structure</th><th colspan="2">Manchuria (continuous ranges)</th><th colspan="2">Solomon Islands (discontinuous ranges)</th></tr>
<tr><th>Number</th><th>Percent</th><th>Number</th><th>Percent</th></tr>
<tr><td>Widespread, uniform species</td><td>15</td><td rowspan="2">69</td><td>1</td><td rowspan="2">24</td></tr>
<tr><td>Minor geographic variation</td><td>59</td><td>11</td></tr>
<tr><td>Species with isolates nearing species level</td><td>1</td><td rowspan="2">3</td><td>17</td><td rowspan="2">52</td></tr>
<tr><td>Groups of semispecies or allopatric species</td><td>2</td><td>9</td></tr>
<tr><td>Species with ordinary subspecies</td><td></td><td>28</td><td></td><td>24</td></tr>
</table>

the passerine birds of continental Manchuria, 69 percent of the species show clinal variation and only 3 percent are strong isolates, while in the Solomon Islands only 24 percent show clinal variation and 52 percent are composed of strong isolates (Table 13.1). The superspecies of lizards *Lacerta muralis* in the western Mediterranean consists of three semispecies: *L. bocagei* (Iberian Peninsula), *L. pityuensis* (Pityusas Islands), and *L. lilfordi* (Balearic Islands). The continental semispecies *bocagei* has only three slight subspecies in the immense area of the mainland of Spain; there are 37 subspecies of *pityuensis* and 13 subspecies of *lilfordi* on the respective islands, in spite of the fact that these lizards are absent on the main islands of the Balearic Islands (Majorca and Minorca), apparently having been exterminated by the lizard snake *Macroprotodon*. Yet every island rock nearby has its own race except for a few very small bare rocks (Eisentraut 1949). The same difference between a "continental" and an "insular" pattern of variation is found wherever natural (geographical-ecological) barriers break up the continuity of populations.

Differences between Central and Peripheral Populations

Naturalists have long been aware of differences between central and peripheral populations of a species. Discussions of this subject have, however, been almost invariably confused by a failure to distinguish between various superficially similar but unrelated phenomena. Matthew (1915), for instance, discussed at length the persistence of primitive genera, families, and orders at certain peripheral, isolated localities such as New Zealand, Tasmania, Madagascar, and Ceylon. Although the groups to which these taxa belong have become extinct elsewhere, Matthew broadened his observation to the generalization: "At any one time the most advanced stages should be nearest the center of dispersal, the most conservative stages farthest from it." However, the zoogeographic phenomenon of the survival of primitive higher taxa has nothing to do with infraspecific geographic variation. Indeed, one can make a generalization concerning infraspecific variation that is precisely the opposite of Matthew's: the "original" phenotype of a species is usually found in the main body or central part of a species range, while the peripheral populations, particularly the peripherally isolated populations, may deviate secondarily in various ways.

A second source of difficulty involves likewise a confusion of different levels of taxonomy, namely, confusion of genes within a population and strains within a species. Vavilov's (1926, 1951) "centers of diversification" of cultivated plants are not areas in which populations show a maximum genetic variation; rather, they are geographic areas in which the greatest number of distinct cultivated strains are found (or originated). That certain areas (Transcaucasia and northeastern Iran, among others) are such outstanding reservoirs of cultivated varieties can in part be explained by the length of time during which the species had been cultivated in these areas, and in part by the abundant opportunities for isolation in the agricultural oases of these mountainous or semiarid regions. The causal factor is, thus, the same isolation that is responsible for the evolution of the rich indigenous fauna of the Hawaiian Islands. Vavilov himself was fully aware of this. He emphasized that "extremely interesting" deviates from the average type of the species are found "on the periphery of the areas occupied by a given plant and in places of natural isolation, such as islands and isolated mountain regions" (Vavilov 1951:47), and *not* at the center of the species range.

A third area of confusion surrounds the term "variation." When an author says a species is more variable either in the central part of the range or along the periphery, he should specify whether he means the species as a whole or a given local population. A species as a whole may be more variable peripherally than centrally, because it has formed many divergent isolates along its periphery, even though every local population in each isolate has far less genetic variability than any local population in the center of the species range (see below). Unless otherwise indicated, I always use the terms "high variability" and "low variability" with reference to a single local population.

A fourth difficulty consists of confusing the "movement" of genes in a species with that of characters. As far as genes are concerned, it is evident that: all populations of a species actively exchange genes with each other, directly or indirectly, unless such exchange is prevented by dispersal barriers; furthermore, as a result of population surplus the more successful populations will exert a greater "gene pressure" (and will consequently manifest greater population mobility) than the less successful populations; the populations nearer the center of the species range usually live under more optimal ecological conditions and are therefore more successful; finally, the success of such gene flow depends on the ability of the "alien" genes to compete with the "local" genes. The more deviant the environmental conditions are, as is the case in most peripheral areas, the less likely it is that the alien genes will survive for any length of time. There is, however, no evidence that "characters," as such, move from one part of the species range to another.

A species may evolve a specially adapted population in any ecologically "marginal" area, whether this is in the center of the species range or at its periphery. The polytopic origin of similar populations along the periphery of a species range can be interpreted as an independent response to equivalent selection pressures. Voous (1955) shows that when species of birds from humid Venezuela colonize arid islands in the West Indies (Fig. 13.4), they evolve independently similar phenotypes (paleness and so forth). It would be a great mistake to conclude from the convergent similarity of these populations that they are necessarily derived from each other.

Characteristics of Central and Peripheral Populations

The populations near the center of the species range are usually completely contiguous; they also show a relatively high population density (per unit area) and greater individual variation than is the average for populations of the species. Peripheral populations tend to have opposite values for each of the three characteristics (frequent isolation, low population density, low individual variation). This broad generalization, long accepted by naturalists, is based on general observations but lacks, so far, detailed

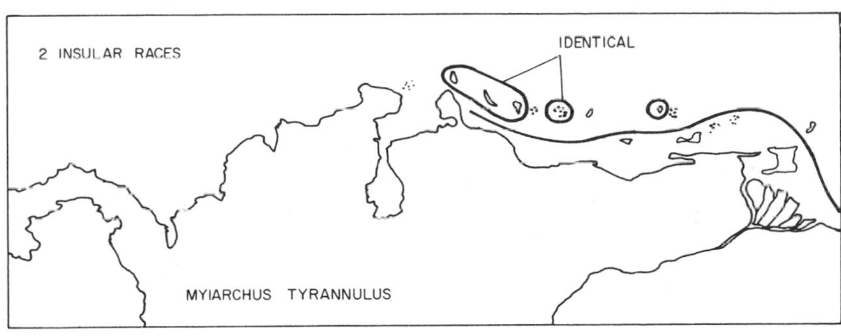

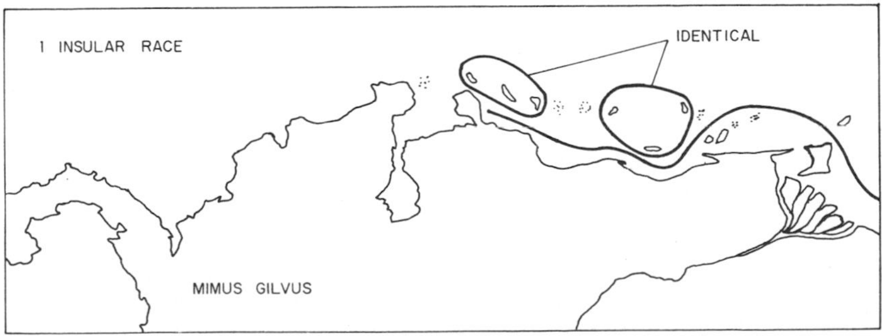

Fig. 13.4. Indistinguishable subspecies may evolve on different islands off the coast of Venezuela, even though independently derived from the mainland subspecies. Top, the tyrant flycatcher *Myiarchus tyrannulus;* bottom, the mockingbird *Mimus gilvus.* (From Voous 1955.)

quantitative support. It is substantiated by the variation observed in polymorph species. A study of such species reveals almost invariably that the degree of polymorphism decreases toward the border of the species and that many of the peripheral populations are monomorphic. In most mimetic butterflies there is a decrease in the number of mimetic forms per population toward the periphery of the species range. All peripheral populations of the highly polymorphic moth *Zygaena ephialtes* are monomorphic.

The best available evidence, however, comes from an analysis of chromosomal polymorphism in *Drosophila*. The widespread and very common tropical American species *D. willistoni* has over 40 different gene arrangements in its variable three pairs of chromosomes. Most inversions occur throughout the range of the species but are absent in a few peripheral populations. The inversional polymorphism is highest in several areas in Brazil, but drops off toward the south (Argentina, Chile), east (easternmost Brazil), and north (northern Central America, Florida, West Indies). At the most isolated point of the range (St. Kitts Island, West Indies) only two inversions are found. Analysis of *D. robusta* reveals a similar pattern (Fig. 13.5). Structural homozygosity has now been found at peripheral iocalities, for instance, southern Florida, in several species of *Drosophila* that elsewhere are polymorphic (Carson 1965).

The reason for the reduction of structural heterozygosity in the peripheral populations of species of *Drosophila* is not entirely clear. It may be favored by selection for two quite different reasons. One is that the gene arrangements may have an ecotypic function (Mayr 1945). This suggestion has been elaborated and broadly documented by Dobzhansky (1951) and his collaborators. There is a great ecological difference between center and periphery of the species range. A species usually finds itself at its ecological optimum near the center of its range. Here, the physical environment is so favorable that the species can, so to speak, make ecological expeiments and occupy various subniches that would be unsuitable under the more adverse conditions at the periphery of the range. A great diversity of gene arrangements is thereby favored, according to Ludwig's theorem. In peripheral or otherwise ecologically marginal areas the ecological leeway of the species is drastically reduced and only a single ecological variant may be able to survive. This hypothesis is supported by the observation that the greatest number of gene arrangements are, on the whole, found in the ecologically most versatile species. The number of arrangements is reduced in *D. willistoni* in areas with many competing species or with adverse conditions (da Cunha and Dobzhansky 1954).

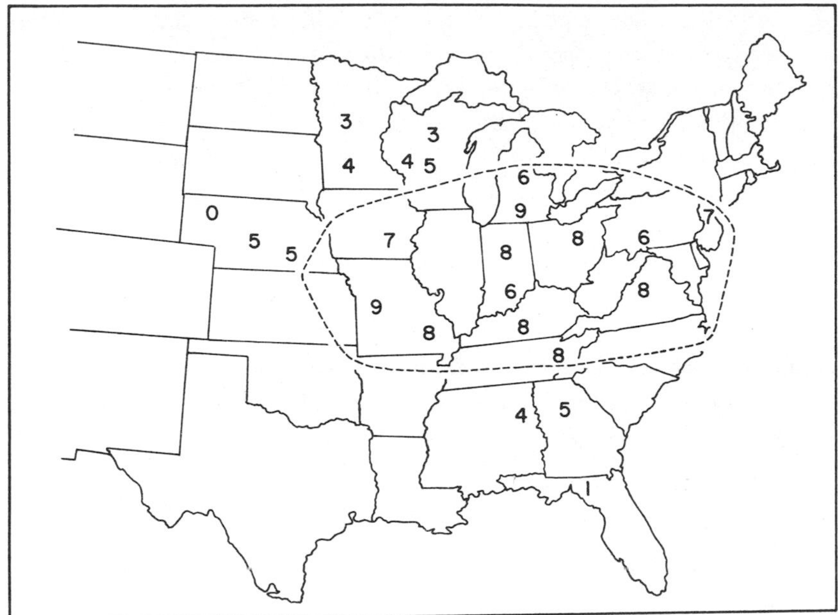

Fig. 13.5. Number of gene arrangements (in addition to standard) in 24 populations of *Drosophila robusta*. The central populations have 6–9 arrangements, the peripheral populations 0–5. (After Carson 1958b.)

A second possible advantage of the extensive chromosomal homozygosity on the periphery of the species range is that it increases the opportunity for recombination (Carson 1959). In the central populations many genes are locked up in nonrecombining inversions. This has two advantages: it prevents the breakup of these coadapted portions of the genome through recombination and it facilitates the building up of heterotic chromosomal polymorphisms. It may be advantageous for marginal populations that live under not only more severe but also more fluctuating conditions to have a greater length of chromosome available for free recombination. This method of increasing genetic variability may compensate for the reduction in gene flow in these peripheral populations.

The major reason for chromosomal uniformity in peripheral areas is presumably that only a limited number of genotypes are able to cope with the one-sided conditions near the species border. This conclusion is supported by the observation (Chapter 11) that genic polymorphism likewise is greatest in central populations and is often replaced by monomorphism in peripheral populations. It is probable that a monomorphism of peripheral populations is a consequence of the genetic cohesion of species. These marginal populations share the homeostatic system, the epigenotype, of the species as a whole. They are under the severe handicap of having to remain coadapted with the gene pool of the species as a whole while adapting to local conditions. The basic gene complex of the species (with all the species-specific canalizations and feedbacks) functions optimally in the area for which it had evolved by selection, usually somewhere near the center. Here it is in balance with the environment and here it can afford much superimposed genetic variation and experimentation in niche invasion. Toward the periphery this basic genotype of the species is less and less appropriate and the leeway of genetic variation that it permits is increasingly narrowed until much uniformity is reached. These peripheral populations face the problems described in the discussion of "species border" (Chapter 17).

The genetic differences between central and peripheral populations can be described as follows. The total amount of gene flow is reduced in peripheral populations and near the periphery gene flow becomes increasingly one-way outward. Many of the peripheral populations, particularly the more isolated ones, are established by a single fertilized female or a small group of founders that carry only a fraction of the total genetic variability of the species. Contiguous central populations, on the other hand, are in the midst of a stream of multidirectional gene flow and harbor at all times a large store of freshly added immigrant genes. Environmental conditions are marginal near the species border, selection is severe, and only a limited number of genotypes is able to survive these drastic conditions. Reduction of gene flow and increased selection pressure combined deplete the genetic variability of the peripheral populations. Lowered variability permits, if it does not favor, a shift into different ecological niches; for the selection pressure at the periphery is not only more severe, but also different. Central populations, being in the area ecologically most favorable for the species, tend to build up large populations whose size is mainly controlled by density-dependent factors. Genes adapting for such density-dependent factors accumulate in such populations. Low-density populations from near the tolerance limits of the species are selected mainly for adaptation to density-independent factors.

Comparative Systematics and Species Structure

There are many ways in which species may differ from each other in their population structure: they may be phenotypically uniform (monotypic) or they may show geographic variation; the species population may be more or less continuous or it may be fragmented into isolates; there may or may not be a central-peripheral differentiation, to mention only a few of the points made in the preceding sections. Comparison of the patterns of population structure found in different species of mammals, birds, insects, snails, and other organisms is one of the tasks of comparative systematics. This field of research is still very new, and much more needs to be known about the geographic variation in the various groups of animals before such investigations can be placed on a quantitative basis and used for broad biological generalizations.

As a first approach to a study of intraspecific variability one may analyze the presence and frequency of subspecies in various groups of animals. The number of subspecies correlates, by definition, with the degree of geographic variability and depends on a number of previously discussed factors (Chapter 11). Degree of variability may differ quite strongly in families belonging to the same order. For instance, among the North American wood warblers (Parulidae) only 20 (40.8 percent) of the 49 species are polytypic, while among the buntings (Emberizidae) 31 (72.1 percent) of the 43 North American species are polytypic. The difference is real and not an artifact of different taxonomic standards. Of the species of passerine birds in the New Guinea area 79.6 percent are polytypic, while only 67.8 percent of the North American passerines are polytypic. Among the 25 species of *Carabus* beetles from central Europe, 80 percent are polytypic, while in certain well-known genera of buprestid beetles not a single species is considered polytypic.

There are still large groups of animals in which all species are listed under binomials. It would be interesting to know to what extent such listing is due to a real lack of geographic variation of the phenotype and to what extent to insufficient taxonomic analysis. Sibling species, of course, are nearly always monotypic.

Classifying species as monotypic or polytypic is a first step in a quantitative analysis of phenotypic variation. Another way is to analyze the subdivisions of polytypic species: What is the average number of subspecies per species in various groups of animals and what is their average geographic range? There are believed to be about 28,500 subspecies of birds in a total of 8,600 species, an average of 3.3 subspecies per species. It is unlikely that this average will be raised materially (let us say above 3.7) even after further splitting. The average differs from family to family: 79 species of swallows (Hirundinidae) have an average of 2.6 subspecies, while 70 species of cuckoo shrikes (Campephagidae) average 4.6 subspecies and

75 species of larks (Alaudidae) 5.1 subspecies. The total number of subspecies is, however, much higher in a few species. The North American Song Sparrow (*Passerella melodia*), for instance, has some 30 subspecies. Species of birds with 20 or 30 well-defined geographic races are not rare in highly insular areas (such as the Indo-Australian region), the extreme apparently being the Golden Whistler (*Pachycephala pectoralis*), with over 70 races. The average range of subspecies is very much smaller in many groups of mammals. Sixty-one species of rodents have an average of 2.75 subspecies in the single state of Utah. The pocket gophers (*Thomomys*) are noted for the high number of subspecies distinguished by taxonomists. Of the two species in Utah, one has 11 subspecies, the other 24.

Even smaller appear to be the ranges of recognizably distinct populations in many groups of invertebrates. Nearly every isolated oak grove in Mexico seems to have an endemic population of gall wasps (*Cynips*). Nearly every stream along the Gulf Coast of Florida has its endemic crayfish (*Cambarus*). Flightless carabid and tenebrionid beetles tend to form very local subspecies. Even more localized are populations in many genera of tropical land snails: the Hawaiian species *Achatinella mustelina* from Oahu has 26 subspecies and 60 additional microgeographic races in an area 20 by 5 miles in size; similar distribution patterns have been found in *Partula* and *Cerion* (see Fig. 2.1).

We can conclude that virtually all species are composed of numerous local populations, each adapted to the local environment, yet sharing much of their genetic system with the other conspecific populations and retaining contact with them through gene flow. Each species has its own characteristic population structure, and it is one of the tasks of comparative systematics to determine the differences in population structure of various kinds of animals.

14 · Kinds of Species

The unity of the biological species in animals has been stressed so far in all chapters of this book. This unity is apparent even when one discusses different aspects of the species: the biological species is usually also a morphological species (except for a few sibling species); the multidimensional polytypic species is nondimensional at any given place or time. Indeed, species of animals on the whole are a singularly uniform phenomenon. Terminologies previously proposed to differentiate between different kinds of zoological species were in most cases not based on contrasting biological properties of species; they resulted instead from efforts to resolve difficulties in the application of the species concept to specimens or samples (see Chapter 2). Paleontologists might speak of paleospecies, fossil species, or chronospecies, while practicing taxonomists might distinguish morphospecies and biospecies, and students of asexual organisms recognize agamospecies. Yet, depending on the criteria applied, a sample of the same population might be called a morphospecies by one author and a biospecies or a paleospecies by another.

In spite of this essential unity of the zoological species, animal species may differ from each other in various significant characteristics. To distinguish kinds of species in a meaningful way, one must use biological attributes as classifying criteria, such as population structure, genetic system, or mode of reproduction.

Almost any property of a species might be used in an attempt to classify kinds of species. A list with no claim of completeness is given in Table 14.1. This tabulation could be expanded almost ad infinitum. Every species can be classified theoretically into one of the alternatives of each of the numbered categories. There is thus an enormous amount of overlap between the various classifying criteria. The important question is which of the criteria are most important for the biologist, and particularly the evolutionist. Very little is known so far about the amount of correlation between the different sets of criteria. Certain correlations are obvious, such as that

Table 14.1. Classifying criteria for kinds of species.

Criterion	Kind of species
1. System of reproduction	Biparental sexual reproduction Self-fertilizing hermaphroditism Parthenogenesis Reproduction by fission or vegetatively
2. Degree of intra- and interspecific fertility	Cenospecies
3. Presence or absence of hybridization	Occasional interspecific hybrids Occasional introgression With allopatric hybrid zones Sympatric hybrid swarms Amphiploidy
4. Variation in chromosome number or pattern	Variable chromosome numbers (dysploidy) Polyploidy With more or less extensive structural heterozygosity
5. Difference in origin	Gradual by geographic speciation Gradual by sympatric speciation Instantaneous by polyploidy or macromutation By fusion
6. Structure of species	Monotypic Polytypic
7. Size of populations	Constant Highly, often cyclically, fluctuating
8. Sequence of generations	Rapid Annual Slow, a single generation extending over several to many years
9. Amount of gene flow	Essentially panmictic With numerous geographical isolates Largely inbreeding
10. Pattern of distribution	Cosmopolitan Widespread Insular Relict
11. Environmental tolerance	Euryecous Stenecous
12. Rate of evolution	Slow or stagnant Rapid
13. Phenotypic plasticity	Sibling species Polymorphic species

panmictic or sibling species will usually also be monotypic. But rather little is known so far about the correlation between environmental tolerance, mating system, population structure, and rate of evolution. What ecological properties are correlated with specific genetic systems? It will not be possible to answer such questions until the systematics of animals, particularly of the lower invertebrates, has advanced a great deal further. Without the development of a field of comparative systematics there can be no firm basis for the study of comparative evolution. Until better data are available, only a tentative discussion can be attempted, to deal particularly with those sets of factors believed to affect the evolutionary potentialities of species. Among these the genetic systems, along with ecological and behavioral factors, seem most important.

Genetic Systems

The totality of genetic factors affecting the population structure of a species and its evolutionary potential is sometimes designated its *genetic system.* White (1954:366), for instance, writes:

> Under the general term *genetic system* we include the mode of reproduction of the species (bisexual, thelytokous, haplodiploid, etc.), its population dynamics (population size, sex ratio, vagility, extent of panmixia or inbreeding, etc.), its chromosome cycle (meiosis normal in both sexes or anomalous in one or both), its recombination index, presence or absence of various forms of genetic or cytological polymorphism in the natural population, and, in brief, all those characteristics which determine its hereditary behavior over periods of time sufficient for evolutionary changes to occur.

This broad definition embraces most factors listed in Table 14.1 as determining kind of species and several additional ones. Most listed mechanisms are merely alternative means to achieve the same end, namely, regulation of the balance between inbreeding and outbreeding. The factors that determine the degree of genetic difference of the zygote-forming gametes have been combined under the term *breeding system* (Darlington and Mather 1949). This controls genetic variability, population structure, and, ultimately, evolutionary change. Stebbins (1950, 1960), Grant (1963, 1964), and other authors have pointed out that the amount of outbreeding depends not only on genetic and chromosomal factors of sexuality, ploidy, and recombination index, but also on such more or less ecological factors as dispersal facility, size of population, stability of population (versus fluctuations), length of life, overlap of generations, number of offspring, differences between larval and adult ecology, and so forth.

Variation in Chromosome Number

The occurrence of several different chromosome numbers within a single species may be explained in different ways (for superb surveys of this topic see White 1954, 1957a). It is most frequently caused by the presence of supernumeraries or by Robertsonian fusions or fissions. Geographic variation in chromosome number is not infrequent in animals. In the rodent *Gerbillus pyramidum* (Wahrman and Zahavi 1955) chromosome number varies from 40 (Algeria) to 52 (coastal plain of Israel) and 66 (the Negev and adjacent parts of Egypt). The significance of this pronounced variation of closely related populations is not diminished even if one raises these isolates to the rank of full species, as one probably should (Chapter 15). Chromosome number is comparatively constant in most groups of animals but shows great variation in others, for instance in butterflies. The number of chromosomes sets a lower limit to the number of linkage groups, but the actual number is controlled by other chromosomal factors, such as chiasma localization.

Mode of Reproduction

The biological function of sex, long disputed, is to speed up the production of a vast variety of different genotypes, as first pointed out by Weismann (1902). Mutation can assume this function only when, as in microorganisms, the generations follow each other much more rapidly than the changes of the environment to which the organism must remain adapted. The essence of sexual reproduction, thus, is the combining of the genetic factors of two different parent individuals (or cells) into numerous unique zygotes. Any reproduction that does not involve genetic recombination is, biologically speaking, asexual, whether the new individual is produced vegetatively (by fission or budding) or from an unfertilized egg cell.

Modes of reproduction other than orthodox sexuality are rarer in animals than in plants. Vegetative reproduction, such as budding in certain sessile colonial marine and fresh-water organisms or simple fission in certain protozoans, turbellarians, and annelids, is usually only a temporary condition. It normally alternates with sexual stages or generations. However, in certain species no sexual stage has ever been definitely demonstrated. The two most common deviations from bisexual (= gonochoristic) reproduction among animals are hermaphroditism and parthenogenesis.

Hermaphroditism. Hermaphroditism, that type of sexuality in which a single individual produces both male and female gametes, is very widespread in the animal kingdom; it does not necessarily result in close inbreeding. In most hermaphroditic species there are numerous mechanisms that reduce or completely eliminate the chances for self-fertilization. Of these, the most important is successive hermaphroditism, a condition in which the gonads at any one time produce only male or only female gametes, one set prior to the other (termed protandry if male gametes are produced first, proterogyny if female gametes are produced first). Another such mechanism is reciprocal fertilization, best documented in the pulmonate snails. Considering how widespread hermaphroditism is among the lower animals, it is surprising how few cases are known of obligatorily self-fertilizing hermaphrodites. Self-fertilization, in most cases, seems to be subsidiary to cross (reciprocal) fertilization. Except in the few cases of obligatory self-fertilizing hermaphroditism, there is no evidence that hermaphroditism affects species structure. It would by no means lead necessarily to closer inbreeding than does the separation of sexes in different individuals. Rather, its significance seems to be that it permits an increase in general productivity. The production of eggs, on the whole, requires much greater metabolic resources than the production of spermatozoa, and this seems to be the reason that in many hermaphroditic species smaller, younger individuals produce male gametes, and larger, older individuals produce female gametes. Ghiselin (1969) has recently surveyed the occurrence and evolution of hermaphroditism.

The technical term in zoology for the separation of the sexes (in different individuals) is *gonochorism*. It has been suggested that the botanical terms dioecy and monoecy be substituted for the zoologists' hermaphroditism and gonochorism. Such transfer does not seem justified, since the equivalence is not strict. Dioecy and monoecy refer to sporophytes, while gonochorism and hermaphroditism are phenomena relating to gamete-producing individuals.

Cross-fertilizing hermaphrodites are not known to differ in species structure from gonochoristic forms. Self-fertilization not only increases the amount of inbreeding, but also permits single individuals to become the founders of new populations. It might therefore lead to a change in evolutionary potential. Unfortunately, however, good comparative taxonomic studies of self-fertilizing hermaphrodites are not available and it is not known to what extent this mode of reproduction affects species structure and speciation.

Parthenogenesis. Parthenogenesis denotes the development of offspring from egg cells not fertilized by male gametes. It occurs in two forms, of which only female diploid parthenogenesis (thelytoky) is of interest to us. The other type—the production of haploid males from unfertilized eggs (arrhenotoky)—is a form of sex determination, and its main effect on the genetic system is that it eliminates in the males in each generation all deficiencies, homozygous lethals, and other factors that are inviable in hemizygous condition. This male haploidy has arisen only about seven times in the whole history of the metazoa, five times in the insects, once in the mites, and once in the rotifers. Thelytoky, simply called parthenogenesis in the following account, has arisen repeatedly in most larger phyla of animals. Either it occurs as an optional or seasonal condition in otherwise sexually reproducing animals, or it is complete, males being entirely unknown. White (1954, 1970) gives an excellent analysis of the cytological and evolutionary aspects of parthenogenesis.

From the cytogenetic viewpoint we may distinguish two types of parthenogenesis which, with White, we may designate the *meiotic* and the *ameiotic* type. Meiosis is entirely suppressed in the ameiotic type, and, since the maturation divisions in the egg are like any mitotic division, the daughters will have a genetic constitution identical with that of the mother, except for an occasional genic or chromosomal mutation. Since there is no recombination, and only dominant mutations will be exposed to selection, there will be an accumulation of recessive mutations and of structural rearrangements, leading to ever-increasing heterozygosity. Furthermore, since the pairing of the chromosomes is eliminated together with the meiosis, there is no longer any mechanical barrier to the establishment of various chromosomal irregularities, including polyploidy. As a result polyploidy is very widespread in groups of animals with ameiotic parthenogenesis (see Chapter 15).

In the meiotic type of parthenogenesis chromosomal reduction occurs during meiosis but is compensated for by restoration of the diploid chromosome number at some subsequent stage of the life cycle. There are three or four alternate ways of achieving such restoration. Either the first meiotic division is abortive (even though preceded by pairing and crossing over) and the second is a simple mitosis, or meiosis is complete, but two complementary ones of the four pronuclei fuse (automixis). One would expect a steady loss of heterozygosity through crossing over, but this apparently does not necessarily happen. It is prevented by the fusion of the unlike pronuclei (or the abortion of the reduction division) and, more important, by continued selection in favor of heterozygotes. Among meiotic parthenogenetic groups, polyploidy has similarly arisen several times.

The phenotypic variation of completely parthenogenetic animals has been critically compared only a few times with that of bisexual animals. Since each parthenogenetic individual, and the clone to which it gives rise, remain permanently independent of all related clones, one would expect, owing to mutation, a steady genetic divergence between clones, ultimately resulting in high variability in the parthenogenetic "species." Some parthenogenetic species of British sawflies, *Mesoneura opaca* and *Eutomostethus ephippion*, are indeed highly polymorphic. Other parthenogenetic species seem to have no more variation than sexual species. The reason may be that the vast majority of mutations is recessive and not able to become homozygous in the absence of recombination. The probability of phenotypic variation is even smaller in tetraploids, where each new allele has to compete with three wild-type alleles.

Complete parthenogenesis is usually limited to an occasional species or genus in a broad taxonomic group. In the vertebrates, for instance, it has been reported in nature for several teleost fishes, for instance poeciliids and *Coregonus*, also for reptiles (*Lacerta saxicola* complex and *Cnemidophorus*). Studies of nematodes, psychid moths, lumbricids, simuliids, *Drosophila*, and other groups of animals have shown how quickly parthenogenesis can be acquired. However, the only example of a large group of metazoans known to me of which all members show complete, ameiotic parthenogenesis is the rotifer order Bdelloidea (Chapter 15). The absence of a complete continuum of clones between these taxonomic entities in this order can perhaps be explained by secondary elimination of all clones that are competitively inferior. Perhaps this is also the explanation for other cases of discontinuities between closely related parthenogenetic species, such as the white-fringed weevils or the New Zealand phasmids of the genus *Acanthoxyla*.

The evolutionary importance of parthenogenesis is that it permits instantaneous speciation (see Chapter 15). However, by abandoning genetic recombination, it generally gains only a short-term advantage, and, with the apparent exception of the bdelloids, virtually every case of parthenogenesis in the animal kingdom has all the earmarks of recency. Though superficially appearing a "more primitive" type of reproduction, parthenogenesis in recent animals is evidently in all cases derived from sexual reproduction. Its greatest advantage seems to be that it permits a single colonist to establish a new population in an area not previously occupied by the species.

Asexuality among animals. It is sometimes claimed that half or more of the kinds of animals reproduce by self-fertilization or asexually. A study of the literature does not support such claims. Cross fertilization seems to occur regularly even among the protozoans, although it usually alternates with many generations of simple fission or selfing. Mobility, which permits mates to seek each other actively, gives animals much greater efficiency in fertilization than is possible among plants. That is why gonochorism, combined with normal sexuality and protection against hybridization, is by far the most frequent breeding system in animals.

Ecology

The relation of an animal to its environment is an important factor in the determination of the amount of outbreeding and other components of the species structure. Since every species differs in its ecology from every other species, one might say that every species is a different "kind of species." However, some environmental factors affect population structure more than others and are therefore more important for the evolutionary potential of species.

The relation between an organism and the environment in which it lives seems sufficiently important to serve as the basis of a classification of kinds of species. There are perhaps five major types of ecological specialization:

(1) Specialization for a very narrow niche;

(2) Broad tolerance of every individual of the species so that it can tolerate environmental extremes within the species range;

(3) Polymorphism: the presence of several morphs in the population adapted to particular subniches (Ludwig effect);

(4) Ecotypic variation: the formation within each geographic area of numerous localized populations that specialize in ecologically different subareas, and are not obliterated by gene exchange with adjacent, differently specialized populations;

(5) Geographic polytypicism: the formation of geographic races in response to the environmental variation over the major part of the species range.

It is not certain that these five categories can be sharply delimited. Indeed, it is possible that in certain species some of these categories overlap, for instance 2 and 4, or 3 and 4, or 1 and 5. With the possible exception of polymorphism in gene arrangements, nothing at all is known about the genetic systems that permit a species to cope with its ecological needs in such a way that it will be classified in one or another of the abovementioned categories. Extreme specialization is typical for many insects and virtually unknown among the higher plants. Ecotypic variation is common among plants, particularly in specialized inbreeders, but is rare in animals. The role of physiological-ecological polymorphism, found in *Cepaea* and *Drosophila*, is still very poorly understood.

Behavior

Various components of behavior affect the amount of outbreeding and other population properties of species that are of evolutionary significance. The whole field of behavioral isolating mechanisms might be mentioned here as well as behavior-controlled food and habitat selection. Some of these points have been treated in a recent volume on behavior and evolution (Roe and Simpson 1958). Here I shall single out for discussion merely two behavioral elements, parental care and mobility.

Parental Care and Sequence of Generations

The amount of parental care exercised by a species is of great evolutionary importance. An organism that is poorly protected against the physical environment, or whose young are poorly protected, or that is faced by vast fluctuations in its potential resources will benefit by a vast production of zygotes. Much of the mortality will be nonselective, and intraspecific competition will tend to be limited to certain stages in the life cycle. Conversely, the more an organism is protected from the vicissitudes of the environment and the more independent it is of the environment, the greater becomes the role of competition among conspecific individuals. Under these circumstances it is of greater selective advantage to produce well-equipped offspring than many offspring. If there is much competition for females and a long period of parental care, there will be a high selective premium on an increased life span. That again increases intraspecific competition and tends to depress the number of offspring even further. Instead of a survival rate (to the adult stage) of less than one in a million, as is found in so many species of marine invertebrates, there may be a survival rate as high as one out of five or ten. There is a tremendous difference between nematodes and other small invertebrates, with a new generation every few days, and some large mammals and birds, in which only a single offspring is produced every second year and 5 to 10 years are required to reach maturity. Even these two extremes of genetic systems among animals are not as distinct as are among plants the long-lived trees with their immense number of seeds and the short-lived microorganisms with a new generation every 10 minutes. Species with a slow turnover of generations and a small number of offspring require special mechanisms for the maintenance of high genetic variability, such as an increase in the number of chromosomes and perhaps in the rate of mutation.

Mobility

Mobility gives animals an exceedingly flexible means of adjusting the amount of outbreeding. In contradistinction to plants, dispersal in animals is only partially passive. The average distance of dispersal during the dispersal stage in the life cycle of an individual animal is in part controlled by behavioral characteristics. Sedentary habits, territory occupancy, and philopatry (aided by homing ability) tend to reduce the amount of random dispersal. Restlessness, induced by various influences, may lead to long-distance dispersal of a smaller or larger proportion of the population. In animals mobility and dispersal are the most important devices for the prevention of close inbreeding (see Chapter 18).

Animal System and Kinds of Species

It would seem to be interesting and important to have precise quantitative data on the relative frequency of the various kinds of species in the different phyla and classes of the animal kingdom. Unfortunately such a quantitative analysis is not yet possible, since we know too little about the taxonomy of most groups of animals, particularly the lower animals, where most of the more interesting and more aberrant kinds of species occur. In some of the better-known groups, such as insects and vertebrates, different patterns of variation have been found in related genera and families. In some families more than 80 percent of the species are polytypic, while in other groups nearly every species is monotypic. The latter seems to be the case, for instance, in many Microlepidoptera and other insects that feed on a single host plant. The genera are large and ill-defined in some families (for example, weevils, solitary bees, some dipterans); other families have large numbers of clear-cut, often monotypic, genera, for instance, the longhorn beetles (Cerambycidae), or, among plants, the milkweed family (Asclepiadaceae). It would be interesting to know the reasons for such differences. Different chromosomal characteristics have been suspected as the cause, but their responsibility for such differences has never been demonstrated. Much evidence indicates that ecological factors are very important.

The Meaning of "Kinds of Species"

This survey of the reproductive, chromosomal, ecological, and behavioral characteristics by which kinds of species differ from each other confirms the opinion of many authors, including Stebbins (1950), Dobzhansky (1951), and White (1954), that they range between two extremes, inbreeding and outbreeding. The average amount of difference between the gametes that produce the zygotes of the next generation depends largely on the amount of outbreeding. Furthermore, this factor is far more easily changed than are the various chromosomal mechanisms (discussed in Chapter 9) that regulate amount of recombination within a gene pool. It appears that the range in the degree of outbreeding is far greater in plants than in animals, at least in the higher animals. Among plants both extremes, complete inbreeding (= selfing) and extreme outbreeding (= hybridization with other species), are common. In the higher animals both extremes are rare, and the normal alternative lies between moderate and increased outbreeding. To put it in another way, the size of the deme is the only variable. The larger the deme, other things being equal, the greater the probable genetic difference of fusing gametes. The factors that determine the degree of outbreeding (= the size of the deme) thus determine the average genetic difference of the gametes.

Outbreeders and inbreeders differ from each other in numerous ways. The entire breeding system of outbreeders is so organized as to accumulate and preserve genetic variation giving a maximum of ecological plasticity and evolutionary flexibility, but at a price—the production of many inferior recombinants. An outbreeder may also be so well buffered that it stagnates evolutionarily. At the other end is the extreme inbreeder which has found a lucky genotypic combination that permits it to flourish in a specialized environmental situation, but again at a price—an inability to cope with a sudden change of the environment. A species thus has the choice between optimal contemporary fitness combined with considerable evolutionary vulnerability and maximal evolutionary flexibility combined with the wasteful production of inferior genotypes. No species can combine the two advantages into a single system. Every species makes its own particular compromise between the two extremes and every species has its own set of devices for achieving this compromise. To provide for more flexibility, devices exist in many evolutionary lines that permit the increase or decrease of the degree of outbreeding according to need. Outbreeding appears to be the original condition in animals and one can assert that any more extreme form of inbreeding (including the various forms of asexuality) is a derived condition.

Outbreeding, that is, genetic flexibility, as stated by Stebbins (1950), is favored by large, structurally complex, slow-growing organisms that have low numbers of offspring and live in a generalized environment. Inbreeding, that is, genetic fixity, is favored by small, structurally simple, fast-growing organisms that have large numbers of offspring and are more or less adapted to special situations. Most animals are essentially outbreeders, most microorganisms essentially inbreeders.

The various chromosomal, reproductive, and ecological factors are not assorted randomly but are adjusted to each other to form a single breeding system. It is difficult to follow the pathway by which the various characteristics are brought together and the success of an outbreeder is achieved. It is much easier to understand the selective advantage of many of the inbreeding devices because, as in the case of parthenogenesis, they nearly always result in a speed-up of reproductive rate. *Daphnia* in a lake in spring, or an aphid when the new foliage grows in spring, or a bacterium successfully invading a new host, all must multiply as rapidly as possible to keep ahead of competitors or a deterioration of the environment. A temporary suspension of sexual reproduction in these forms has two advantages: a successful genotype can be perpetuated unchanged, without being broken up by recombination; and, more important, fecundity is doubled. Instead of half the zygotes being "wasted" on males, which are not able to reproduce by themselves, all zygotes are fertile egg-producing females. The temporary abandonment of sexuality is characteristic of organisms that invade temporarily vacant niches (lakes and vegetation in spring). The return to sexual reproduction takes place promptly when the habitat is filled and conditions begin to deteriorate.

15 · Multiplication of Species

One of the most spectacular aspects of nature is its diversity. This diversity has the very special property that it is not continuous, but consists of discrete units, species. To explain the origin of these species is one of the great problems in the field of evolution. The local naturalist knew no answer to it: in fact, the mystery seemed to deepen the more he studied the relation of species to each other. At a given locality each species is separated by a bridgeless gap from every other species, as was shown in Chapters 2–5. Not only the naturalists but also the early geneticists failed to solve this problem, the geneticists because their thinking was dominated by typological concepts and they therefore attempted to solve the puzzle with the help of hypotheses of instantaneous speciation involving a single individual. The eventual solution came from a very different direction, namely, from the study of the population structure of species, as shown in Chapters 7–13. A more precise formulation of the problem of speciation was what eventually permitted its solution.

Speciation means the formation of species ("specification," as Darwin called it). In retrospect it is evident that much of the past argument on modes of speciation was due to differences in species concepts (Chapter 2). To be able to discuss the many theories of speciation that have been proposed in the past, it will be useful to tabulate the potential modes of speciation (Table 15.1). This table is based on three major criteria and their alternatives: (1) addition of new, reproductively isolated populations (*III*), or not (*I, II*); (2) instantaneous origin through individuals (*IIIA*) or gradual origin through populations (*IIIB*); and (3) geographic isolation of the speciating populations, or not (*IIIB1* vs. *IIIB2,3*). Combining these alternatives in various ways results in twelve potential modes of speciation.

The true meaning of the term "origin of species" was understood only rather recently. The pre-Darwinian evolutionists were quite vague on this issue, and even Darwin himself seems to have considered "origin of species" the same as "evolution" (Mayr 1959a). He confused two essentially different problems, subsuming them under the single heading "origin of species."

Table 15.1. Potential modes of origin of species.

I. Transformation of species (phyletic speciation)
   1. Autogenous transformation (owing to mutation, selection, etc.)
   2. Allogenous transformation (owing to introgression from other species)
II. Reduction in number of species (fusion of two species)
III. Multiplication of species (true speciation)
   A. Instantaneous speciation (through individuals)
   1. Genetically
      (a) Single mutation in asexual "species"
      (b) Macrogenesis
   2. Cytologically, in partially or wholly sexual species
      (a) Chromosomal mutation (translocation, etc.)
      (b) Autopolyploidy
      (c) Amphiploidy
   B. Gradual speciation (through populations)
   1. Sympatric speciation
   2. Semigeographic speciation (see Chapter 17)
   3. Geographic speciation
      (a) Isolation of a colony, followed by acquisition of isolating mechanisms
      (b) Extinction of the intermediate links in a chain of populations of which the terminal ones had already acquired reproductive isolation

I2, II, and IIIA2(c) may lead to reticulate evolution.

One of these, the one Darwin was primarily interested in, is evolutionary change as such, also referred to under the terms "phyletic evolution," "modification in time," or "descent with modification" (Darwin). Evolutionary transformation of a species does not necessarily lead to multiplication of species. An isolated population on an island, for instance, might change in the course of time from species *a* through *b* and *c* into species *d* without ever splitting into several species (Fig. 15.1). In the end there will be only one species on the island, just as at the beginning. The essential aspect of this type of evolution is the continuous genetic and evolutionary change within the populations composing the species, without the development of reproductive isolation between populations of the species, and consequently without its breaking up into several species. When a paleontologist speaks of speciation, he usually has this type of phyletic evolution in mind; and Darwin, under the somewhat misleading heading *On the Origin of Species,* primarily discussed descent with modification. More recently it has become increasingly customary to restrict the term "speciation" to the process that leads to a multiplication of species through branching of phyletic lines.

The opposite of the splitting of a single species into several daughter species is the complete fusion of two species. Since the product of such a process would be an entirely new unit, truly a new biological species, it is legitimate to consider fusion a form of speciation, the reverse of a multiplication of species. Since species are defined as reproductively isolated populations, the fusion of two species is, on the whole, a logical contradiction. Yet occasionally a previously existing reproductive isolation breaks down and two previously distinct sympatric species merge. This happens most easily in species in which the isolating mechanisms are primarily ecological. How frequently two species fuse is very much in question. Most cases cited in the literature allow different interpretations. Usually they are actually cases of an interbreeding of subspecies, typologically defined as species. Perhaps the nearest approach to a real fusion of two species of birds is the case of the towhees *Pipilo erythrophthalmus* and

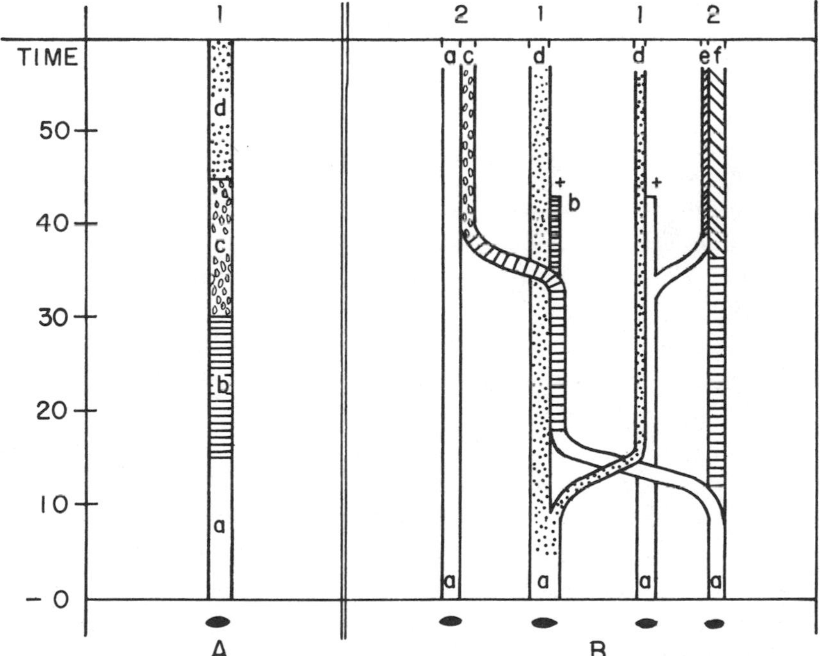

Fig. 15.1. *A* (on the left) designates a strongly isolated island on which a species *a* changes in the course of geological time through *b* and *c* to *d*. *B* (on the right) indicates an archipelago with four islands on which an originally monotypic species *a* breaks up into five currently living species (*a*, *c*, *d*, *e*, *f*) through geographic speciation and cross colonization. (From Mayr 1949.)

P. ocai described in Chapter 6. This case and several others discussed in that chapter (p. 74) prove that breakdown of isolation and subsequent fusion is a distinct possibility.

Speciation in its restricted modern sense, however, means the splitting of a single species into several, that is, the multiplication of species. The problem of the origin of discontinuities between species is made more precise by recalling the meaning of the word "species." A species is a reproductively isolated population. The problem of the multiplication of species, then, is to explain how a natural population is divided into several that are reproductively isolated, or, more generally, how to explain the origin of a natural population that is reproductively isolated from preexisting species. It appears at present that there are nine potential modes of multiplication of species (Table 15.1, *III*). The normal mode by which such multiplication takes place is the process called geographic speciation, discussed in Chapter 16. Its genetic aspects will be discussed in Chapter 17 and its ecological aspects in Chapter 18. All other potential modes are discussed in the present chapter.

Instantaneous Speciation

Instantaneous speciation may be defined as *the production of a single individual (or the offspring of a single mating) that is reproductively isolated from the species to which the parental stock belongs and that is reproductively and ecologically capable of establishing a new species population.* This is what one might call the naive concept of speciation. It was the prevailing concept among the early workers in the field. If one were to ask a lay person how a new species originates, he might say that a species suddenly throws off one individual or several individuals which then form the ancestral stock of a new species. The early Mendelians maintained similar views (Chapter 17). Even at present there are still some adherents of instantaneous speciation. On the whole, however, the reasoning has prevailed that species are populations and that new species populations are not normally founded by single individuals. To what extent founding of species by an individual might, nevertheless, be possible is discussed in the succeeding sections.

Instantaneous Speciation through Ordinary Mutation

Mutation is a phenomenon of such relatively high frequency that in a higher animal with more than 10,000 gene loci almost every individual will be the carrier of a new mutation. Such mutations merely increase the heterozygosity of a population; they do not lead to the production of new species. Any mutation drastically affecting reproductive behavior or ecology will be selected against if it lowers viability, or will displace the original allele if it is of higher viability. In neither case will there be any origin of discontinuities. It is evident that ordinary mutations cannot produce new species in sexually reproducing species.

Certain phenomena among rotifers, cladocerans, and nematodes suggest the occasional occurrence of asexual speciation. Among the most challenging problems is that of speciation in the rotifer order Bdelloidea. In this entire large order not a single male has ever been found, despite much searching and the analysis of large samples. A secondary extinction of all sexual species appears improbable. It is therefore reasonable to assume that the ancestral species itself was parthenogenetic. This ancestral species evolved into over 200 species, about 20 genera, and 4 families. If my assumption is valid, all this took place without the benefit of increased genetic variation generated by sexual reproduction. The discontinuities within this order are presumably the result of natural selection. Among the many clones produced by mutation some have gene combinations that are favored by selection over others. The extermination of inferior genotypes then results in the origin of discontinuities.

Vegetative reproduction (by growth and splitting) might offer a favorable condition for instantaneous speciation. This form of reproduction, however, occurs only in some of the lowest groups of animals, such as sponges, coelenterates (hydroids), turbellarians, and bryozoans, groups in which the taxonomy is as yet too uncertain for a study of speciation. When the form of branching is an important taxonomic character, for instance in hydroids and graptolites, a single somatic mutation may in one step produce a considerable alteration of the phenotype. Such a mutation, of course, does not produce a new species, because the gametes carrying the new mutation are not reproductively isolated from those carrying the "parental" growth gene. Changes of the phenotype are often saltational, as sometimes in polymorphism and as in the case of chromosomal inversions and translocations. The reorganization of the gene pool, required for successful speciation, is (except in the case of polyploidy) never saltational.

Speciation by Macrogenesis

The sudden origin of new species, new higher taxa, or quite generally of new types by some sort of saltation has been termed *macrogenesis*. Before the nature and structure of species was fully known and before the genetics of speciation was as well understood as it is today, the hypothesis of macrogenesis was one of the most widely accepted theories of speciation and of the origin of higher categories. It is the logical consequence of a typological interpretation of taxa. The proponents of macrogenesis considered the individual the real unit of evolution and believed that an individual could undergo a viable major genetic reconstruction. Among modern biologists who supported theories of macrogenesis to a greater or lesser extent are Goldschmidt (1940) and Schindewolf (1936, 1950). The theory of macrogenesis can never be proved, since it is obviously impossible to witness the occurrence of a major jump, particularly one that achieves simultaneously reproductive isolation and ecological compatibility. To support their theory, the adherents of macrogenesis rely primarily on the claim that gradual speciation and the gradual origin of higher taxa are impossible. As far as speciation is concerned, this claim will be refuted in Chapter 16, and as far as the origin of higher taxa is concerned, in Chapter 19.

The origin of new types. The believers in macrogenesis consider the chief support of their thesis to be the "fact" that all new types appear in the fossil record suddenly and abruptly. These types are not connected with the ancestral types by intermediates, they claim, and cannot be derived from them by gradual evolution. Rensch (1960) and Simpson (1953) have shown how erroneous this claim is. As our knowledge of the fossil forms has improved, it has been possible to demonstrate in one case after another that one "type" can be derived from a previously existing one. The fossil record is admittedly very incomplete, particularly among the soft-bodied early invertebrates. However, among organisms with hard parts that date from the geological periods in which the fossil record is more complete, so-called missing links have been found time after time. For instance, intermediates between most of the major classes of vertebrates have been found in the 100 years since Darwin. This is quite remarkable in view of the fact that an estimation of the total number of species that must have existed at one time and of the fossil species already described by paleontologists reveals that only about one out of 5000 species has so far been discovered.

Some saltational postulates are based on the assumption of essentially invariant evolutionary rates. Since birds, for example, do not show any significant evolutionary change from the Cretaceous to the present, the postulate of a uniform rate of evolution leads by backward extrapolation to an absurdly early date for the origin of birds. Taking the case of bats, the first known bat from the Eocene being hardly different from modern bats, Simpson points out that by this system of extrapolation the origin of the mammals would antedate the origin of the world. All we know about evolutionary rates shows clearly how unequal they are and, more specifically, with what rapidity the valley between one adaptive peak and another one is crossed. He who postulates essentially steady evolutionary rates cannot help but arrive at absurd conclusions.

The mechanism of macrogenesis. Goldschmidt is the only recent adherent of macrogenesis who specifies how he envisions speciation and the origin of higher taxa through macrogenesis. It is evident, however, from the writings of other representatives of this school that their thinking is similar. Essentially they all agree that the production of a new type by a complete genetic reconstruction or by a major "systemic mutation" is the crucial event. Such an event will produce a "hopeful monster," as Goldschmidt called it, which will become the ancestor of a new evolutionary lineage. The occurrence of genetic monstrosities by mutation, for instance the homeotic mutants in *Drosophila*, is well substantiated, but they are such evident freaks that these monsters can be designated only as "hopeless." They are so utterly unbalanced that they would not have the slightest chance of escaping elimination through stabilizing selection. Giving a thrush the wings of a falcon does not make it a better flier. Indeed, having all the other equipment of a thrush, it would probably hardly be able to fly at all. It is a general rule, of which every geneticist and breeder can give numerous examples, that the more drastically a mutation affects the phenotype, the more likely it is to reduce fitness. To believe that such a drastic mutation would produce a viable new type, capable of occupying a new adaptive zone, is equivalent to believing in miracles.

Individuals are members of populations. Goldschmidt completely ignores the relation between the freak individual and his parental population. The finding of a suitable mate for the "hopeless monster" and the establishment of reproductive isolation from the normal members of the parental population seem to me insurmountable difficulties.

The objections to the theory of macrogenesis are so numerous (Mayr 1963:435–439), and evidence to support this theory so singularly lacking, that it would be contrary to the scientific principle of parsimony (Occam's razor) to entertain any longer any theory of evolution by saltation except those that are well substantiated by the evidence, such as polyploidy, hybridization, shift in sexuality, and chromosomal rearrangements, discussed below and in Chapter 17.

Speciation through Major Chromosomal Changes

There was a widespread belief among early cytogeneticists that chromosomal rearrangement was the essential step in speciation. Proposed as an alternative to geographic speciation, the chromosomal speciation hypothesis is not valid. However, there is good evidence that the building up of isolating mechanisms in geographical isolates may be facilitated by chromosomal reorganization. The problem will be discussed in Chapter 17.

Polyploidy

The only unequivocally established mode of instantaneous speciation is polyploidy. *Polyploidy is a multiplication of the normal chromosome number.* If, for instance, the normal diploid ($= 2n$) chromosome number of a species is 14, all multiples of 7 higher than 14 are polyploids. Individuals with 3 chromosome sets ($3n$) are called triploids; with $4n$, tetraploids; with $6n$, hexaploids; with $8n$, octoploids; and so forth.

Two types of polyploids can be distinguished, each having different significance in evolution: *autopolyploids* and *allopolyploids* (or amphiploids). An autopolyploid arises when more than two haploid chromosome sets of a single species participate in the formation of the zygote; an allopolyploid when the chromosome sets of two species are combined. Normally an allopolyploid arises through the doubling of the chromosomes in a hybrid. Too little is yet known about animal polyploids to determine which of them are autopolyploids and which allopolyploids.

Polyploidy is very widespread among plants and is an important mechanism of speciation in the plant kingdom (Stebbins 1950, Grant 1963). Probably more than one third of all species of plants have arisen by polyploidy, although the phenomenon is rare in some groups of plants, for instance the conifers.

Polyploidy is much rarer in animals than in plants. Why this is so is still a somewhat controversial question (see White 1954 for an exhaustive treatment; see also Mayr 1963:440–449). An imbalance in sex determination is unquestionably the most important factor. Sex is determined in most animals by a balance of sex factors on the sex chromosomes and autosomes. Polyploid parents with the sex chromosomes *XXXX* in the females and *XXYY* in the males would produce enough *XXXY* individuals in their offspring to be severely discriminated against by selection. The frequency with which polyploidy has arisen in many groups of unrelated parthenogenetic species of animals (and its rarity in all other groups of animals) supports the sex-determination imbalance theory.

Polyploidy seems to occur in all animal groups with permanent parthenogenesis. In a few specialized groups it has become the principal method of speciation. In the lumbricid earthworms, for instance, up to 70 percent of the species found in a given area may have arisen by polyploidy. In the turbellarians and in certain groups of weevils, the percentage of polyploids is likewise appreciable. There seems little doubt (Suomalainen 1950) that parthenogenesis preceded the origin of polyploidy in all these cases. See White 1954 and Basrur and Rothfels 1959 for further detail.

Polyploidy poses considerable difficulties for the taxonomist. An autopolyploid may be virtually indistinguishable from the parental diploid in every character except its chromosome number. Such a form is often referred to as a "polyploid race." For instance, in the psychid moth *Solenobia* (Seiler 1961) three "races" occur, a sexual diploid, a parthenogenetic diploid, and a parthenogenetic tetraploid. The sexual form is found mainly near the edge of the last glaciation and near the nunataks along the northern slope of the Alps. The tetraploids are found in the upper valleys of the northern Alps and along the entire slope of the southern Alps. The Swiss Jura is inhabited almost exclusively by the parthenogenetic diploids. When the polyploid is reproductively isolated from its diploid parents and differs from them in ecology and distribution, it is usually best to consider it a separate species. Such a case is a typical illustration of the frequent conflict between a morphological and a biological species concept.

Variation of chromosome number in related species. Numerous cases have been reported in the literature in which related species have chromosome numbers that are multiples or near multiples of each other or of the haploid number, for instance, 20, 30, 40, 51, or 22 and 44, or 24, 45, 90, 135, 190. It is the existence of such series that has fostered a strong belief in the frequency of polyploidy in animals. However, further analysis has almost invariably shown either that there are all sorts of intermediate chromosome numbers or that the total amount of DNA is the same despite the highly unequal chromosome numbers. It is obvious in most of these cases that processes other than polyploidy are responsible for the changes in chromosome number. Most important among these are chromosomal fusions and fissions.

It is evident that changes in chromosome number may play a role, and in some cases a decisive role, in speciation, since closely related species often differ considerably in chromosome number. What this role might be is treated in Chapter 17 (pp. 310–319).

Summary. On the whole, speciation through polyploidy is very rare among animals. Only in parthenogenetic groups is its occurrence at all frequent. A great deal of cytological work remains to be done before it can be stated whether and to what extent polyploidy occurs in sexually reproducing groups of animals such as crustaceans, phasmids, mantids, Lepidoptera, and fishes. Even if polyploidy should be confirmed in some of these groups, one would still be justified in saying that in animals this mechanism is exceptional, in contrast to the situation in plants. There is no evidence that whole species groups or genera owe their origin to polyploidy, as is frequently the case with plants.

Gradual Speciation: Sympatric Speciation

Gradual speciation is the gradual divergence of populations until they have reached the level of specific distinctness. Two modes of gradual speciation have been postulated, one involving geographical separation of the diverging populations (geographic speciation, Chapter 16), the other involving divergence without geographic separation (sympatric speciation).

The Concept of Sympatric Speciation

The majority of authors until fairly recently considered *sympatric speciation,* that is, *speciation without geographic isolation,* to be the prevailing mode of speciation. Such speciation is based on two postulates: (a) the establishment of new populations of a species in different ecological niches within the normal cruising range of the individuals of the parental population; (b) the reproductive isolation of the founders of the new population from individuals of the parental population. Gene flow between daughter and parental population is postulated to be inhibited by intrinsic rather than extrinsic factors. A rapid process of species formation is implied in most schemes of sympatric speciation.

The concept of sympatric speciation is far older than that of geographic speciation and goes back to pre-Darwinian days. Darwin was rather vague on the subject and made no clear distinction between speciation through individuals and speciation through populations. In some of his statements he seems to give due recognition to the need for geographic isolation while in others he seems to ignore the geographical element altogether. Since

Darwin's time, many authors have presented detailed theories of sympatric speciation, none of which explain the origin of the decisive difference between the incipient species. (For literature on the history of the many attempts to prove sympatric speciation see Mayr 1963:450.) What is rather disappointing in this perennial controversy is that the same old arguments are cited again and again in favor of sympatric speciation, no matter how decisively they seem to have been disproved previously. Few of the authors proposing these arguments are aware of the extensive prior literature in the field. In the last analysis, most of the schemes make arbitrary postulates that at once endow the speciating individuals with all the attributes of a full species. They attempt thus to bypass the real problem of speciation. To judge from past experience, the problem of sympatric speciation will continue to be a controversial issue. It is important, therefore, to analyze the various premises and postulates of the theory of sympatric speciation in considerable detail and to determine whether or not there is evidence suggesting that populations could acquire reproductive isolation without geographic separation.

Two basic points must be clearly stated at the outset. The first is that the theories of sympatric and geographic speciation agree in their emphasis on the importance of ecological factors in speciation. They differ in the sequence in which the steps of the speciation process follow each other. The theory of geographic speciation lets an extrinsic event separate the single gene pool into several gene pools, with the ecological factors playing their major role after the populations have become geographically separated. According to the theory of sympatric speciation, the splitting of the gene pool is itself caused by ecological factors, and any spatial isolation of the populations formed thereby is a secondary, later phenomenon.

The second point concerns the definition of *sympatric speciation*. In order to avoid circular reasoning it must be defined as *the origin of isolating mechanisms within the dispersal area of the offspring of a single deme*. The size of this area is determined, for instance, in marine organisms by the dispersal of the larval stages. In most insects it is determined in the adult state by the more mobile sex. Since there are normally a great many ecological niches within the dispersal area of a deme, niche specialization is impossible without continued new pollution by immigrants in every generation.

Reasons for postulating sympatric speciation. Numerous biological phenomena suggest, on first sight, the occurrence of sympatric speciation. One is the abundance of finely adapted sympatric species in local faunas. For instance, there are several hundred (perhaps 500) species of bees of the genus *Perdita* in North America. All are oligolectic, confining their visits to the flowers of a single species or of a group of closely allied species. Many species of *Perdita* may be found at a single locality but never together because they occur on different plants or in some cases at different times of the year. How could all these species have arisen by geographic speciation? This is a question that at first seems difficult to answer.

A similar ecological specificity characterizes the species of many other genera of animals. All such species are perfectly adapted to the particular environment in which they occur. Could such a perfect fitting of species into their ecological niches have evolved as a purely accidental by-product of genetic changes accumulated during geographic isolation?

It required a full understanding of the genetics of geographic speciation, only recently acquired, before it could be demonstrated that these phenomena are entirely consistent with the theory of geographic speciation. Ecological difference can be acquired in geographic isolation and can be increased (by "character displacement") after the secondary establishment of sympatry (see Chapter 17).

Biological races. A necessary corollary of any theory of gradual speciation is that there should exist in nature some "forms" or "varieties" or "populations" that are *incipient species.* The supporters of sympatric speciation have always cited the occurrence of *biological races* as examples of such incipient species. Most of the earlier supporters of sympatric speciation adhered to a morphological species definition and assumed that the acquisition of ecological or other nonmorphological species characters preceded the acquisition of morphological species characters. Any discrete population characterized by nonmorphological characters was accordingly interpreted as a biological race and incipient sympatric species.

Kinds of animals that show no (or only slight) structural differences, although clearly separable by biological characters, are called biological races. A closer analysis reveals that a very heterogeneous assemblage of phenomena falls under this definition. With the help of two classifying criteria (locality and population) they can be sorted into three groups (Table 15.2). A full analysis of the various categories of biological races is given by Mayr (1963:453–460).

Morphs, clones, geographic races with distinctive physiological or ecological properties (Fig. 15.2), semispecies, and polyploids obviously do not qualify as incipient species under the concept of sympatric speciation. Most so-called biological races are now known to be sibling species, as discussed

Table 15.2. Phenomena listed as biological races.

Occurrence	Population	
	Same	Different
Sympatric	Morphs Clones Host "races"	Polyploids Ecological races Host races Sibling species
Allopatric		Geographic races Semispecies Sibling species

in Chapter 3. The only biological races that, at least in principle, might qualify as incipient sympatric species are seasonal races (see p. 271) and host races.

Host races. In many species of animals, particularly nematodes and insects, temporary strains may develop on specific host plants. In spite of the enormous literature on these so-called "host races," we are still far from understanding this phenomenon and its evolutionary implications. As an introduction to the literature see the reviews of Dethier (1954) and Thorsteinson (1960), which include abundant references to the earlier literature.

A study of this literature permits the following generalizations:

(a) Preference for a given species of host plants may and nearly always does have a double basis—conditioning (including larval conditioning) and a genetic predisposition.

(b) Nearly all species with host races concentrate in a given district on one host; yet they also have the ability to establish themselves on a variety of other host plants, particularly under crowded conditions, and may have different preferred host plants in different districts.

(c) Local populations of insects often have considerable genetic variability with respect to host specificity. Any monophagous or oligophagous species of insect will come in contact, during its dispersal phase, with numerous plant species other than its normal host. If the species has the appropriate genetic constitution, it will establish itself on the new host and this, according to the Ludwig theorem, leads to an expansion of the food niche of the species. Numerous such cases of an acquisition of new host plants by insect species have been recorded (Andrewartha and Birch 1954).

(d) During the dispersal phase of the species (which in these insects usually coincides with the adult stage and the period of reproduction) a

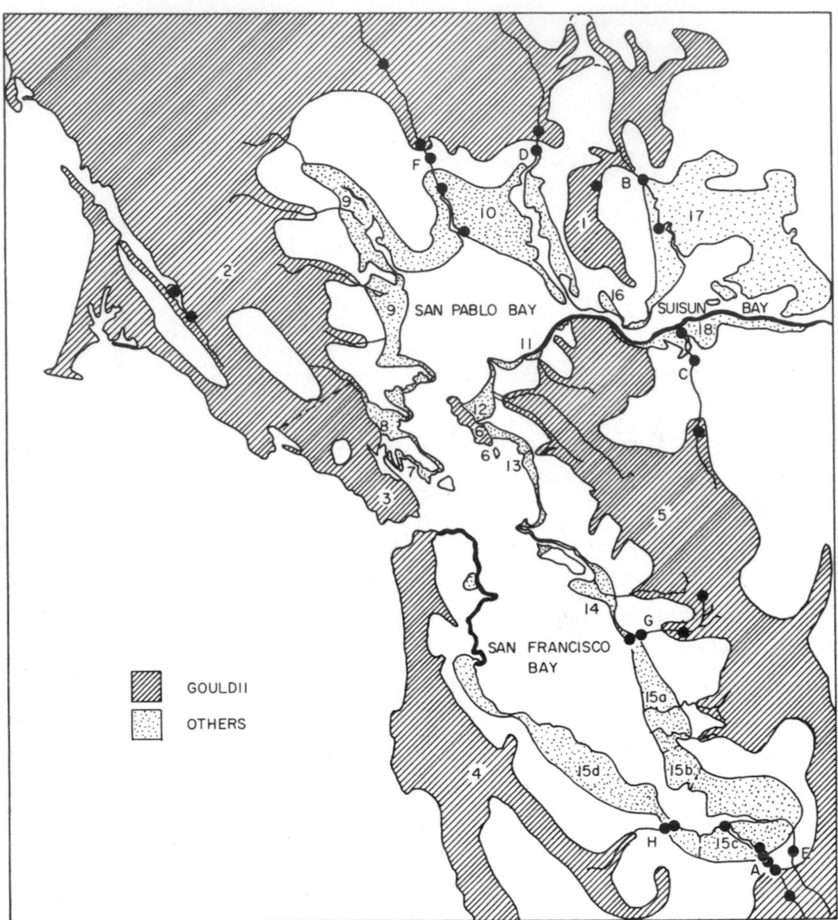

Fig. 15.2. Geographic races of the Song Sparrow (*Passerella melodia*) in the San Francisco Bay region with habitat preferences. The upland race *gouldii* (populations 1–5) is largely separated by unsuitable habitat from the three salt-marsh races *samuelis*, *pusillula*, and *maxillaris* (6–18), as are the latter from each other. There is evidence for gene flow along certain creeks marked *A–H*. (For details consult Marshall 1948.)

lesser or greater mixing of local strains occurs. This is limited to the males if the females are completely restricted to the plant on which they feed as larvae and if mating takes place on the host species. A dispersal phase is, of course, necessary in every species to permit its spread and to permit utilization of new sites.

(e) If such mixing is prevented artificially and an inbred strain is selected experimentally on a single one of several original hosts, such a strain may become progressively less tolerant ecologically until a stage is reached when it may be difficult to reestablish it on any of the other original host species.

(f) Mortality occurs whenever a strain is established on a new host. The more unusual the host, the heavier the initial mortality.

Among the many kinds of so-called biological races, it is the host race that has the best claim to the designation "biological race." That host races with minimal geographic isolation occur is well established. To what extent host preference is correlated with incipient reproductive isolation between host races is largely unknown. The evidence in favor of regarding the host race as an incipient species, in a process of sympatric speciation, is discussed below.

Reputed Cases of Sympatric Speciation

One of the arguments used in favor of sympatric speciation is the existence of certain situations in nature that, it is claimed, can be accounted for by sympatric speciation but not by geographic speciation. These situations include speciation in sibling species (Chapter 3), monophagous species groups, parasites, and species swarms, and the instantaneous splitting of fossil lineages (for further discussion see Mayr 1963).

Monophagous species groups. In many groups of insects there are genera with many species, each of which appears to be limited to a single host. Genera with essentially monophagous species have been described for Microlepidoptera, solitary bees, buprestid beetles (*Acmaeodera, Agrilus*), chrysomelid beetles (*Calligrapha, Arthrochlamys*), leaf-mining Diptera, and other groups. Two phenomena, in particular, seem to suggest a mode of speciation in these food specialists that is different from speciation in most other animals. The first is that monophagous insects often, if not usually, belong to large genera. It is claimed that sympatric speciation permits more rapid and more frequent speciation than geographic speciation. This is not, however, a convincing argument. It would seem far more plausible to ascribe the greater number of congeneric species to the vastly increased number of available niches and the reduction of competition. Likewise, high food specificity should greatly enhance the efficiency of ecogeographical barriers. Such enhancement has been substantiated in the case of flower constancy among desert bees.

It is sometimes argued that monophagy and geographic speciation are incompatible concepts. Indeed, it is asked, how can a monophagous species speciate geographically and thereby become attached to a different host? Without doubt this poses a serious problem. However, monophagy is rarely as rigid as is sometimes claimed. There may be subsidiary host species in the entire range or in part of it. Events crucial for speciation in monophagous groups, as in all speciating animals, are apt to take place in peripherally isolated populations, and these have not yet been studied adequately in even a single one of the groups of monophagous species. As a working hypothesis one might assume that a subsidiary host may become the primary host in such an isolated population and may offer more favorable conditions under the changed ecological situation of the marginal environment (Fig. 15.3). The shift from one host to another will set up an increased selection pressure that will result in a rapid genetic alteration of the population. The amount of genetic variability appears to be comparatively low in a species that has been selected for life strictly on a single host. Host specificity is thus an ideal prerequisite for rapid speciation. In due time such an isolated population may not only become perfectly adapted to the new host, but may also acquire reproductive isolation from the parental population as a by-product of the geographical isolation and the genetic changes in the population. As soon as reproductive isolation is achieved the newly evolved species can reinvade the range of the parental species and live side by side with it. The situation in the leaf beetles described by Brown (1956) indicates that this is a feasible interpretation of speciation in monophagous groups. The same model has been applied successfully to oligolectic bees (Linsley and MacSwain 1958).

The fact that much speciation in monophagous insects is presumably geographic does not preclude the possibility of occasional sympatric speciation by way of host races (see above). Unfortunately, very few cases of strongly differentiated host races have been sufficiently well analyzed to settle the argument, not even the races of the well-studied Codling Moth (*Carpocapsa*) on apples and walnuts. Perhaps the best-studied case is that of the Apple Maggot (*Rhagoletis pomonella*). This species is unknown as an apple pest in Europe. It was first reported infecting apple trees in North America in 1866, more than 200 years after the first apple trees were grown there. Native hawthorn (*Crataegus*) seems to have been the original host of *R. pomonella*, although the species occurs also on other American Rosaceae. In the 100 years since apples first became infested, a tendency toward an earlier breeding season developed in the apple-infesting strain (as compared to the *Crataegus* form) as well as an average difference in

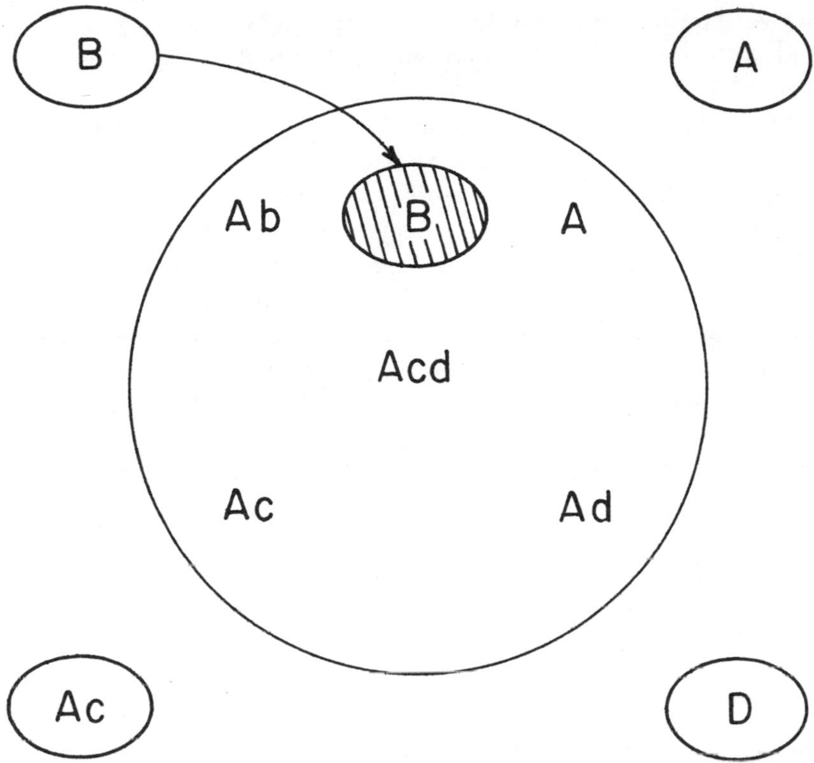

Fig. 15.3. Primary hosts (capital letters) and subsidiary hosts (small letters) of an essentially host-specific species. The large circle indicates the main range of the species; the outlying small circles, the peripheral isolates. New primary hosts (*B, D*) are acquired in some peripheral isolates, facilitating subsequent speciation. Reinvasion from the isolate that is host-specific for *B* will establish a new species (hatched area) if reproductive isolation was acquired during the geographic isolation.

ovipositor length. It is not known how much gene flow there still is between the two host races or how widely separated were the apple orchards and hawthorn stands in which these host races developed. There is, however, a definite possibility of incipient sympatric speciation in this and several other similar cases (Bush 1969).

Parasites. Among parasites there are also many situations that at first sight appear difficult to reconcile with geographic speciation. Two types of distribution of parasites, in particular, might appear challenging. One is the occurrence of different closely related species of parasites in or on different body parts of the same host, and the second is the coexistence of several related species on different host species in the same geographical area. Both, however, can be interpreted as the result of geographic speciation.

Clay (1949) has shown this for speciation in feather lice (Mallophaga). The essential factor here as always is the interruption of gene flow by extrinsic barriers. Related congeneric species of feather lice often occur in different plumage regions (head, body, wing) of the same individual bird. Clay points out that these sympatric species cannot be attributed to "ecological speciation"; rather they are the result of double colonizations, the reproductive isolation having been acquired during previous geographical or host isolation. The zone of contact (and consequently potential gene flow) between lice specialized for distinct body regions is much too great to permit speciation on a single host.

The case of the human head and body louse (*Pediculus*) is not yet fully understood. These lice presumably originated as ecotypes in different human races. Well-clothed human races, like the Eskimos, would seem a prerequisite for the development of the body louse; scantily clad races with much head hair, such as the Melanesians, for development of the head louse. It is conceivable, and indeed probable, that the coexistence in many human races of head and body lice is a case of secondary overlap.

Speciation in internal parasites, with their much more complex life cycles, appears to be strictly allopatric in most cases. The same species of parasite may have different hosts or intermediate hosts in different regions; and if the isolation is sufficiently complete, it will first allow for the development of a geographical race and eventually of a separate species.

Species swarms. Special situations occur in various parts of the world where a considerable number of closely related species, a so-called species swarm, is confined to a narrowly circumscribed area, no close relatives occurring elsewhere. Species swarms occur with particular frequency in the so-called ancient fresh-water lakes, such as Lake Baikal in Siberia (Kohzov 1963), and also in more recent lakes, such as Lake Lanao in the Philippines. Other lakes with such species swarms are Victoria, Tanganyika and Nyasa in Africa, Ochrid in the Balkans, and Titicaca in South America. Leading specialists ascribed the origin of such species swarms to a mode of speciation different from geographic speciation. Rensch (1933:38) dissented from that interpretation and suggested that five factors contribute to the richness of the fauna of these lakes: preservation of relict types, multiple invasions from adjacent river systems, fusion of temporarily separated lakes, fusion of lake basins, and geographic barriers within the lakes. All these factors are contributory, but recent researches indicate that the localization of populations by extrinsic barriers within the lakes is by far the most important factor. The most remarkable aspect of this speciation is that much of it must have proceeded at an inconceivably rapid rate (see Chapter 18). The recognition of this aspect of lake speciation has led to essential unanimity in the interpretation of these species swarms as the product of geographic speciation (Brooks 1950; Fryer 1960; Lowe-McConnell 1969).

The barriers set up by the ice during the various Pleistocene glaciations resulted in the temporary isolation in Eurasia of various lakes or ocean basins and permitted speciation. Svärdson (1961) has attempted to reconstruct the pathway of speciation for pairs of sibling species in the following groups: sculpin (*Myoxocephalus*), smelt (*Osmerus*), cisco (*Coregonus albula*), whitefish (*Coregonus oxyrhynchus*), herring (*Clupea*), and stickleback (*Gasterosteus*) (Table 15.3). Like the species swarms in lakes, these sibling species developed during a remarkably short period of isolation. The current coexistence of two sibling species in the same body of water is due to reinvasion after the lifting of the geographical barrier.

One can summarize the problem of species swarms in lakes by saying that the available evidence is consistent with the hypothesis of a spatial separation of populations corresponding to allopatric speciation in terrestrial animals. However, since such separation in different water masses may result in adaptation to different water temperatures, to different breeding seasons, and to different food niches, it presumably sets up very powerful selection pressures leading to a more rapid divergence of these populations than is usual in terrestrial animals. When secondary contact of such previously isolated populations is established, selection will be intensified by competition among close relatives. Finally, the presence of many previously unoccupied ecological niches, particularly in the older

Table 15.3. Speciation in Eurasian fish (after Svärdson 1961).

Species pair	Location of isolate	Duration of isolation (years)
Sculpin, smelt	Onega Valley Ice Lake	60,000
Cisco	South Scandinavia	120,000
Herring	Mediterranean	120,000
Whitefish, stickleback	Great Siberian Ice Lake	100,000

lakes, encourages a high degree of adaptive radiation that is quite unparalleled among terrestrial animals except in a few isolated archipelagos such as the Hawaiian Islands.

Instantaneous splitting of fossil lineages. Paleontologists have described a considerable number of cases in which, supposedly, a single lineage suddenly split into two clearly separated species. The Steinheim snails (*Planorbis multiformis*), the ammonite *Kosmoceras* (Brinkmann 1929), and the echinoid *Micraster* are such cases. The situation of the Steinheim snails has not been reexamined in recent years, but in the case of *Kosmoceras* and *Micraster* it is now evident that the original interpretation was not correct. Whenever a second species appeared suddenly, it owed its origin not to sudden speciation but to immigration from elsewhere. Every case of so-called sympatric splitting in *Kosmoceras* is preceded by a break in the strata. For *Micraster* Nichols (1959) has shown that *senonensis* invaded the range of *M. cortestudinarium* after the end of the *Holaster planus* zone and did not split off from it. The exposures accessible to the paleontologist are usually localized, and it is in most cases quite impossible to determine the place of origin of a new invader that had originated elsewhere in a localized geographical isolate.

A reanalysis of five reputed proofs of sympatric speciation—sibling species (Chapter 3), speciation in monophagous groups, speciation in parasites, species swarms in lakes, and the instantaneous splitting of fossil lineages—thus shows that four are consistent with the theory of geographic speciation, but that a process essentially amounting to sympatric speciation may occur in host-specific plant feeders.

Models for Sympatric Speciation: Assumptions

Those who believe in the frequent occurrence of sympatric speciation have enumerated a number of factors that—they postulate—would permit the segregation of a single population into several species without the help of geographic isolation. The detailed models that they have worked out are based, in part, on unsupported assumptions, which must be discussed before the models can be adequately analyzed.

Homogamy. According to this concept the most similar individuals of a population tend to mate with each other. Mayr (1947) discussed this concept and its history and showed that the observed evidence contradicts its validity. Homogamy had to be postulated under the pre-Mendelian theory of blending inheritance because without it random mating would soon have led to a complete elimination of all genetic variability. The assumption that monogamy would lead to homogamy is not supported by facts. Monogamy is the rule among birds, but there is no evidence for homogamy except for some nonrandom mating in the case of the Snow Goose and the Blue Goose. These two kinds of geese have now been clearly shown to belong to a single species (*Anser coerulescens*), the two major color types differing primarily in a single gene, semidominant for blue (Cooke and Cooch 1968). Nevertheless there is evidence for nonrandom mating, the males (who apparently exercise the choice) preferring, owing to imprinting, a mate of the phenotype of their parents (? their mother). There is, however, no evidence that the imprinting is complete and there is considerable variation among the heterozygotes; hence it would be improbable in this case for the assortative mating to lead to speciation. Such a possibility exists, of course, for all species with parental care and imprinting, as shown by Seiger (1967) and others in theoretical models, but there is no evidence that the rigorous assumptions demanded by these models are met with in nature. All cases of incipient and recently completed speciation in the estrildid finches, a bird family that is highly subject to imprinting, are fully consistent with geographical speciation.

A mild form of homogamy occurs in many animals with highly variable adult size. The mates in pairs of such species are on the average more similar in size than would be expected if mating were totally random. The same individual, however, as it continues to grow year after year, passes into larger and larger size classes. Most students of polymorphic species have confirmed the randomness of pairings, in spite of the sometimes striking visible differences of the sex partners. Indeed, in *Panaxia dominula* Sheppard (1952b) found a slight preference for the unlike partner and the same has been described for morphs in the White-throated Sparrow (*Zonotrichia albicollis*). Finally, in *Drosophila* the rare genotype often has increased mating success (Chapter 9). Positive assortative mating can have very little importance as an evolutionary process unless it is exceedingly intense.

Conditioning. Several authors have recently revived the well-known suggestion that the establishment of a new sympatric species population in a new niche might be achieved through conditioning. Thorpe (1945), in particular, has presented convincing evidence for the powerful effect of conditioning in shifting insects from one type of food to another. Yet Mayr (1947) has shown that complete isolation of the two populations was never achieved in any of these conditioning experiments. Preference for a type of food may be raised from 35 to 67 percent or lowered from 20 to 8 percent, but these changes do not decrease gene flow between the two subpopulations to the point where speciation is possible.

In all carefully conducted experiments of this sort it was possible to demonstrate a high mortality on the new host or on the new food. It would seem impossible that a reproductively isolated new host population could become established within the cruising range of the parental population in the face of such high counterselection and with the help of such a slight advantage through conditioning. It is far more likely that the ecological amplitude of the species would be enlarged, according to the Ludwig theorem, by the incorporation of an additional kind of host into the niche of the species.

Preadaptation and niche selection. Most hypotheses of sympatric ecological speciation postulate that dispersing individuals search actively for that particular niche to which they are best adapted on the basis of their particular genotype. There is some evidence for the validity of this assumption, since a given species usually has well-defined species-specific habitat preferences and these are not necessarily identical for all the various genotypes of a species (Chapters 9 and 18). Most habitat selection is, however, species-specific only in a very generalized way and is also often affected by nongenetic influences (conditioning, and so forth). Only in an exceptional case will an individual search out that particular subniche for which it is specifically preadapted by its genetic constitution. The concept of sympatric speciation by preadaptation is quite typological in making the assumption that a single gene preadapts an individual for a new niche. Indeed, it would require a veritable systemic mutation to achieve the simultaneous appearance of a genetic preference for a new niche, a special adaptedness for this niche, and a preference for mates with a similar niche preference. The known facts do not support these assumptions.

Two Proposed Models for Sympatric Speciation

Among the various models of sympatric speciation proposed in recent decades, two deserve particular attention.

(*1*) *Speciation by disruptive selection.* Fisher (1930) pointed out that the mutationist interpretation of the origin of mimetic polymorphism in *Papilio polytes* is highly improbable, it being far more likely that the striking discontinuities between the mimetic morphs are gradually acquired by natural selection. This assumption has since been largely substantiated through genetic analysis. Such acquisition through selection of several distinct phenotypes in a population has recently been designated *disruptive selection* (Mather 1955) *or diversifying selection* (Dobzhansky). Muller (1940) was apparently the first to suggest that the accumulation of different sets of specific modifiers might lead to sympatric speciation. In recent years Thoday and his collaborators have studied various aspects of disruptive selection. They found, for instance, that when simultaneous selection for high and low bristle number was carried on in a population of *Drosophila melanogaster*, it responded with a strongly bimodal distribution within 12 generations of selection (Thoday and Gibson 1962). Attempts by several other laboratories to achieve similar results have not been successful so far. When the selection was phenotypically successful, one of the resulting morphs was of lowered viability and could be maintained only with difficulty in competition with the more successful morph. Even though it seems conceivable that the experimental conditions could be improved to the point where the sublines would behave to each other like two species, such conditions would be so different from anything one might find in nature as to be irrelevant to the problem of sympatric speciation. Likewise, a selection pressure that would permit the survival of only two opposite extremes is unlikely under the variable conditions of the natural environment. See Wallace (1968a:397) for a perceptive discussion of the role of disruptive selection in speciation.

A polymorphism established by diversifying selection is unlikely to lead to speciation. A species would lose all the advantages of improved utilization of the environment acquired through adaptive polymorphism if it were to split into a series of narrowly specialized species. K. Jordan pointed out long ago that selection of different physiological varieties within a population can have only two outcomes, either polymorphism or extinction of the inferior types. The case of mimetic polymorphism, perhaps the most frequent and best-analyzed product of diversifying selection, fully substantiates Jordan's conclusion: even the most powerful selection in favor of a complete discontinuity of phenotypes does not result in speciation.

Jordan's conclusion is well illustrated by the North American butterfly *Limenitis arthemis*. In the northern part of its range, it lacks iridescence and it has large red-orange spots and, like most related species, a broad disruptive wing bar (*arthemis* morph). Further south, where its range coincides with that of the strongly distasteful swallow tail *Battus philenor*, it looks exactly like its model: it has acquired iridescence and it lacks the red-orange wing spots and the wing bar (*astyanax* morph). A highly variable population bridging the extremes is found in a narrow zone between *arthemis* and *astyanax*, from Minnesota to New England. This highly variable population was originally considered a hybrid belt between two semispecies, but is now interpreted as a zone of relaxed selection in an area where the model (*B. philenor*) is rare (Platt and Brower 1968) (Fig. 15.4). There is no evidence that this diversifying selection represents incipient speciation.

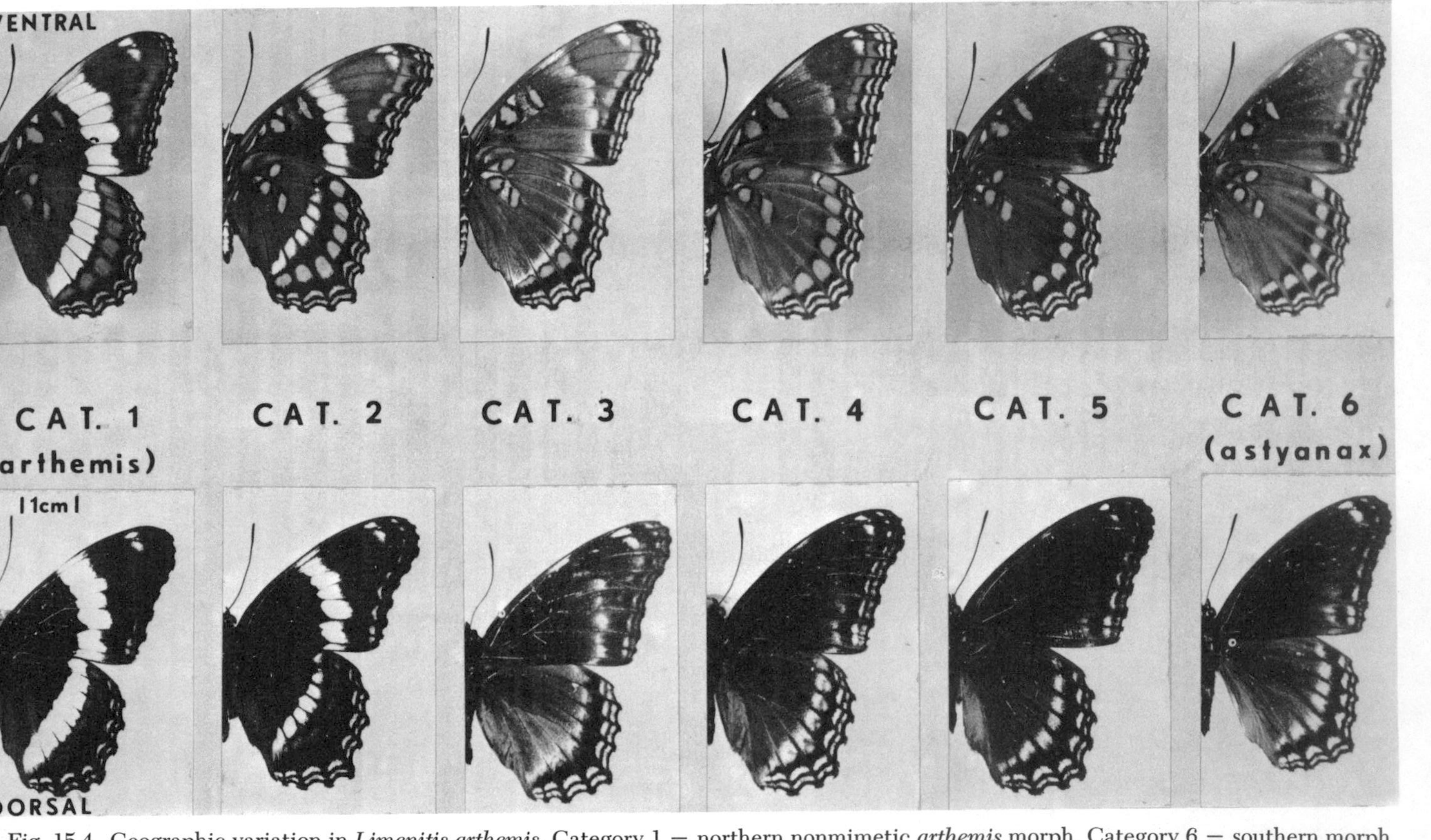

Fig. 15.4. Geographic variation in *Limenitis arthemis*. Category 1 = northern nonmimetic *arthemis* morph, Category 6 = southern morph (*astyanax*) mimicking black swallow-tails. Categories 2-5 = intermediate types near northern edge of black swallow-tail distribution. (From Platt and Brower 1968.)

(2) *Speciation by seasonal isolation.* It has been postulated by various authors that a species with a very long breeding season might be sympatrically split into two if the genetic continuity between the earliest and the last breeders of the year could somehow be interrupted. The occurrence of *seasonal races* is often cited as evidence for such a process of speciation, in spite of the grave objections raised by Jordan (1905). The term "seasonal race" is confusing, since at least four different kinds of breeding behavior have been lumped under this heading: (a) the succession of generations within a single year, the later generations being direct descendants of the earlier generations; (b) in fresh-water and marine animals the occurrence of subpopulations of a species that breed at distinctly different water temperatures but that may interbreed at intermediate temperatures; (c) the occurrence of a spring and a summer or fall generation when each year's spring generation is the descendant of the previous year's spring generation and the summer or fall generation a descendant of the previous year's summer or fall generation; (d) so-called biological races, which differ in their breeding season, but are actually sympatric sibling species.

Of these four kinds of "seasonal races" only those listed under *c* are serious contenders for the rank of incipient sympatric species. Unfortunately, not a single case has been well analyzed in which different seasonal races of the same species are believed not to interbreed. In many cases such seasonal races seem to be the product of a secondary overlap of populations that had originated as geographic races and had adapted their season of reproduction to local conditions. The various broods and sibling species of the periodical cicada (*Magicicada*) have evidently originated in this manner (Lloyd and Dybas 1966; Alexander and Moore 1962). Many cases of "seasonal races" recorded in the literature are undoubtedly sibling species. The occurrence of seasonal races and of sibling species that differ primarily in their seasonal activities has led to the proposal of several models of speciation by seasonal isolation. Such a model might postulate the following conditions:

Let there be a species with a single annual generation, but with a prolonged breeding season, which lasts from spring to fall. Let an event take place, like a killing off by climate or a newly invading competitor, that leads to the extinction of the midseason breeders. As a result, only two sets of individuals remain, early-season breeders and late-season breeders. Both kinds are henceforth reproductively isolated and can accumulate genetic differences.

Several of these assumptions are so unrealistic that the operation of this model in nature would seem improbable. First, the existence, in an area with pronounced seasons, of a species with a very long breeding season but only one generation per year is altogether unlikely. Prolongation of the breeding season is normally achieved by having several successive generations or broods per year. Second, if there were such a species it would have a wide geographic range and it is improbable that the extermination factor would affect all the populations simultaneously and in a like manner. The midseason gap would then be filled by gene immigration. Segregation by recombination from the gene reservoir of the spring and fall breeders would tend to fill the midseason gap even without immigration. Finally, it is exceedingly unlikely that an exterminating factor would single out a segment in the middle of the curve. Competitors as well as climatic factors are far more likely to eliminate the extreme members of the population, that is, the earliest or the latest breeders.

The third of these three objections is successfully met by Alexander and Bigelow (1960) in their attempt to explain the separation of two field crickets (*Gryllus*) by allochronic speciation. These two very similar sibling species differ most conspicuously in their breeding season and overwintering method (Fig. 15.5). The Northern Spring Field Cricket (*G. veletis*) breeds from May to July and overwinters in a late nymphal stage. The Northern Fall Field Cricket (*G. pennsylvanicus*) breeds from July to October and overwinters in the egg stage. Since the early instars are not cold resistant, Alexander and Bigelow propose that the ancestral species was split into two by the killing off of the early nymphs during a Pleistocene cold period. Yet their model fails to answer the other two objections raised above. To me, it would seem far more reasonable to assume that the range of the ancestral species was fractionated into several geographical isolates (during one of the glaciations?), in one of which fall breeding proved more adaptive, in the other, spring breeding. Otherwise the two species remained very much the same, so that they could occupy essentially the same region during their post-Pleistocene range expansion. The assumption of Alexander and Bigelow that all populations of the parental species would be subjected, in an identical manner, to an elimination of midseason nymphs seems unrealistic. The brevity of the breeding season in crickets and grasshoppers precludes the existence of a univoltine (one generation per year) continuous breeding season. It is rather probable that the difference in breeding cycle of *G. pennsylvanicus* and *G. veletis* was not yet complete when they first met, after emerging from their geographic isolation. If so, competition eliminated any tendencies toward spring breeding in *pennsylvanicus* and

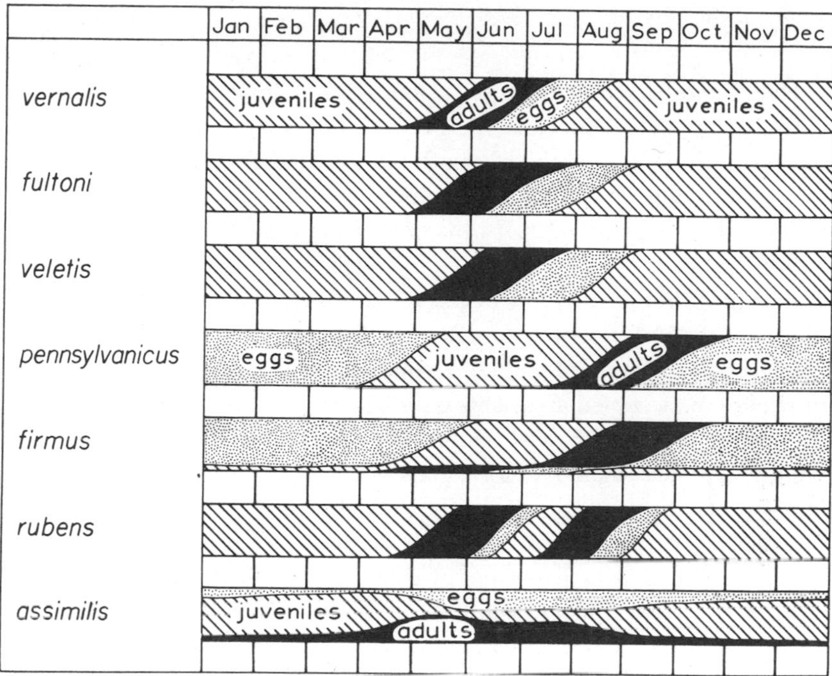

Fig. 15.5 Diagrams of the life cycles of the seven field crickets (*Gryllus* spp.) in eastern North America, illustrating four of the seven general kinds of life cycles known in crickets. (From Alexander 1968.)

toward fall breeding in *veletis*. Related species, not in competition with a close relative, are rarely as narrowly specialized.

It is interesting to note how many species of insects are bivoltine (two generations per year) in part of their geographic range and univoltine in another, without this situation leading to speciation. A change in life cycle, therefore, does not necessarily result in species formation. All cases of closely related species of insects that differ in their life cycle (seasonality) can be explained by geographic speciation. However, such an explanation does not refute the possibility of allochronic speciation, along the lines of the Alexander-Bigelow model. The evidence and arguments in favor of this model have recently been comprehensively summarized by Alexander (1968).

One can summarize this discussion of models of sympatric speciation by saying that all of them show serious, if not fatal, weaknesses. In not a single case is the sympatric model clearly superior to an explanation of the same natural phenomenon through geographic speciation.

Difficulties of Hypotheses of Sympatric Speciation

Hypotheses of sympatric speciation are usually proposed in order to eliminate so-called difficulties of the theory of geographic speciation. However, the authors of these hypotheses overlook the fact that they create many more difficulties than they remove.

Neglect of dispersal. Dispersal is one of the basic properties of organic nature, yet it is conspicuously neglected in most schemes of sympatric speciation. The life cycle of every species includes a dispersal phase. In most insects with wingless larvae it is the adult stage. In most marine animals it is the larval stage. Dispersal permits the species to expand into previously unoccupied areas and insufficiently filled niches. What is often overlooked is that dispersal results in a great deal of intermingling of populations. This includes individuals not only of different geographical, but also of different ecological origin. Such dispersal occurs also in host-specific species and in those in which mating takes place on a host plant. As shown above, it is this dispersal stage that permits the interchange of individuals between incompletely conditioned populations.

Ecological plasticity and polymorphism. Most species show much more ecological variability than is consistent with the hypotheses of ecological speciation. This variability is exhibited both by individuals belonging to a single population and by the various populations of a species. Some of the variation may have a genetic basis; much of it, however, is evidently nongenetic. Where a species has a preferred food plant, some individuals are usually also found on other plant species, either exceptionally or regularly. If a species is rigidly tied to one host plant in part of the species range, it may be more plastic in other parts. Every population of a species lives in an environment that is different from that of the other populations, and its genotype is continuously adjusted by selection to this specific local environment. Occupation of multiple niches and shifts from niches in one part of the species range to other niches in different areas are the normal situation in most species of animals. If this led to species formation without geographic isolation, we would have a vastly greater number of species than we actually have. To interpret the ecological variability of species as a mechanism of sympatric speciation is to misunderstand the function of this variability. Ecological variation has been misinterpreted in the same way as chromosomal variation. Both are adaptive mechanisms rather than speciation devices. Both, however, may facilitate and accelerate speciation, when superimposed on geographic speciation.

Genetic difficulties. By far the most decisive argument against hypotheses of sympatric speciation comes from the field of genetics. That zygotes in sexually reproducing organisms are diploid and that sexual reproduction maintains the genetic cohesion of every local population present serious obstacles to all hypotheses of sympatric speciation.

Bateson and other early Mendelians always looked for the singular genetic event that would differentiate a new cross-sterile line from the parental one. However, a single genic change cannot accomplish such differentiation in a diploid organism. This can be illustrated by a simple genetic model. A new mutation in a diploid is always heterozygous in the beginning because it occurs on only one of the two equivalent chromosome sets. For instance, if, in a homozygous *aa* population, a dominant mutation *A* were to make its carriers sterile with *a*, every carrier of the gene *A* would automatically be sterile (*Aa*). If the production of genotype *Aa* were to result not in sterility but merely in reproductive isolation from *aa*, it would still doom the carriers of *A* to extinction—since they would have no potential mates unless *A* were subject to a highly improbable process of mass mutation or unless they adopted an asexual mode of reproduction. Even then it would segregate 25 percent of *aa* individuals per generation, which would provide for continuous gene flow from the daughter to the parental species. These are only a few of the most obvious objections to the idea that a single mutation could produce new species. Additional, more general, objections are raised in Chapter 10 and 17.

For successful speciation, a minimum of two complementary genes is necessary (Dobzhansky 1937). This process in its simplest form is presented in Fig. 15.6. Let us assume that a uniform population, homozygous for *a* and *b*, is separated by geographic isolation into two populations, 1 and 2. Let us postulate that *AA* individuals are reproductively isolated from *BB* individuals. If a mutation *A* occurs in population 1, it will result in the production of an individually variable population with the three genotypes *Aabb*, *aabb*, and *AAbb*. Let us assume that natural selection favors *AAbb*, so that population 1 eventually becomes homozygous for this genetic constitution. While this happens in population 1, population 2 will go through a parallel transformation, with the *b* locus mutating to *B*, resulting in the genotype *aaBB*. If the geographic isolation between the two populations should then break down, they could now intermingle without interbreeding because, according to our initial postulate, *AA* is reproductively isolated from *BB*. From Fig. 15.6 it is quite evident that in the intermediate stage, when each population consisted of three genotypes, only one of the

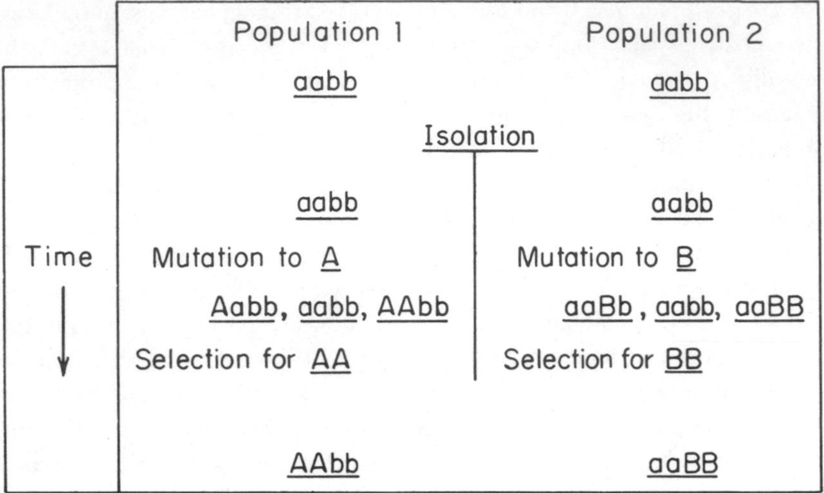

Fig. 15.6. Speciation through the acquisition of two complementary factors *A* and *B* from a parental population *aabb*.

possible nine combinations of these six genotypes would be reproductively isolated. Such almost complete panmixia would, of course, prevent speciation.

Our model is grossly simplified, since we started with a homozygous population, at least as far as the crucial loci are concerned. Actually, natural populations are highly heterozygous and composed of a genetic environment that favors heterozygosity. A bridge of heterozygosity thus exists not merely at one or two loci but at scores or hundreds of them. There is no reason to doubt that the genes involved in the control of the isolating mechanisms are at the same time part of the total genetic cohesion of the population (Chapter 10). Our second simplification is that we let the reproductive isolation be determined by only two loci. Actually, as shown in Chapters 5 and 6, reproductive isolation usually has a highly complex basis, and many mutational steps or other genetic rearrangements are required to complete reproductive isolation. All these considerations help to make clear why geographic isolation is ordinarily a prerequisite for successful speciation.

Conclusions

The discussion of sympatric ecological speciation permits us to conclude that the hypothesis is neither necessary nor supported by irrefutable facts. It tends to overlook the fact that speciation is a problem of populations, not of individuals, and it minimizes the difficulties raised by dispersal and recombination of genes during sexual reproduction. The essential component of speciation, that of the genetic repatterning of populations, can take place only if these populations are temporarily protected from the disturbing inflow of alien genes. Such protection can best be provided by extrinsic factors, namely, by spatial isolation. It appears that such spatial isolation is normally effected by geographical barriers. The possibility is not entirely ruled out that forms with exceedingly specialized ecological requirements or life cycles may diverge genetically without benefit of geographic isolation, but the reported cases are also open to other interpretations.

The discussions in this chapter, in particular the discussions on instantaneous and sympatric speciation, have shown that nearly all the potential modes of speciation (Table 15.1) suggested in the past are improbable and that reputed instances of speciation without geographic isolation are quite consistent with the theory of geographic speciation, indeed are more easily understood through this interpretation than any other. Most distribution patterns of closely related species cannot be readily explained except through geographic speciation. It is thus clear that geographic speciation is of overwhelming importance in explaining the problem of multiplication of species. Chapter 16 will therefore be entirely devoted to this important process.

16 · Geographic Speciation

That geographic speciation is the almost exclusive mode of speciation among animals, and most likely the prevailing mode even in plants, is now quite generally accepted. And yet this thesis was vigorously contested as recently as 30 years ago and such a distinguished biologist as Goldschmidt never accepted it. The theory of geographic speciation is one of the key theories of evolutionary biology and it would seem appropriate to present in considerable detail the proofs of its validity.

The problem of speciation is one that illustrates particularly well the basically different modes of thinking of the functional biologist, who is primarily concerned with mechanisms, and the evolutionist, who is concerned with natural populations. Although evolutionists have stressed since 1869 that the population structure of species is the key to the problem of speciation, most publications of the geneticists were, as late as the 1930's and 1940's, still displaying a complete preoccupation with the mechanisms. How little the geneticists before Dobzhansky (1937) understood the problem of speciation is exemplified by R. A. Fisher's choice of epigraph for the first chapter of his book *Genetical Theory of Natural Selection,* 1930; he quoted with evident approval W. Bateson's statement "As Samuel Butler so truly said: 'To me it seems that the "Origin of Variation," whatever it is, is the only true "Origin of Species." ' " In other words, mutation is the only true Origin of Species.

Actually, the basic problem of speciation consists in explaining the origin of the gaps between sympatric species. These gaps, as we saw in Chapter 5, are clear-cut, well defined, and maintained by isolating mechanisms. As long as species were considered static, typological, and nondimensional, it was virtually impossible to solve the problem of the crossing of these gaps. Hence, they were termed, with good reason, "bridgeless gaps" by Turesson, Goldschmidt, and others. During much of the nineteenth century, when typological concepts prevailed, no other mode of the origin of species seemed possible than that by sudden jumps, macromutations.

A far more plausible solution of the previously seemingly insoluble problem of the bridgeless gaps was made possible by a fundamental change in certain concepts of systematics, a change consisting in the expansion of the species from the nondimensional and typological to the multidimensional and polytypic species, as described in Chapter 12. The work of numerous taxonomists finally culminated in the theory of geographic speciation, which states that in sexually reproducing animals *a new species develops when a population that is geographically isolated from the other populations of its parental species acquires during this period of isolation characters that promote or guarantee reproductive isolation after the external barriers break down* (Mayr 1942). A detailed history of the gradual development of the theory of geographic speciation is given by Mayr (1963:482–488).

A historical survey shows the venerable age of the theory of geographic speciation. Since it has only recently come to the attention of botanists, paleontologists, and geneticists, it is sometimes erroneously considered a "new" theory. That is quite wrong. The early statements by Leopold von Buch go back 145 years and in its full development by M. Wagner the theory is more than 100 years old; since then it has been tested and retested by four generations of systematists. It is one of the most interesting examples of a theory that was found strictly empirically and for which the "how?" was answered about 100 years before the "why?"

Evidence for Geographic Speciation

Every careful modern generic revision provides new evidence for geographic speciation. Yet taxonomic evidence is inferential and a few saltationists still refuse to accept it. Speciation is a slow historical process and, except in the case of polyploidy, it can never be observed directly by an individual observer. Is there any method by which slow past events can be reconstructed and "proved"?

Yes, there is, and this method is well tested because the evolutionist is not alone in his predicament. Any scientist who must interpret past events, like the archeologist, the historian, or the geologist, or who, like the cytologist, studies dynamic processes that can be observed only at definite fixed stages, like the stages of cell division, faces the same difficulty and solves it with the help of the same method. This method consists in the reconstruction of an essentially continuous series by arranging fixed stages in the correct chronological sequence.

Speciation being a slow process, it should be possible to find natural populations in all stages of "becoming species." The question then is: "Through what stages does a population pass that is in the process of becoming a separate species?" Darwin and other early evolutionists were quite aware of this approach and looked for *incipient species.* A "variety," declared Darwin, is such an incipient species, but his study of varieties was not very fruitful, since he made no distinction between variant individuals and variant populations. It is more rewarding to start with the assumption that speciation is a population phenomenon and to look for populations that are incipient species or that appear to have just completed the process of speciation ("new species"). Do such populations show a characteristic pattern of distribution, and, if so, is this pattern consistent with the theory of geographic speciation? There are three sets of phenomena that give us information on this question: (a) levels of speciation, (b) geographic variation of species characters, and (c) borderline cases and distribution patterns.

Levels of Speciation

Numerous species groups, recently analyzed throughout their area of distribution, have been found to consist of populations that represent every stage of divergence, up to recently completed speciation. As an example I quote findings in the genus *Drosophila* (Patterson and Stone 1952:548):

> We have been particularly intent on demonstrating different degrees of divergence and the correlated difference in factors involved in the several available strains [of *Drosophila*]. These range from partial strain isolation (*repleta, peninsularis*), to subspecific divergence (*texana-americana, fulvimacula-flavorepleta,* to closely related species (*virilis-texana, mojavensis-arizonensis*), to distantly related species in the same species group which would not cross, and to members of different species groups and subgenera which do not ordinarily even show interest or attempt to mate in *Drosophila.* We have been able to show the presence of some isolation between separate subspecies or even strains. We do believe that this evidence favors the view that ordinary species differences have come about through the accumulation of steps, through allopatric strains and subspecies to species.

Similar situations have been described for numerous genera of invertebrates and vertebrates. A particularly well-known case is that of the platyfishes of Central America (*Xiphophorus maculatus* species group), excellently analyzed by Gordon and associates (Fig. 16.1). The northernmost species (*X. couchianus*) is restricted to a single river system. To the south follows *X. variatus,* which has three well-defined subspecies in the states

Fig. 16.1. Distribution of the species and subspecies comprising the platyfish superspecies *Xiphophorus maculatus*. (From Rosen 1960.)

of Tamaulipas, San Luis Potosi, and Vera Cruz, Mexico, among which the subspecies *xiphidium* is so distinct that it has long been considered a full species. Southernmost is *X. maculatus,* famous for the polymorphism of its spotting patterns, which occurs in numerous stream systems from Vera Cruz to British Honduras. The frequencies of the micro- and macromelanophore patterns differ from stream to stream and population crosses have revealed other genetic differences, for instance in the sex-determining mechanisms. Here then we have a series of related, allopatric populations showing every stage from the local genetic race, to the ordinary subspecies, to the almost specifically distinct subspecies (*xiphidium*), to the full species (*couchianus*).

The findings in other recently analyzed species groups fully confirm those reported for *Drosophila* and *Xiphophorus*. All findings agree that in every actively evolving genus there are populations that are hardly different from each other, others that are as different as subspecies, others that have almost reached species level, and finally still others that are full species. Sometimes these last are still allopatric; in other cases the most distinct ones may already have been able to overlap the ranges of their closest relatives. To find the complete hierarchy of successive levels of speciation, without looking at a single fossil, is most impressive. The fact that these more and more distinct populations remain allopatric until they have reached species level is one of the most convincing proofs of geographic speciation.

Two postulates can be derived from the theory of geographic speciation, that there should be geographic variation in species characters and that the pattern of distribution of natural populations should reflect the pathway of geographic speciation. Any fact that substantiates these postulates would be further proof of geographic speciation.

Geographic Variation of Species Characters

One can summarize the evidence presented in Chapter 11, "Geographic Variation," by saying that any character ever described as distinguishing species is also known to be subject to geographic variation. Of special interest are those species characters (discussed in Chapters 4 and 5) that facilitate sympatry of related species, that is, isolating mechanisms and ecological compatibility factors. If speciation is geographic, these two sets of characters should be as much subject to geographic variation as morphological characters. This conslusion is indeed supported by abundant evidence.

Ecological properties. In most species where the populations are contiguous and gene flow is unimpeded there is little geographic variation in niche requirements. Ecological shifts are more characteristic of geographic isolates. Yet most widespread species that have been carefully studied have been found to contain geographically representative populations that differ from each other to a lesser or greater extent in their ecology. In the Island Thrush (*Turdus poliocephalus*) the populations on Bougainville and Kulambangra in the Solomon Islands live in the high mountain forest above altitudes of 6000 feet. On nearby Rennell Island the species occurs in the lowlands and may feed among coral boulders along the seashore. Wilson's Warbler (*Wilsonia pusilla*) in North American occurs in Canadian sphagnum bogs in the east, while it is found in chaparral and other exposed hot and rather arid localities in the west. For other cases see Mayr 1963:492. Geographic variation in habitat occupation is by no means restricted to mammals and birds. Certain species of mosquitoes, such as *Anopheles pseudopunctipennis, A. bellator,* and *A. labranchiae,* enter houses in part of their geographic range but fail to do so in others. It is easy to see that such shifts preadapt the populations for eventual sympatry. On the basis of available observations I would hazard the guess that ecological exclusion (Chapter 4) of incipient species is in most cases well established before secondary contact is established.

Isolating mechanisms. The isolating mechanism whose geographic variation is best documented is the sterility barrier. The well-studied genus *Drosophila* supplies many illustrations of partial sterility among geographically distant races of the same species (Patterson and Stone 1952). Reduced fertility has also been found in crosses between geographically distant populations of numerous other terrestrial, fresh-water, and marine species. Although the greatest reduction in fertility usually occurs among the most distant populations, the amount of sterility is not completely correlated with distance. Rather distant populations are sometimes more fertile with each other than populations less far removed.

Ethological barriers are the most important isolating mechanisms in animals. Unfortunately the evidence for the existence of such barriers between geographical races of a species is scanty. Incipient sexual isolation in the Japanese newt *Triturus pyrrhogaster* has been established for races from the islands Amamioshima, Kyushu, and Honshu. There is even nearly complete isolation between two populations on Honshu Island. The hybrids are, however, perfectly viable. The breakdown of the courtship between males and females belonging to different races seems to be due to the inadequacy of the mutual chemical stimuli. The earliest work on geographical variation in isolating mechanisms among essentially olfactory animals, reacting to chemical stimuli, was done by breeders of moths. Standfuss (1896:107) reports of the Tiger Moth (*Panaxia dominula*) that when freshly hatched females of the Italian subspecies *persona* [*italica*] were exposed near Zurich, Switzerland, only a few males of the native subspecies *dominula* appeared, even though they appeared in large numbers at a nearby place where freshly hatched females of native *dominula* were released.

The evidence for the geographic variation of ethological isolation based on visual stimuli is almost entirely inferential. For instance, the plumes of birds of paradise (Paradisaeidae) vary geographically to such a striking extent that one would expect differences in the courtship poses and in the responses of the females, particularly in the genera *Astrapia*, *Parotia*, and *Paradisaea*. An analysis of such differences would be particularly interesting, since all stages of speciation are represented in these genera. Behavioral differences between geographic races of birds do occur, but it is rarely known to what extent they have the potential to effect reproductive isolation.

Gompertz (1968) found that hybrids bred in captivity between the British and the Indian races of *Parus major* were ignored by British birds when released among them. The hybrids acted (and were treated) like individuals of a different species. Because British and Indian birds differ both in color and in vocalization, it cannot be determined which character played the more important role in the sexual isolation.

Geographic variation of song, an important isolating mechanism in birds, is widespread. For instance, the Pacific warbler (*Acrocephalus*) has on Guam a song so beautiful that the bird is called the nightingale warbler; on other islands the song is undistinguished, and on Pitcairn it seems to have been lost altogether. The nuptial song is the most important isolating mechanism in many orthopterans. In this group also, geographic variation of song has been described (for example, Cantrall 1943), but in view of the many as yet undescribed sibling species of grasshoppers none of the recorded cases seemed to be unequivocally established. Several cases of geographic variation in the calls of anurans have been reported, and completion of speciation by this process is probable in at least one case (Blair and Littlejohn 1960).

Ultimately, all differences in fertility, structure, ecology, and behavior between geographical races of a species are nothing but a reflection of an overall difference in their genetic constitution. The more barriers to gene flow between populations and the greater the differences in their environments, the greater will be the genetic rebuilding and the greater the probability of changes in the components of isolating mechanisms. That pronounced genetic differences among geographic races may exist has now been abundantly established for *Drosophila*, amphibians, lepidopterans, and other animals (Chapter 11; Ford 1964).

Borderline Cases and Distribution Patterns

The geographic isolate is the key unit in the process of geographic speciation. We should therefore be able to find isolates in every stage of speciation. New species, furthermore, should often still be either allopatric to the species from which they have diverged or just barely overlapping their range. It should be possible to find isolates that have some of the attributes of species but lack others (borderline cases). Finally, there should be segments of species which consist of former isolates that have failed to acquire fully efficient isolating mechanisms before rejoining the parental species along zones of secondary hybridization. And indeed all these evidences of geographic speciation are found abundantly in the taxonomically better-known animals. We can classify the various kinds of borderline cases into seven categories.

(*1*) *Peripheral isolates.* The most distinct isolates of a species are nearly always situated along the periphery of the species range. Most polytypic species in well-analyzed groups of animals have such peripheral isolates. They are almost invariably a source of disagreement among taxonomists, some of whom consider them "still" subspecies, others "already" species. In the Spangled Drongo (*Dicrurus hottentottus*) there are almost a dozen peripheral subspecies that are considered full species by some authors (Fig. 16.2). One of these isolates (*D. megarhynchus* of New Ireland) is so aberrant

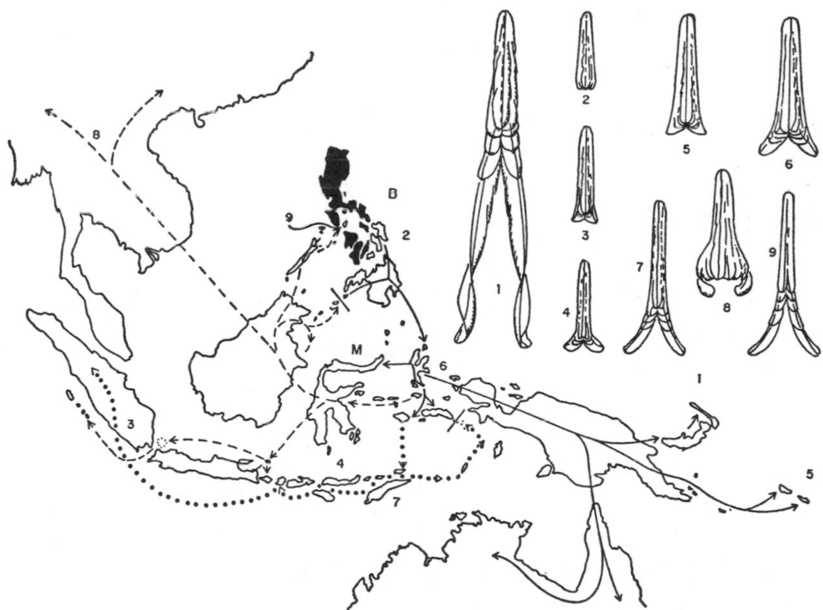

Fig. 16.2. Branches of the polytypic species *Dicrurus hottentottus.* Solid line, oldest group; dotted line, more recent group; broken line, most recent. *B* = closely related species *D. balicassius; M* = *D. montanus*, the product of a double invasion of Celebes by *hottentottus.* The figures indicate the ranges of the nine forms, the tails of which are shown in the insert. The tails of 4 and 6 are typical for the species; the tails of the peripheral forms 1–3, 5, 7–9 are aberrant and specialized in various directions. (From Mayr and Vaurie 1948.)

that until recently it was always considered a separate genus. A rich list of such cases is presented by Mayr (1963:496–499; see also Blair 1965). Every island or island group that is situated in front of a continent, like Britain, Ceylon, Formosa, the Ryukyus, Japan, and Tasmania has given rise to peripheral isolates.

The majority of isolates are "borderline" cases, that is, they have some but not all attributes of new species, with isolating mechanisms more or less incompletely developed. There are far more such cases in existence than one would guess from a casual study of the systematic literature. The rigidity of zoological nomenclature forces the taxonomist to record borderline forms either as subspecies or as species. An outsider would never realize how many interesting cases of evolutionary intermediacy are concealed by the seeming definiteness of the species and subspecies designations.

(2) *Superspecies. A superspecies consists of a monophyletic group of entirely or essentially allopatric species that are morphologically too different to be included in a single species.* The principal feature of the superspecies is that geographically it presents essentially the picture of a polytypic species, but that the allopatric populations are so different morphologically or otherwise that reproductive isolation between them can be assumed. In some cases this has subsequently been confirmed by breeding or mate-selection experiments. The superspecies is an interesting stage of evolution and a particularly convincing illustration of the geographical nature of speciation.

Superspecies constitute a regular and sometimes rather high percentage of every fauna. To substantiate this statement numerically is impossible at the present time because, except for birds, the taxonomy is not sufficiently mature to allow us to determine unequivocally which groups of allopatric species taxa are polytypic species and which others superspecies. In birds, certainly, superspecies are very common and many instances have been cited by Mayr (1942, 1963), for instance the birds of paradise of the genus *Astrapia* (Fig. 16.3). Almost one third of the species of Australian birds belong to superspecies (Keast 1961) and so do 185 (30.5 percent) of the 607 breeding species of North American birds and over 50 percent of the African birds (Hall MS). There is much to indicate that superspecies might be even more common in animals with lower dispersal facilities than birds. For example, among tropical sea urchins almost every genus contains a superspecies. They seem to be widespread also in mammals, fish, mollusks, and certain groups of insects.

The frequency of superspecies in a group of organisms is largely a

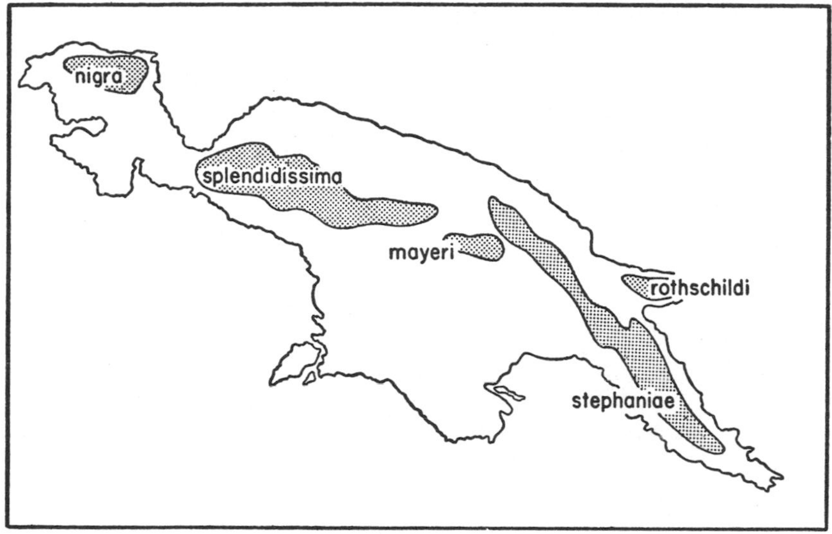

Fig. 16.3. A superspecies of paradise magpies (*Astrapia*) in the mountains of New Guinea. Each of the component species has been described as a separate genus. Hybridization has been recorded in the zone of contact between *A. mayeri* and *A. stephaniae.*

function of the physiographic features of the area it occupies. The Indo-Australian archipelago offers an ideal situation for isolation and has resulted in the production of superspecies in almost every kind of organism that occurs there. Benthonic deep-sea organisms are restricted to more or less isolated deep-sea basins and trenches and thus give rise to groups of allopatric species. Superspecies among continental terrestrial animals are most common in groups in which highly specific habitat or edaphic requirements favor a fractioning of the area of distribution, such as montane or cave species (Barr 1967).

(3) *Semispecies.* This term designates natural populations that have acquired some but not all properties of species. It is employed when an incipient species has not yet fully completed the process of speciation. Gene exchange is still possible among semispecies (although usually prevented by geographic isolation), but it is not as free among them as among ordinary conspecific populations. The existence of semispecies has been established for *Drosophila* and for a number of other well-studied genera of animals. The beach mice (*Peromyscus polionotus*), formerly considered to constitute an ordinary polytypic species, were shown by Bowen (1968) to consist of

two semispecies, which are apparently partially sympatric. The species of which superspecies are composed are generally considered semispecies.

(4) *Secondary contact zones and incomplete speciation.* When a geographical isolate reestablishes contact with the parental species (owing to a breakdown of the extrinsic barrier) before its isolating mechanisms have been perfected, a hybrid zone will develop in the contact area (Chapter 13). Hybridization may be either random or much restricted, as in the case of the *Sphyrapicus* sapsuckers described by Howell (1952). Acquisition of different habitat preferences may reduce the width of the hybrid belt or prevent hybridization altogether (Chapter 6).

(5) *Partial overlap.* The invasion of the geographic range of a parental or sister species by a newly formed species is conclusive proof of completed speciation. The general distribution pattern of the joint descendants of the parental species is often still largely allopatric, except for smaller or larger areas of marginal overlap. An almost infinite number of such cases are recorded in the taxonomic literature (Mayr 1963). During the Pleistocene a population of the kingfisher species *Tanysiptera hydrocharis* (Fig. 16.4) was isolated on an island that ran from the Aru Islands to the mouth of the Fly River and was separated from the mainland form *galatea* by a branch of the ocean. When this strait was filled by alluvial debris from the mountains of New Guinea, dry land joined the island with the mainland of New Guinea, and *galatea* was able to invade the range of *hydrocharis,* where the two species now live side by side without interbreeding and without obvious ecological competition. In lakes and oceans, likewise, the fusion of previously isolated seas and embayments often leads to an overlap between newly evolved species.

The Pied Wagtail of Europe and Asia, *Motacilla alba,* has developed two essentially allopatric species on the periphery of its range. However, the Indian *M. maderaspatensis* has invaded the range of *M. alba* in Kashmir, and *M. alba* has invaded the range of the Japanese *M. grandis* in Hokkaido and northern Honshu. There is no interbreeding in these areas of overlap. Such marginal overlaps in otherwise allopatric groups of species are of widespread occurrence in marine organisms. It is not claimed that each one of these overlaps proves recently completed speciation; yet this is certainly true for the majority of such cases. Marginal overlaps often indicate the pathway of geographical speciation.

(6) *Multiple invasions.* Well-isolated areas, such as islands, mountain tops, or caves, are sometimes inhabited by two or more species of a widespread species group that elsewhere is represented at a given place by only a single species. In the past such situations were often interpreted as cases of

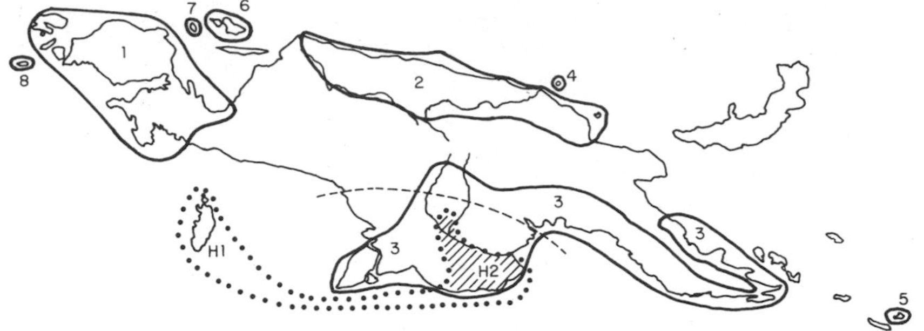

Fig. 16.4. Distribution of the kingfishers of the *Tanysiptera hydrocharis-galatea* group in the New Guinea region. The three mainland forms (1, 2, 3) are barely distinguishable. The island populations (4–8) are strikingly distinct, most of them originally described as separate species. The Aru Islands (H_1) and South New Guinea (H_2) originally formed an island, on which the form *hydrocharis* (*H*) differentiated. When South New Guinea became attached to the main island (along the broken line) the southeast New Guinea subspecies (3) invaded the area. The fact that it does not interbreed with *hydrocharis* (*H*) proves that the latter had become a species during its isolation. (From Mayr 1954.)

sympatric speciation. It is now evident that a different interpretation is the correct one: the occurrence of two or more species in an isolated habitat is the result of multiple invasions. The isolated locality was first colonized by a single group of immigrants from the parental population on the mainland or an adjacent island; this isolated population then diverged genetically from the parental population to such an extent that it had acquired isolating mechanisms as well as ecological compatibility before the second set of colonists reached the island. In exceptional cases there may be even a third colonization, as with the white-eyes of Norfolk Island (*Zosterops albogularis, Z. tenuirostris,* and *Z. lateralis norfolkiensis*).

There is hardly a well-isolated island on which such double invasions have not been reported. As regards birds, they are known in the Canary Islands, Samoa, Australia, Tasmania, Kauai, Ceylon, Andamans, Luzon, Celebes, Flores, Kulambangra, Comoros, Tristan da Cunha, Juan Fernandez, Norfolk Island, Antipodes, and elsewhere. There is no known case, however, in such easily accessible recent continental islands as Great Britain. Double invasions are known not only in birds but also in lizards, snails, butterflies, beetles, *Drosophila,* and other insects. They not only occur on real islands, but may also occur in any isolated location, such as mountains, caves, and fresh-water lakes. Faunal interchange between humid southeastern and southwestern Australia across an arid corridor has resulted in numerous multiple invasions (Fig. 16.5).

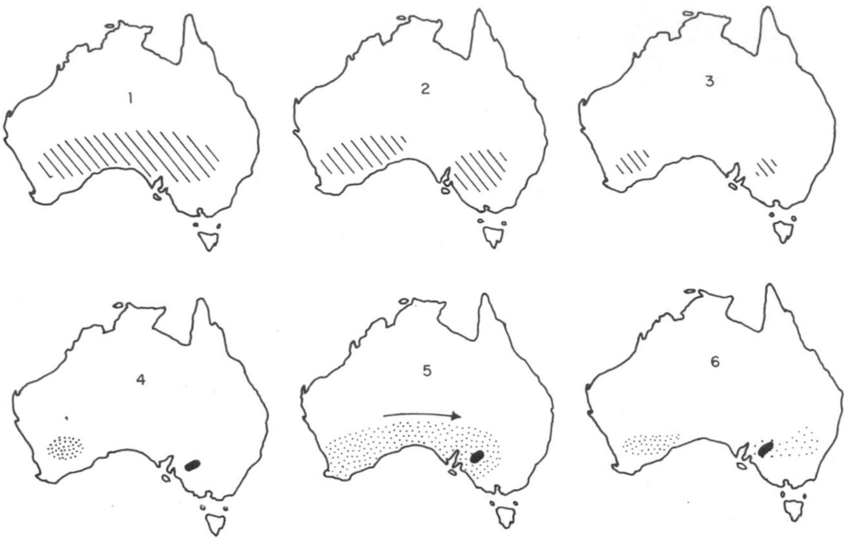

Fig. 16.5. Successive stages (1–6) in the speciation of the Australian mallee thickheads (*Pachycephala*). The present distribution is shown in map 6 with *P. rufogularis* indicated by the black area and the eastern and western areas of *P. inornata* by stippled areas. The range expansions and contractions of these birds are correlated with vegetational changes caused by shifts of rainfall belts. (From Keast 1961, which see for further details.)

Archipelago speciation. It is only a small step from double invasions to the spectacular speciation that has occurred among the birds of archipelagoes like the Galapagos Islands and the Hawaiian Islands. In these archipelagos a single pair or ancestral flock has many times given rise to a vigorous new phyletic line. By colonizing one island after another, and recolonizing the islands from which they have come after reaching species level, the fauna of such an archipelago may become progressively enriched. There is little doubt that all of Darwin's finches had such a monophyletic origin (Lack 1947); and, likewise, the entire highly diversified family of Hawaiian honey creepers (Drepanididae), with 22 species and 45 subspecies, has descended from not more than 1 or 2 species of ancestral immigrants. Even more amazing is a similar archipelago speciation in certain groups of insects, described by Zimmerman (1948). About 4300 species of endemic Hawaiian insects have already been described (1969), but it is estimated that there may be 6500 species. These have descended from only about 250 original colonizations. The number of derived species is astonishingly large in some of the groups. For example, from only one or two original colonizations by *Drosophila* an estimated 600–700 endemic

Hawaiian species have arisen, possibly within less than 5 million years. Similarly, single colonizations by snails, true bugs, moths, beetles, and wasps have resulted in the development of large clusters of descendants, each 100–500 species (according to E. Zimmerman). Speciation by inhabitants of fresh-water lakes often corresponds to archipelago speciation and accounts for some of the species swarms found in ancient lakes (Brooks 1950).

(7) *Circular overlaps.* The perfect demonstration of speciation is presented by the situation in which a chain of intergrading subspecies forms a loop or overlapping circle whose terminal links have become sympatric without interbreeding, even though they are connected by a complete chain of intergrading or interbreeding populations. Among the many proofs of geographic speciation, circular overlaps have always rightly been considered a particularly convincing one. Circular overlaps can obviously develop only where there are highly exceptional constellations of geographical factors. It is therefore rather surprising how relatively common they are. Nine cases were described by Mayr in 1942, and more than a dozen additional cases have been documented in the ensuing years (see Mayr 1963:510–512).

Additional information has tended to show that matters are not quite as diagrammatic as originally believed. This is true, for instance, for the classical case of circular overlap, in which two sympatric species of European gulls, *Larus argentatus* and *L. fuscus*, are considered the terminal links of a chain of subspecies circling the north temperate region (Fig. 16.6). There is, however, no complete continuity of populations and some of the gaps in the series of populations between western Europe and eastern Asia are very pronounced. It is apparent that the range of *argentatus* was split during part of the Pleistocene into a number of refuges. The yellow-footed *cachinnans* group evolved in the Aralo-Caspian region and later gave rise to the Atlantic *fuscus* group. A pink-footed group (*vegae* and relatives) evolved on the Pacific coast of Asia and subsequently gave rise to the closely related typical *argentatus* in North America. The transatlantic invasion of *argentatus* into Europe is a comparatively recent event. For a somewhat different interpretation, see Barth 1968. Where *vegae* and *cachinnans* meet (*mongolicus*) they exchange genes, likewise where *cachinnans* and *argentatus* meet in the north Baltic. However, where *argentatus* and *fuscus* meet along the coasts of Europe they live unmixed side by side, hybridizing only quite rarely. The behavioral and ecological differences that permit the two forms to coexist like good species have been described by Goethe (1955).

During the Pleistocene the *Larus argentatus* group was separated into refuges in North America as well as in Europe and Asia. The eastern

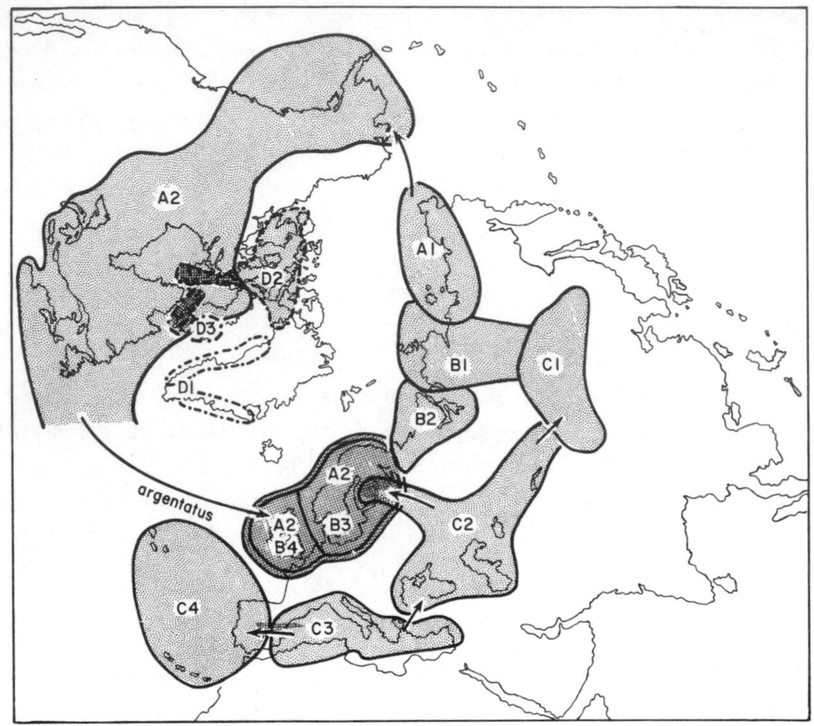

Fig. 16.6. Circular overlap in gulls of the *Larus argentatus* group. The subspecies of *A*, *B*, and *C* evolved in Pleistocene refuges; *D* evolved in North America into a separate species (*L. glaucoides*). When *A* expanded, post-Pleistocene, probably from a north Pacific refuge (? Yukon, ? Alaska, ? Kamtchatka), it spread across all of North America and into western Europe (*argentatus*). Here it became sympatric with *fuscus* (*B*3, *B*4), the westernmost of a chain of Eurasian populations.

American isolates became the species *L. glaucoides* and *L. thayeri*, which are characterized by the purplish-red color of the orbital rings (not orange as in sympatric *argentatus*). When their range was invaded by *L. argentatus*, during the retreat of the ice, hybridization did not occur, but habitat exclusion did. *L. glaucoides* and *thayeri* are essentially cliff nesters along the seashore, while *L. argentatus* more often nests inland and at less precipitous locations. Where the two species nevertheless coexist in the same colonies, they do not interbreed owing to ethological isolating mechanisms (Smith 1966).

It is immaterial whether these instances of circular overlap present themselves in the simplest and most diagrammatic manner or whether they are somewhat more complicated, as indicated by the recent reinvestigation of *Larus argentatus*. In either case the process of geographic speciation can be followed step by step. A more dramatic demonstration of geographic speciation than cases of circular overlap cannot be imagined (see Fig. 16.7).

Reconstruction of the history of speciation. By carefully plotting the current geographic ranges of related species and by evaluating all kinds of borderline cases (see above), it is sometimes possible, in taxonomically well-known groups, to reconstruct the probable path of speciation. This was done, for instance, by Mengel (1964) for some North American warblers, by Hall and Moreau (1970) for the birds of Africa, and by Haffer (1967, 1969) for various South American birds. It has also been done for lizards (Gorman and Atkins 1969), mammals, and marine species. (See also Chapter 18.)

The Universality of Geographic Speciation

The widespread occurrence of geographic speciation is no longer seriously questioned by anyone. The problem now has shifted to the question whether there are also other processes of speciation among sexually repro-

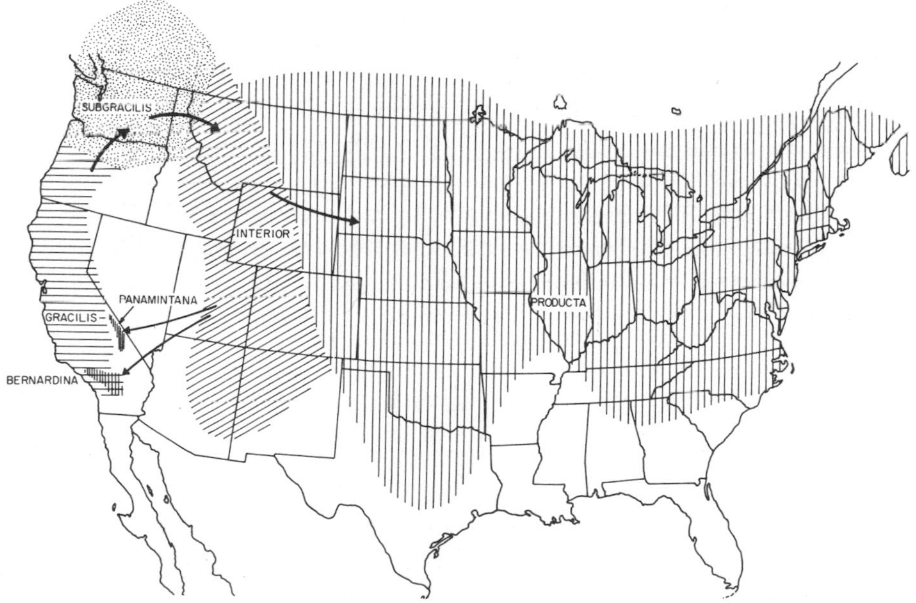

Fig. 16.7. Circular overlap in the bee *Hoplitis* (*Alcidamea*) *producta.* The subspecies *interior,* derived from *gracilis* via *subgracilis,* has twice reinvaded the range of *gracilis* across the desert barrier. The derived populations, *panamintana* and *bernardina,* now coexist with *gracilis* without interbreeding. (From Michener 1947.)

ducing animals. In Chapter 15 I examined the feasibility of alternative modes of speciation. Here I shall try a more positive approach. If it is true that species originate geographically, an abundant supply of incipient species must be present because only a fraction of all incipient species will ever move up to the rank of full species. Most habitats are saturated with species at any given time, and there is room only for as many new species as can find niches for themselves. Most isolates are ephemeral; they become extinct before they have had an opportunity to function as full species. Most ecological shifts in isolates are failures. When discussing the evolutionary role of peripheral isolates it is of cardinal importance to ask the right questions. To ask whether peripheral isolates frequently (or usually) produce new species and evolutionary novelties is not at all the same as asking whether new species and evolutionary novelties are usually produced by peripheral isolates. The answer to the first is no, to the second, yes. The reason is that peripheral isolates are produced 50 or 100 or 500 times as frequently as new species; hence most peripheral isolates do *not* evolve into new species, but *when* a new species evolves, it almost invariably evolves from a peripheral isolate.

The crucial question concerning geographical isolates is, therefore: Are there enough geographical isolates in all stages of divergence to replace extinct species and fill vacant niches and adaptive zones? This question has been answered in Chapter 13 in the affirmative. What is now needed is a systematic analysis of many families in different orders and classes of animals, along the lines of the analysis prepared by me for a few avian families (Table 16.1). In the Dicaeidae, for instance, there are 3.82 isolates per species, more than there are subspecies. The data gathered in this table indicate that there are indeed more than enough isolates available to jump into the gaps caused either by the extinction of formerly existing species

Table 16.1. Frequency of various components of species structure in four families of birds.

Family	Superspecies		Species		Isolates		
	Superspecies	Semispecies	Polytypic	Monotypic	Strong	Subspecific	Weak
Dicaeidae	6	20	37	18	11	86	113
Dicruridae	2	5	11	8	10	37	44
Paradisaeidae	6	19	27	13	1	39	78
Ptilonorhynchidae	2	2	9	7	4	8	9

or by the opening up of new niches. Keast (1961) comes to the same conclusion in his analysis of geographical isolates on the Australian continent. A study of the number of geographical isolates in other kinds of organisms is highly desirable.

To summarize: innumerable aspects of the geographic variation of species, of distributional patterns, and of semispecies give evidence for the widespread occurrence of geographic speciation. Indeed, all the known phenomena with a bearing on this subject are consistent with the theory of geographic speciation.

17 · The Genetics of Speciation

The evidence presented in the two preceding chapters supports the conclusion that speciation in sexually reproducing higher animals normally takes place by way of geographic isolation. Although naturalists had asserted this fact for nearly 100 years, until recently it remained a strictly empirical finding, never receiving a causal explanation, never leading to a genuine theory of speciation. When Moritz Wagner proclaimed in 1868 that geographic isolation was a necessary prerequisite for species formation, he failed to submit any real proof. Worse than that, some of his early explanations were so obviously absurd that they endangered his thesis as a whole and were in part responsible for its cool reception.

Before a theory of geographic speciation could become acceptable, two prerequisites were necessary: (1) a far greater understanding of genetics, particularly of the particulate nature of inheritance, had to be achieved; and (2) population thinking had to become general, replacing the formerly predominant typological thinking. As recently as the 1930's geneticists were still primarily concerned with mutation and selection, that is, with evolutionary change as such. Only after the problems posed by evolutionary change had been essentially solved were population geneticists ready to work out a genetic theory of speciation. Only then was it possible to search for the genetic reasons that necessitate geographic isolation during speciation.

The Genetic Problem of Speciation

The essence of speciation, as we now realize, is the production of two well-integrated gene complexes from a single parental one. All early attempts to explain the genetics of speciation missed this essential point, since they were concerned entirely with the problem of the origin of difference. To be sure, the differences between species are due to mutation and selection, but demonstrating that does not explain how species split.

Most of the genetic variability produced by mutation is almost immediately removed by normalizing selection. Any mutation that would initiate an incipient difference in reproductive behavior among individuals of a population would be particularly vulnerable to normalizing selection. And yet it is precisely such differences in reproductive behavior and physiology that are needed to build up isolating mechanisms. How can they be accumulated without being at once removed by normalizing selection?

The real problem of speciation is not how differences are produced but rather what enables populations to escape from the cohesion of the gene complex and establish their independent identity. No one will comprehend how formidable this problem is who does not understand the power of the cohesive forces in a coadapted gene pool. It was the object of Chapter 10 to establish this point.

It is now evident that there is only one situation in which a gene pool can be completely reconstituted genetically (with reference to a parental population) while all of its elements remain well integrated and coadapted: spatial isolation. Most students of the speciation of sexual animals from M. Wagner on realized clearly the indispensability of this condition, but they based their conclusion strictly on empirical findings. Why isolation was needed remained a puzzle until the genetics of integrated gene complexes had replaced the old "beanbag" genetics.

Gene Flow and Genetic Cohesion

To what extent and under what circumstances are the different populations of a species held together by cohesive forces? Conversely, under what conditions can this cohesion be broken? The most important question of all is: What role does gene flow play in the maintenance of genetic cohesion among populations? Is it advisable to classify populations according to the amount of gene flow to which they are exposed?

Closed and open populations. All selection experiments, whether on mice, *Drosophila,* or domestic animals, have involved *closed populations* with a negligible genetic input. Population size was far too small in all these cases for mutation to have played a major role unless it was artificially induced. The response of the selected stocks, no matter how spectacular it was, must have been largely due to recombination of the initial gene complement, a souce of variation which is bound to dry up eventually. Natural populations (except the most rigidly isolated ones), in contradistinction, are *open populations* with a steady input through gene immigration. This difference in genetic input causes a number of fundamental differences between closed and open populations (Mayr 1955). In the open system, available genetic variation is not only infinitely greater but also of a different kind. The large and continuous influx of alien genes into every local population, as well as the diversity of the environment in space and time, will never permit the gene complex to reach complete stability. The response to selection in an open system is very different from that in a closed one. It is therefore not admissible to apply automatically the findings made on closed laboratory populations to natural populations. This limitation must be kept in mind when one wants to construct models of species structure.

In all widespread, successful species of relatively mobile sexual organisms, there seems to be sufficient gene flow to maintain great similarity in the gene pools of all local populations. Population geneticists, who work all their lives with closed populations in which all genetic input is due to mutation, tend to underestimate the magnitude of genetic input in open populations. To be sure, it is immaterial for certain aspects of evolution whether mutation or immigration is responsible for new genes in a population. Yet it would be a great mistake to lump these two sources of variation together in calculations of their effect, because they are of totally different orders of magnitude. I estimate that genetic change per generation due to mutation in a local population rarely exceeds 10^{-5} per locus, while the exchange due to normal gene flow is at least as high as 10^{-3} to 10^{-4} for open populations that are normal components of species. With an effective local breeding population often as low as 2×10 and usually not higher than 3×10^2, this difference in order of magnitude becomes of vital importance. For example, if one assumes a population of 200 individuals, each with 100,000 loci mutating at the rate of 1 in 10,000, a total of 2,000 mutants for the 20 million loci would almost certainly be a maximum figure. Dispersal, however, brings new individuals into the population; they might constitute 40 percent of it, that is, 8 million genes, of which perhaps 200,000 might be new for the population. If one assumes that there will be the same amount of duplication among mutated and immigrant genes, gene replacement by immigration would be 100 times that by mutation in this deme.

The natural landscape is only rarely so uniform that a species (or a part of it) consists of a single contiguous population. Far more frequently, indeed normally, a species is subdivided into numerous partially isolated local populations (Fig. 17.1). Consequently, it is of paramount importance to determine the order of magnitude of the partial isolation. Where the gaps between suitable habitats are smaller than the normal dispersal range of

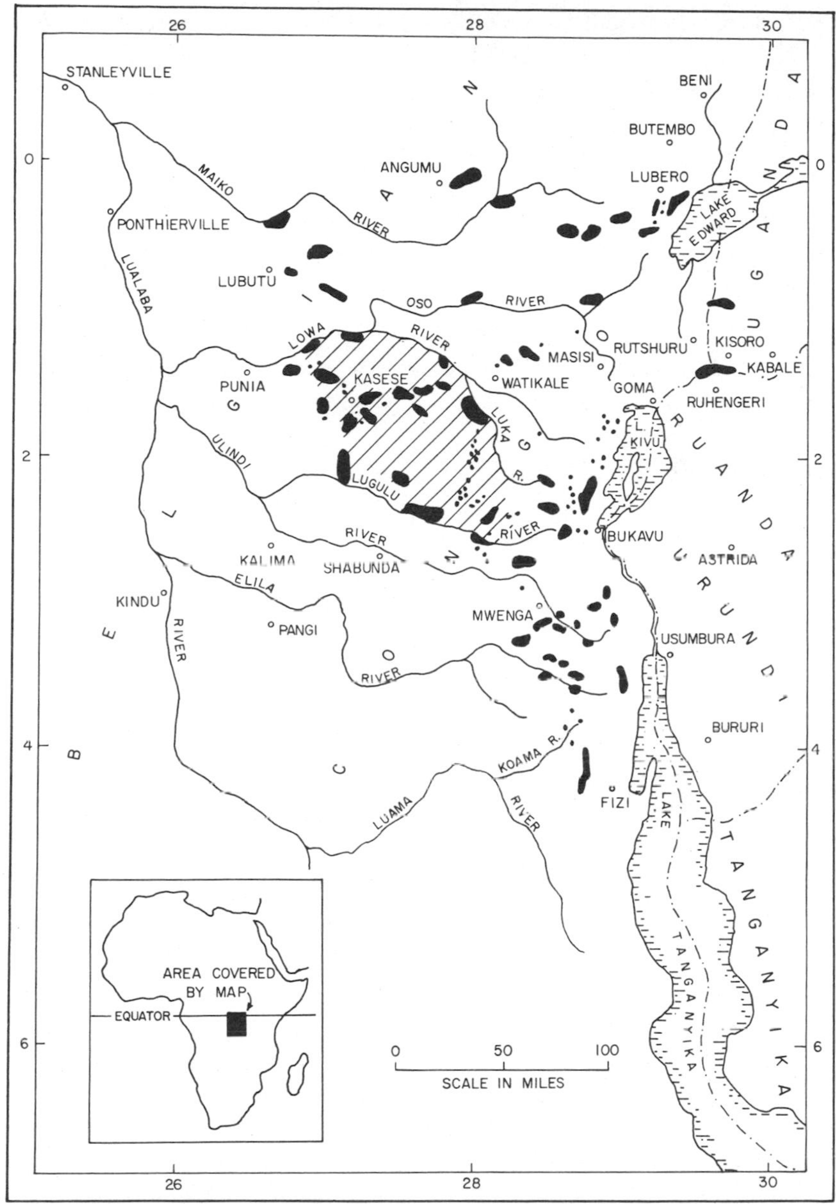

Fig. 17.1. Actual distribution of the Mountain Gorilla in East Africa. Each of the black areas indicates the location and approximate shape of sixty gorilla areas ranging in size from about 10 to about 100 square miles each. The hatching marks a central region of fairly continuous but sparse population. (From Emlen and Schaller 1960.)

individuals, there will be about as much gene flow as if the range were continuous. In view of the steady selection in favor of genes that coadapt easily with immigrant genes, there may well be nearly as much cohesion in this type of a partially isolated system as there is in a panmictic one. The genuinely sharp break is not between the panmictic and the partially isolated system, but between the partially and the virtually fully isolated system. The importance of complete isolation becomes evident as soon as the extensive epistatic effects of genes are properly taken into consideration. As a consequence, a population cannot change drastically so long as it is exposed to the normalizing effects of gene flow.

How efficient is gene flow? One arrives at rather astronomical figures when one tries to calculate the time it would take for a gene to percolate from one end of the range of a widespread species to the other; the figures are especially high if the organisms in question have poor dispersal facilities. It is evident that gene flow alone is not enough to overcome entirely the local effects of mutation and selection. What, then, is responsible for the uniformity of species? I advanced the thesis (1963:523) that—in addition to gene flow—the cohesion of a species is due to the fact that all of its populations share the same homeostatic systems and that this species-wide system of canalization provides great stability: "It is a limited number of highly successful epigenetic systems and homeostatic devices which is responsible for the severe restraints on genetic and phenotypic change" displayed by every species. The total epigenotype governs the range of normalizing selection, which will maintain such epigenetic systems in the face of the centrifugal tendencies of local gene pools. Recent studies (for instance, Lewontin 1967a) have provided evidence for the existence of such genotypic systems.

The adjustment of local populations to the local environment through race or ecotype formation has been stressed so much in previous chapters (Chapters 9, 11, 12), that it would now seem important to stress the basic uniformity of most continuously distributed species. A good example of this is the New Guinea kingfisher *Tanysiptera galatea* (see Fig. 16.4), which displays no significant geographic variation in the vast area of that island, with its strong climatic contrasts. Yet each of the adjacent islands inhabited by this species has a markedly differentiated race even though these islands are in the same climatic zone as the neighboring mainland. Every taxonomist can cite dozens of similar cases. There is, for instance, the butterfly *Maniola jurtina*, which has been studied so intensely by E. B. Ford and his associates:

> One of the most striking features of the *M. jurtina* females is their remarkable uniformity across most of southern England [except Cornwall]. This area includes some of the greatest variations in temperature, rainfall, and geology to be found in Britain. Evidence from Cornwall and the Scillies shows that the spotting is capable of marked variation. The fact that it is so stable elsewhere [= across most of southern England] indicates not only that natural selection is holding it at an optimum value but also that the species is in some way insensitive to environmental variation in this part of its range (Dowdeswell 1956).

The fact, taken for granted by every taxonomist, that he can identify individuals of a species (unless its range is dissected by geographical isolation) regardless of where in the range of the species they may come from is further illustration of this phenomenon. Physiologists and embryologists, likewise, have published evidence indicating a remarkable uniformity of physiological constants throughout the ranges of most species. The essential genetic unity of species cannot be doubted.

The Species Border

The range of a species is delimited by a line beyond which the selective factors of the environment prevent successful reproduction. This line, called the *species border*, is one of the aspects of the population structure of species that can be understood only by taking gene flow into consideration. Single individuals may appear annually in considerable numbers beyond this line, yet fail to establish themselves permanently. Even if they succeed in founding new colonies, these will sooner or later be eliminated in an adverse season. As a result, the species border, though fluctuating back and forth, remains a dynamically stable line. The species border is one of the most interesting phenomena of evolution and ecology, yet as a scientific problem it has been almost totally ignored, except by Finnish authors, who have studied shifts of the species border as correlates of changing climatic conditions (Chapter 18). In the border region there is a never-ending race between reproductive capacity and mortality due to adverse conditions. Population density is far below the saturation point and the border region is a place in the area of a species where density-dependent factors are likely to be of minor, if not negligible, importance.

The essential stability of the species border, on which the annual and long-term fluctuations are superimposed, would seem to contradict our belief in the power of natural selection. One would expect a few individuals to survive in a zone immediately outside the species border and to form a new local population that would gradually become better adapted under the continuous shaping influence of local selection. One would expect the species range to grow by a process of annual accretion, like the rings of a tree. That this does not happen is particularly astonishing in the frequent cases where conditions beyond the borderline differ only slightly from conditions inside the species border and where no drastic barriers prevent expansion.

The solution to this puzzle is probably that the process of local adaptation by selection is annually disrupted by the immigration of alien genes and gene combinations from the interior of the species range (Mayr 1954). This influx prevents the selection of a new stabilized gene complex adapted to the conditions of the border region. Presumably, the border populations barely maintain themselves, and the new colonists beyond the species border (in mobile species such as birds and insects) come from farther inside the species range, where conditions permit a greater surplus of individuals and the resulting increased population density stimulates emigration of individuals whose gene complex is not adapted for the conditions of the border region.

The Genetic Reconstitution of an Isolated Population

The argument in the preceding section was essentially negative. It demonstrated that contiguous populations of a species are held together by such close ties of genetic cohesion that one can scarcely conceive of this essentially single gene pool being divided into two. I shall now take up the other half of the argument and show what genetic events take place in spatially isolated populations and how these permit the formation of isolating mechanisms while leaving the genetic integration of the gene pool at all times undisturbed.

The isolated population. Let us study the genetic history of a newly founded population that is spatially isolated from the parental population from which it branched off. We shall at first make two simplifying assumptions: (1) that the environments of the two populations are identical, and (2) that the new population is at the beginning completely identical genetically with the parental population. Yet, even so, as soon as they are separated, the two populations will drift apart in their genetic contents, for a number of reasons. The probability is nil that the same mutations will occur in the two populations in the same sequence. Each incorporated mutation changes the genetic background of the population and thus affects the selective value of all subsequent mutations. Furthermore, recombination will produce different genotypes in the two gene pools and thus, since the same gene may have different selective values in different genotypes, will lead to a gradual shifting of gene frequencies. A third factor leading to the divergence of the gene pools is "genetic indeterminacy" (Chapter 8). If several gene combinations have equal selective values (with respect to a given selection pressure), pure chance or some irrelevant pleiotropic effect may decide which of them becomes established in a given gene pool. The changes in gene frequency due to "genetic indeterminacy" will again not be the same in two independent populations. Each divergence of the two gene pools increases the difference in the genetic background of all the genes of the two populations and will thus tend to set up new selection pressures. The drifting apart is thus evidently an accelerating process.

The rate of divergence, however, is even greater than can be accounted for by the stated factors, owing to the fact that neither of the two simplifying assumptions made above is valid. The selection pressures to which the two separated populations are exposed are not the same, since there are no two places on the face of the earth where even the physical environment is quite identical. Every completely isolated population exists in a biotic environment that is different from any other, and this shift of the biotic environment adds another powerful selection pressure. Competition, predation, and other ecological interactions are apt to be entirely different in the new environment. These local conditions exert selection pressures reenforcing the steady change of gene contents and leading to the development of numerous new adjustments.

Founder principle. The second assumption made above, that the new population is at the beginning genetically completely identical with the parental population, is likewise invalid. The founders of a new colony of a species inevitably contain only a small fraction of the total variation of the parental species (the founder principle; see Chapter 8). All subsequent evolution will proceed from this limited original endowment. The importance of this restriction is demonstrated by recent selection experiments in which several parallel lines were exposed to the same selection pressure. Almost invariably the end results were different in the different lines. The smaller the starting populations, the greater the degree of indeterminacy. Ten experimental populations, each descended from only 20 founders, diverged far more from each other than ten other populations founded by 4000 individuals each, all 20 populations having been derived from the same parental population (Fig. 17.2).

The founder principle is an important concept only in the framework of population thinking. For the essentialist any individual of a species has

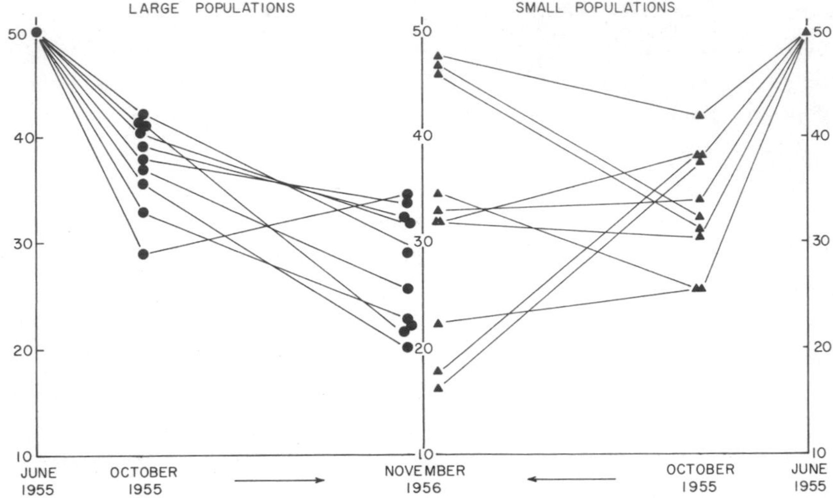

Fig. 17.2. The frequency (percent; vertical scales) of *PP* chromosomes in 20 replicate experimental populations of mixed geographic origin (Texas by California). The populations that have gone through a bottleneck of small population size show far greater variance after 17 months than the continuously large populations. (From Dobzhansky and Pavlovsky 1957.)

"the essential characters" of the species, and if such an individual establishes a new population by colonizing a previously unoccupied area, its descendants will be typical representatives of the species. For the populationist no two individuals are the same and no individual is ever a typical representative of the species. The particular sample that a founder population represents is inevitably different from the mean values of the parental population. It constitutes a genetically unique population, even though the probability is high that the most common genes of the parental population will be represented.

How far a founder population may diverge genetically is well illustrated by the isolated Bogotá (Colombia) population of *Drosophila pseudoobscura.* All over Central America (south to Guatemala) and North America (north to British Columbia) this species shows only modest geographic variation in a set of investigated enzyme genes. The Bogotá population, however, has a strikingly different composition in the frequency of the enzyme alleles (Table 17.1).

It must be remembered that each genotype is a discrete genetic system, constituting an experiment in coadaptation. Occasionally such a genotype is uniquely adapted to be successful in its evolutionary experiment. If this exceptional genotype is a member of a large population, it is bound to be broken up (in the next generation) by recombination. In a small founder population, however, this coadapted system has a good chance of being perpetuated in its descendants and of giving rise to an evolutionary novelty (Chapter 19).

The Chances of Success of a Founder Population

The founders carry such a small reservoir of genetic diversity with them that the population founded by them is highly vulnerable to the dangers of inbreeding (homozygosity). The situation is aggravated in most cases by the ecological uniformity of the insular environment and the resulting one-sidedness of selection. These phenomena are principally responsible for the frequent extinction of island populations. Indeed, extinction under adverse conditions is the fate of most peripheral isolates. The smaller a population the more vulnerable to extinction it seems to be (Fig. 17.3; Mayr 1965a).

Awareness of the frequency of extinction among island species and of the severe inbreeding depression observed by animal breeders might induce one to take a dim view of the prospects of founder populations. However, one must be cautious when generalizing from highly artificial selection experiments. Selection in nature, even in the smallest population, is primarily for overall fitness. The same is true of so-called "unselected" inbred lines in the laboratory. On the whole, loss of genetic variance through inbreeding occurs far more slowly than one might expect.

Table 17.1. Geographic variation of two enzyme genes (*Pterine-8* and *Xanthine dehydrogenase*) on chromosome II of *Drosophila pseudoobscura*. The difference in the frequency of the alleles between the three North American and the isolated South American locality is striking (from Prakash et al. 1969).

Allele	California	Mesa Verde	Texas	Bogotá
		Pterine-8		
.80	.014	.009	.011	.870
.81	.472	.410	.441	.100
.83	.514	.576	.512	.030
.85	—	.005	.035	—
		Xanthine dehydrogenase		
.90	.053	.016	.018	—
.92	.074	.073	.036	—
.99	.263	.300	.232	—
1.00	.600	.580	.661	1.00
1.02	.010	.032	.053	—

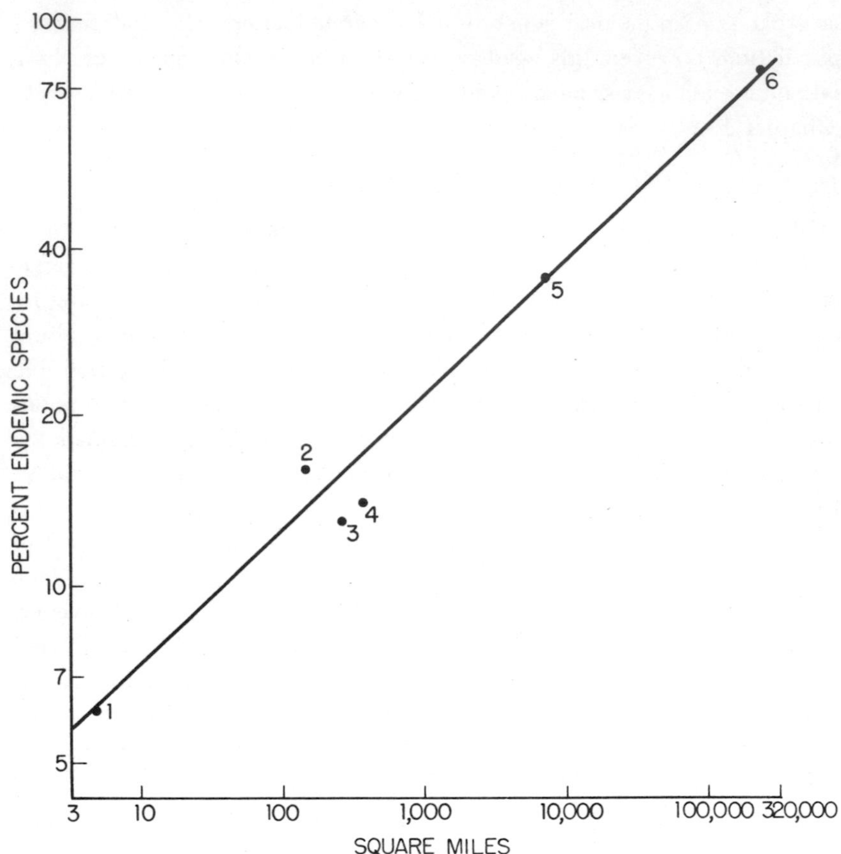

Fig. 17.3. Double logarithmic plotting of area against percentage of endemic species of birds on single well-isolated islands. The smaller the island the more frequently the endemic species become extinct, the more rapid the faunal turnover. 1 = Lord Howe, 2 = Ponape, 3 = Rennell, 4 = Chatham, 5 = New Caledonia, 6 = Madagascar.

There is abundant evidence in the literature for an occasional phenomenally successful founder population (Mayr 1963:530). Most of the mammals and birds successfully introduced into North America, Australia, and New Zealand, and nearly all accidentally introduced insects, were the offspring of a handful of individuals (Elton 1958). Inbreeders are, of course, vastly more successful colonizers than outbreeders.

Sometimes an exceedingly small population maintains itself successfully over a long series of generations. Examples are fish in desert springs and a number of protected mammals and birds that, like the American bison and the European ibex, successfully overcame an extreme reduction in numbers. The millions of golden hamsters (*Mesocricetus auratus*) in the laboratories around the world are supposedly all descendants of a single pregnant female. The establishment of highly successful colonies by single founders is not only feasible, but quite likely; it seems to be the normal method of spreading in many species of animals and plants (Baker and Stebbins 1965).

These observations also have a bearing on the problem of the genetic composition of rare and localized species. One would expect great homozygosity owing to inbreeding. The available information is slight, but indicates that this assumption is not necessarily true. In *Keyacris scurra,* an Australian grasshopper reduced to isolated populations in cemeteries, many of them quite small, White (1957b) found that only a few of the usually polymorphic karyotypes of this species had become homozygous. Clearly, drift had been unable to override an obvious but not yet analyzed selective advantage of the heterozygous condition. There are numerous records in the genetic literature of the tenacity with which genetic variability is maintained in small laboratory populations through scores of generations of inbreeding (Chapter 9).

The Genetic and Biotic Environment of the Founder Population

The founder population is differentiated from the parental population not only by the drastic reduction of the diversity of its gene pool, but also by its exposure to a totally new constellation of environmental factors, biotic as well as genetic. The most important of these is the sudden conversion from an open to a closed—and at that to a *small closed*—population. It is the suddenness and completeness of this shift that is decisive. The new population is at once completely emancipated from the parental population. In an open population there is a steady and rather high input of alien genes. "Little or no thought [has been given in the past] to the effect of these alien genes on the relative viability of the genes of the gene complex into which they were introduced. It appears probable that the frequent introduction of alien genes into a gene pool will lead to the selection of such 'native' genes as are tolerant to combination with these alien genes, that is, which produce viable heterozygotes with a great assortment of alien genes or gene combinations" (Mayr 1954:162) (Fig. 17.4). I have referred to such genes as "jack-of-all-trades" genes or "good mixers." The less inflow of alien genes in a population, the lower the special selective advantage of the good mixers.

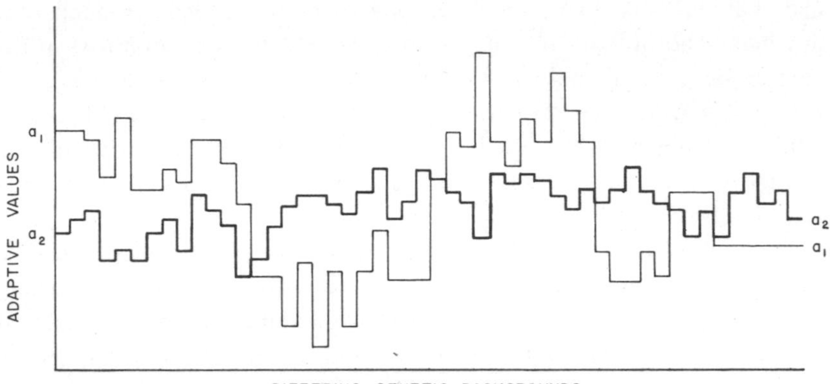

Fig. 17.4. Diagrammatic representation of the changing adaptive value of genes a_1 and a_2 on different genetic backgrounds. Gene a_1 is very good on some backgrounds and very poor on others; gene a_2 fluctuates only slightly around the mean. (From Mayr 1954.)

The effect of the increased homozygosity at some loci on other loci of the founder population is, perhaps, more important.

> As a consequence of their increased frequency in the founder population, homozygotes will be much more exposed to selection and those genes will be favored which are specially viable in homozygous condition. Thus, the "soloist" is now the favorite, rather than the "good mixer." We come thus to the important conclusion that *the mere change of the genetic environment* [and particularly the change from an open to a closed population] *may change the selective value of a gene very considerably.* This change . . . is the most drastic genetic change (except for polyploidy and hybridization) which may occur in a natural population, since it may affect all loci at once. Indeed, it may have the character of a veritable "genetic revolution." Furthermore, this "genetic revolution," released by the isolation of the founder population, may well have the character of a chain reaction. Changes in any locus will in turn affect the selective values of many other loci, until finally the system has reached a new state of equilibrium (Mayr 1954:169).

There is some experimental evidence which supports these conclusions. In several recent selection experiments, it was found—contrary to classical assumptions—that more homozygous stocks sometimes respond phenotypically more strongly to selection pressures (for example, for more bristles in *Drosophila*) than genetically more variable stocks. The genetically more variable populations seem to be richer in balanced systems of regulating genes and epistatic balances and thus better equipped to resist one-sided selection pressures. The depletion of such balancing systems in founder populations may facilitate their phenotypic response to new selection pressures.

There are other factors that may favor a rapid genetic turnover in populations which pass through the bottleneck of reduced population size. One is that the rate at which a gene is replaced by its allele depends on its initial frequency. In large populations the frequency of a new favorable gene will be very low, by definition, and the rate of substitution correspondingly slow (Haldane 1957). In a founder population, however, a gene that is elsewhere rare may start with a relatively high frequency and thus be able to replace its allele very quickly. This may be the most important reason for rapid evolution in small speciating populations.

Haldane (1956) has called attention to an important difference between central and peripherally isolated populations: the role of density-independent mortality. In the central populations of a species, the physical as well as the biotic environment is optimal, and much, if not most, of the mortality is somehow connected with the high population density. In peripheral and peripherally isolated populations, conditions are usually near the minimum for the species. In bad years the populations will be wiped out, or nearly wiped out; in good years they may build up to large numbers (because of the scarcity of species-specific predators and pathogens). There will thus be strong population fluctuations and these will favor genetic turnover. The strength of density-dependent factors in the central part of the species range damps such fluctuations (see Chapter 13).

Consequences of the Genetic Revolution

During a genetic revolution, the speciating population passes from one well-integrated and stable condition through a highly unstable period to another period of balanced integration. Various genetic phenomena characterize the passing through the bottleneck. Most conspicuous among these is a great loss of genetic variability (Fig. 17.5). For this loss there are a number of reasons: (1) the founders represent only a fraction of the variability of the species; (2) owing to inbreeding, more recessives will become homozygous and thus be exposed to selection; (3) owing to the reduced population size, there will be changes in the selective value of alleles and certain alleles will be eliminated (loss of "good mixers"); (4) during the reconstitution of the epigenotypes, many genes will lose the advantage of being part of a balanced system and will be selected against; (5) as long as the new population is small it may lose additional genes through errors of sampling.

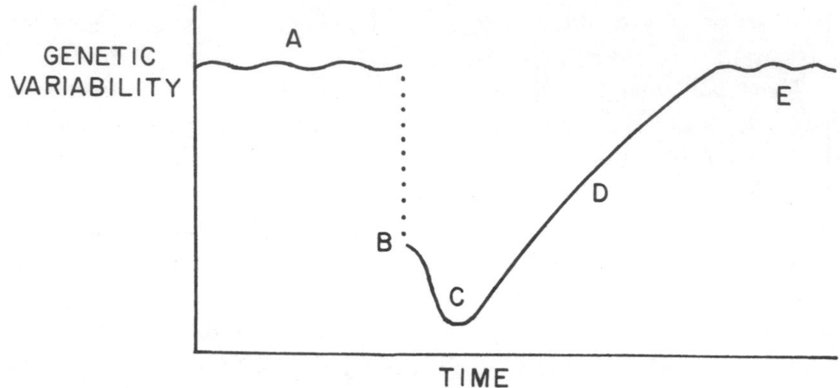

Fig. 17.5. Loss and gradual recovery of genetic variation in a founder population. The founders (*B*) have only a fraction of the genetic variation of the parental population (*A*) and further genes are lost during the ensuing genetic revolution (*B* to *C*). Variation is gradually recovered (*D*) if the population can find a suitable niche, until a new level (*E*) is reached. (From Mayr 1954.)

Not all of this loss of genetic variability is deleterious. It reduces the genetic load quite drastically and gives the surviving population a "clean start." It has been found in natural as well as experimental populations of *Drosophila* (several species) that small populations carry far fewer genes that are deleterious in homozygous condition than large ones. Furthermore, the loss of balancing systems often permits a stronger response to selection pressure.

A genuine genetic revolution is characterized by a breakdown of genetic homeostasis through a loss or a reconstitution of previously existing balancing systems. The population will go through a labile state. The situation is made still more acute by the fact (discussed below) that the population has to cope with new selection pressures owing to the changed physical and biotic environment in the isolate. Those populations that succeed in surviving the genetic revolution, presumably very few, will enter a new state, one characterized by the renewed accumulation of genetic variability and the acquisition of new and usually very different balancing systems.

Chromosomal Rearrangements and Speciation

Closely related species often differ more conspicuously in their karyotype than in their morphology. Among aspects of the karyotype that differ are chromosome number; the number of metacentric or acrocentric chromosomes; the presence and kind of paracentric or pericentric inversions, or of supernumerary chromosomes; and just about every other aspect of chromosomal evolution. It has, therefore, been held by most cytogeneticists that chromosomal reconstruction ("chromosomal mutation") is an important and indeed indispensable component of the speciation process. This belief is based on two assumptions: (1) that the degree of difference displayed by two species requires a speciation process of such drastic dimensions that only chromosomal mutation can qualify, and (2) that reproductive isolation between two species cannot be achieved without chromosomal reorganization. Although it is now known that both of these assumptions are invalid, there is still need for explaining the frequency with which speciation seems to be accompanied by chromosomal rearrangements. The analysis may be facilitated by showing in a diagram (Fig. 17.6) the three possible relationships between the occurrence of new chromosomal rearrangements and the acquisition of reproductive isolation.

(*1*) *Speciation without chromosomal rearrangement.* Numerous cases in various groups of organisms are now known of well-defined and reproductively isolated species that agree completely in their chromosome structure and differ only in their gene contents. Such cases show that speciation can be completed by genic differentiation without a structural repatterning of the chromosomes. To prove this conclusively is difficult in the case of ordinary chromosomes, such as those of vertebrates. The evidence supplied by the giant salivary gland chromosomes of certain Diptera with highly specific banding patterns is, however, irrefutable. In a study of 69 related Hawaiian species of *Drosophila,* Carson and his coworkers (1967, 1970) found 10 groups of species in each of which several related species had the identical highly complex banding pattern in their salivary chromosomes.

		Chromosomal Mutation	
		No	Yes
Reproductive Isolation	No		2
	Yes	1	3

Fig. 17.6. The three possible relationships between chromosomal rearrangements and the acquisition of reproductive isolation.

In one case three species with structurally identical chromosomes coexist in the same habitat without interbreeding. Clearly in these cases the isolating mechanisms were built up without the help of chromosomal mutations. Carson refers to species with an identical sequence of bands on the chromosomes as *homosequential* species. The extreme similarity of the karyotype in various groups of animals, for instance, in certain families of artiodactyls among the mammals and in certain groups of insects, indicates that much speciation and adaptive radiation without restructuring of the chromosomes may also occur in cytologically less favorable groups.

(2) *Chromosomal rearrangement without speciation.* Much, if not most, chromosomal rearrangement, unless deleterious, leads to chromosomal polymorphism rather than to the development of isolating mechanisms. The paracentric inversions of *Drosophila* (Chapter 9) are a well-known example; for others see White (1954, 1969) and other textbooks of cytology. Yet, each species has its own species-specific polymorphism, and only very rarely do even the most closely related species share the same chromosomal polymorphism. This fact underlines the drastic nature of the chromosomal reconstruction during much of speciation.

These two classes of phenomena (1 and 2) prove that there is no necessary correlation between chromosome mutations and speciation, since either can occur without the other. Yet, closely related species are far too often differentiated by a major chromosomal rearrangement to permit considering this fact a pure coincidence.

(3) *Speciation coinciding with a chromosome mutation.* It is obvious that a reconstruction of the karyotype during speciation would not occur so frequently if it did not have a selective advantage. What could this advantage be? It is now becoming evident that there are two advantages, two separate phenomena contributing to the success of chromosomal speciation: (a) chromosomal mutations have the potential to serve as (or contribute to) isolating mechanisms, and (b) the locking up and protection of a particularly favorable gene complement through a chromosomal mutation may create a new supergene, as Wallace (1959a) was perhaps the first to see clearly. Both of these components of chromosomal speciation can subsequently be improved by natural selection, either during a period of segregation in a geographic isolate or during subsequent parapatric speciation (see below) or by both processes.

(*a*) *Chromosomal mutations as potential new isolating mechanisms.* Any change in the structure of the chromosomes is called a chromosome mutation, whether it is an inversion, translocation, duplication, or any other change in the linear sequence of the genes or in the mechanics of the chromosomes (for instance, spindle attachment). For details see textbooks of cytology or recent reviews (for example, White 1969; Benirschke 1969).

Chromosomal mutations (mostly inversions) are estimated to occur at a rate of 1 in 1000. Most of these are sufficiently deleterious to be eliminated at once, that is, before the mutation's carrier can reach reproductive age. Others are capable of giving rise to a system of balanced polymorphism and they will be retained in the population. In addition, there is a third class of rearrangements that appears to reduce the fecundity of the heterozygotes to some extent. It is the incorporation of these chromosome mutations into the gene pool of species and the role they play in speciation that remains a controversial issue to this day. Heterozygotes for the kind of chromosome mutations that seem important in speciation usually encounter the following difficulties during meiosis: (1) meiotic asynapsis (partial failure of chromosome pairing), or (2) malorientation of multivalents at the first metaphase, or both. Both of these difficulties may lead to production of gametes carrying chromosomal deletions or duplications or broken or acentric chromosomes and thus lead to a significant reduction in the fecundity of male hybrids.

Chromosome mutation is an entirely different phenomenon from gene mutation. First of all, it almost always has no effect on the visible phenotype, and is therefore difficult to detect. Second, heterozygotes are not shielded by dominance; indeed the heterozygotes with their meiotic difficulties are the prime target of selection (John and Lewis 1969). Finally, chromosome mutation is not (or at least not primarily) a change in the DNA program, but in the linear sequence (or duplications or deletions).

Breakage and fusion of chromosomes are prominent in chromosomal mutation, but there is still considerable disagreement among experts about the cytological details of these processes. One's interpretation of certain sequences in chromosome phylogeny will depend on whether one assumes that a metacentric chromosome can simply divide by "fission" into two acrocentric ones or one demands that a new centromere has to be acquired by translocation, whether one feels that broken chromosomes can simply "heal" their open ends or one demands that they must acquire a new "telomere" (end piece), also by translocation or inversion. See White (1969) and John and Lewis (1969) for a discussion of some of the opposing views.

The most frequent or at least most conspicuous chromosomal change is the *fusion* of two acrocentric chromosomes into a single metacentric chromosome. The result is a reduction in chromosome number. A trend toward such a reduction is widespread among animal taxa. For instance, the most primitive isopods have 28 haploid chromosomes and this number is independently reduced through fusion to the lower number of 8 in several unconnected lines. It is now clear that most differences in chromosome numbers between closely related species of animals is the result of such fusions.

If only fusions occurred during evolution, soon all species would have only a single pair of chromosomes. Obviously some opposing process must also occur that leads to an increase in chromosome number. The seemingly simplest such process would be the exact reverse of fusion, that is a *fission* (or disassociation) of a metacentric chromosome into two acrocentrics. That such simple fissions can occur is, however, doubted by most cytologists. Yet increases in the chromosome number (not due to polyploidy), whatever the mechanism, are frequent in animal evolution and may well play an important role in speciation.

There seem to be considerable differences in the mechanical aspects (for example, spindle attachment, spindle orientation) of the chromosomes among different groups of animals. Centric fusion, for instance, permits in some organisms the development of perfectly viable polymorphisms, as in the mollusk *Thais*, the shrew *Sorex*, and certain beetles and mantids. On the other hand, no species of *Drosophila* or morabine grasshopper is polymorphic for a fusion. In these groups fusion occurs only in association with speciation.

(*b*) *Chromosomal mutation and the production of new supergenes*. The cyto-mechanic aspects of chromosome mutation have perhaps been overemphasized in the past. If mechanical difficulties during chromosome pairing were the major selection pressure against chromosome mutations, one would expect such mutations to accumulate in species without meiosis. This inference, however, is not substantiated by a study of parthenogenetic species. They do not seem to have a larger number of chromosomal rearrangements than sexual species. Furthermore, the mere fact that so many populations pass successfully through the heterozygous bottleneck of a "deleterious" chromosomal mutation likewise militates against too exclusive a stress of the mechanical aspects. It now appears that gene contents rather than mechanical qualities determine in many cases whether or not a new gene arrangement can establish itself in a population.

Most structural rearrangements of chromosomes inhibit or prevent crossing over in heterozygotes. A new gene arrangement may "lock up" a coadapted gene sequence and, by protecting it from crossing over, create a new supergene (Mayr 1963:536). This eventuality is particularly important in peripheral populations where only an unusual constellation of genes can expect to be successful. Wallace (1959a) pointed out that bearers of such protected chromosomes or chromosome segments, being members of peripheral populations, are better adapted for the marginal environment of the species border than are genotypes from the center of the species range. To him, the most important aspect of the chromosomal reorganization is the protection from disruptive recombination that it affords certain new supergenes. Chromosomal mutation, thus, is an instrument of ecotypic adaptation.

It is evident that these two aspects of chromosomal mutation—production of mechanical incompatibilities and development of new supergenes—reinforce each other. There will be a steady selection for an improvement of the adaptation of the supergene and this will tend to produce an increase in genic heterozygote inferiority. This, in turn, will strengthen the effectiveness of the cyto-mechanic isolating mechanisms.

The reason chromosomal speciation is so controversial is the temporary nature of the transition stage. The duration of the stage in which the original chromosome arrangement and the new rearrangement coexist in the population is too short to be directly observed (except in rare cases). It is so short because the coexistence of both chromosomal types is discriminated against by the lowered fitness of the heterozygous combinations, in contrast to a balanced polymorphism maintained by heterozygote superiority. Interpretation must rely on inference, the best clues being furnished by the distribution pattern of species and incipient species that differ from each other in chromosomal mutations. These patterns suggest that such speciation must occur somewhere at the periphery of the range of the parental species. A study of the chromosomal situation in peripheral populations is, therefore, a prerequisite for a full understanding of chromosomal speciation.

Chromosomal Variation in Peripheral Populations

In all species with polymorphism for chromosome arrangements there is a tendency toward a reduction of this polymorphism in the peripheral populations, some of which may be entirely monomorphic. This chromosomal and genetic uniformity is selected for primarily because only a very small assortment of genotypes is able to cope with the marginal conditions of the species border (see above). As White (1959) has pointed out, however, reduction in the number of heterozygous balances yields a further advantage: it reduces genetic homeostasis and evolutionary inertia in these peripheral populations. They are far better capable of responding to new selection pressures, and consequently, to new evolutionary opportunities than populations from the "dead heart" of the species. They are better situated to utilize new gene combinations that are generated during the genetic revolution than are populations in which the genes are tightly knitted together by numerous balancing mechanisms. Finally, being essentially monomorphic, they are in a better position to incorporate new chromosomal arrangements.

That speciating populations pass through a stage of drastically reduced polymorphism is also supported by the observation that closely related, chromosomally polymorphic species, have only little of this polymorphism in common. For example, the two sibling species *Drosophila pseudoobscura* and *D. persimilis,* both of which are highly polymorphic for inversions of the third chromosome, have only a single inversion in common. The same has been found for many other closely related species of *Drosophila,* for instance, those in the *virilis, repleta,* and *willistoni* (except *paulistorum*) species groups. A karyotypic homozygosity of incipient species of *Drosophila* is further substantiated by the relative chromosomal uniformity of many of the most successful and widespread species of *Drosophila* (Carson 1965). It is apparently this chromosomal monomorphism that has permitted them to acquire a genetic constitution of such flexibility and adaptability that they can feed and breed in the most diverse sites in many climatic zones and geographic regions. *Drosophila simulans, D. melanogaster, D. immigrans,* and *D. m. mercatorum* fit into this category.

The Population Aspects of Chromosomal Speciation

Two kinds of karyotype changes can be distinguished (John and Lewis 1966) that—from the point of view of chromosomal variability of populations—have quite different rates of occurrence and evolutionary potential (neglecting in this context all strictly deleterious chromosome mutations): (1) chromosome mutations, such as paracentric inversions in *Drosophila,* that give rise to polymorphism, thereby demonstrating that they are compatible and perhaps even heterotic with the parent arrangement; and (2) more drastic chromosomal mutations, such as translocations, fusions, fissions, and the like, which usually occur (or at least are successful) only during speciation.

Chromosomal speciation, mediated by the second of these kinds of karyotype changes, apparently occurs only under special circumstances.

This conclusion is indicated by the following facts: most chromosomal sibling species are parapatric; a definite inferiority of heterozygotes is found in zones of secondary hybridization between such species; and intermediate populations (except those produced by secondary hybridization) are absent. The special circumstances required for chromosomal speciation are provided in peripherally isolated populations. Whether established as founder populations or consisting of relict populations isolated owing to climatic fluctuations, such populations can undergo drastic chromosomal reorganization for several reasons: (1) their small population size, which favors errors of sampling; (2) the environmental pressures, which favor unusual genotypes in these populations; and (3) the reduction or cessation of gene flow, which permits an undisturbed reorganization of the isolated gene pool. All these conditions favor the "locking up" of new gene constellations—new supergenes—which are especially favorable in the marginal environment.

It is thus an unusual combination of factors—chromosomal monomorphism, small population size, a rigorous environment, and temporary shielding from gene flow, possibly accompanied by a genetic revolution—that permits a population to pass through a temporary stage of heterozygosity for two somewhat incompatible gene arrangements. The amount of potential incompatibility between the two chromosomal conditions is usually sufficiently great to require a very rapid transition. Therefore, it would seem altogether improbable that this could happen in a large continuous population. Possibly another factor favors the occurrence of chromosomal speciation in marginal populations. Mortality produced by the adversity of the marginal environment takes a very high toll. Should a new rearrangement produce a large number of deleterious segregation products, they may not be fatal for its establishment if the deleterious zygotes can be "charged" to this inevitable high mortality. If a new gene arrangement in homozygous condition is of high selective value, it has a good chance of becoming established (helped by errors of sampling), even if only 2 among 100 or 1000 offspring survive. Such a low survival rate is common among animal and plant species.

As soon as the new gene arrangement is established in homozygous condition and its new supergene (or genes) has become coadapted with the residual genotype, in other words, as soon as the incipient new species has consolidated its new genotype, it can take advantage of its special adaptive potential and begin to expand into unoccupied territory until a new species border is established along the line where its adaptive potential and the resistance of the environment are in equilibrium. Where it comes parapatrically in contact with the parental species, it will establish a narrow secondary hybrid zone (reproductive isolation will rarely be complete at the beginning). If the genotype of the new species is superior to that of the parental species, it will "roll back" the hybrid zone at the expense of the parental species.

On the basis of his studies of morabine grasshoppers in Australia, White (1968) has advanced a slightly different interpretation of chromosomal speciation, at least for that group of organisms. He envisages the chromosomal mutation as "arising at a single point in the area of occupation of a species and spreading out from there on an advancing frontier," forming a narrow zone of polymorphism. He designates this process *stasipatric speciation*. Key (1968) considers such a process improbable. If the heterozygotes are inferior, as White himself infers, it would seem rather unlikely that the new arrangement would advance, considering that it occurs in the frontier zone primarily in heterozygous condition. Instead of interpreting the belt of polymorphism (which is usually only a couple of hundred meters wide) as an advancing frontier of heterozygosity it would seem simpler to consider it a zone of secondary hybridization. In most mammalian genera in which parapatric chromosomal species occur, such belts of secondary hybridization are even narrower, if not altogether absent. In the case of the morabine grasshoppers, an interpretation of past speciation is facilitated if one assumes that the size of the populations pulsated with the fluctuations of the climate. Much remains to be done to answer the many unsolved problems raised by chromosomal speciation. With the steady refinement of the techniques of chromosomal analysis and with the growing interest in the study of the geographic variation of the karyotype, there is every reason to hope that such answers will soon be forthcoming.

Semigeographic speciation. It has often been postulated by evolutionists that genetic differences and eventually isolating mechanisms can be built up in an ecological "tension zone." The occurrence of such *semigeographic speciation* seems unlikely. There is no evidence that reproductive isolation can be acquired in a zone of primary intergradation as long as the two adjacent populations are in broad contact with each other. Gene flow and the cohesion of the gene pool of the species prevent semigeographic speciation, regardless of diversifying selection. The minor ecotypic adaptations on either side of a habitat border can be acquired and strengthened without a disturbance of the basic epigenotype of the species. For a discussion of the feasibility of such speciation see Clarke and Murray (1969).

There is, however, the possibility that speciation can be completed in belts of secondary contact. If hybrids are of sufficiently lowered fitness, there will be a steady elimination of those genes and chromosome arrangements that permit the interbreeding of the two incipient species. Concurrently there will be a steady accumulation of those components of the genotype (previously acquired in geographic isolation) that discourage interbreeding. This process of improving allopatrically acquired isolating mechanisms in a contact zone (a process corresponding to character displacement) can be called *parapatric speciation.* Key (1968) considers that the "tension zones" of morabine grasshoppers illustrate this type of speciation in its nearly final stages.

Obviously not all populations involved in secondary intergradation are in the process of completing speciation. The *Corvus corone-cornix* case (p. 221) proves this conclusively. If the compatibility of the two incipient species is high, that is, if the hybrids are only slightly inferior, there is nothing to stop the continuing process of interbreeding.

Populations with conspicuously different phenotypes meet in many species along a sharp border. Such meetings occur most often in slow-moving organisms like snails. It is still uncertain in all of these cases whether these sharp borders are zones of primary or secondary intergradation (Chapter 13). Even less certain is whether such a phenotypic escarpment can ever evolve into a species border (through semigeographic speciation). The absence of species swarms related to species (like *Cepaea nemoralis*) that are rich in phenotypic escarpments, and everything we know about the cohesion of genetic systems, argues against the possibility of semigeographic speciation with the help of purely genic differentiation.

Speciation in plants is often very similar to that in animals, likewise involving "a rapid process of speciation at the diploid level in peripheral populations" and "catastrophic selection in ecologically marginal populations" (Lewis 1962). A repatterning of the chromosomes is frequent in the founder population that gives rise to a new species.

Speciation and Continuous Ranges

Not all isolates are established by founders. Sometimes they arise through the contraction of a previously continuous species range into isolated pockets. Mutation, recombination, and selection will henceforth be different and independent in the two areas and an increasing genetic divergence is inevitable. How rapidly this will proceed and whether or not it will result in speciation are largely determined by population size. It occurs not infrequently in plants and in highly sedentary animals that the population in such an isolated pocket is drastically decimated by an environmental adversity. Such catastrophic selection (Lewis 1966) may produce the same results as a genetic revolution in a founder population. If aided by chromosomal mutations, it may lead to an almost instantaneous origin of a new species.

Yet, if two separated populations remain large throughout, they will not pass through a genetic revolution and will continue to share the same balancing systems, the same epigenotype. Normalizing selection will tend to eliminate the same deviants in both daughter gene pools, which, although now independent, will continue to act as if they were parts of a single cohesive system. No one knows how long such a "parallel cohesion" can be maintained. The case of the American and Asiatic sycamores (*Platanus*), which have failed to acquire reproductive isolation after millions of years of separation (Stebbins 1950), gives one pause. It appears that the basic epigenotype of a species, its system of developmental canalizations and feedbacks, is often so well integrated that it resists change with remarkable tenacity (Chapter 10). Isolated populations sometimes remain amazingly similar to the parental populations during long periods of complete isolation. In a large population the genetic change toward species level will progress only slowly, unless speciation is favored by environmental conditions and the genetic structure of the isolated population.

Speciation by distance. Speciation by distance is a process that seems far less well established now than it did 20 years ago. It would seem reasonable to assume that gene flow is so slow in highly sedentary organisms that it cannot compensate for the centrifugal forces of genetic changes in all populations. As a consequence one should find abundant evidence for speciation by distance in all widespread, sedentary organisms. However, such evidence is singularly missing in all species with a contiguous distribution of populations. Not even the cases of circular overlap (Chapter 16), often cited as evidence, are conclusive. There are major gaps in nearly all of these chains of populations or at least evidence for the former existence of such gaps. Circular overlaps, therefore, are the product of orthodox geographic speciation.

The rarity of speciation by distance and the slowness of gene flow in most species are two phenomena seemingly in conflict with each other. How can this conflict be resolved? Perhaps, it might be suggested, gene flow is not quite so slow as it appears. Even highly sedentary species have a dispersal stage during which a few individuals scatter unexpectedly far. But this occurrence of long-distance dispersal cannot by itself compensate for the overall slowness of gene flow. Yet gene flow alone cannot prevent abundant speciation by distance. However, it is strongly reenforced by the epistatic systems that the members of a species share (see p. 170), and it is presumably these shared homeostatic mechanisms that ordinarily prevent speciation by distance.

The Genetics of Species Differences

Recent studies on the integration of species-specific gene complexes have considerably changed our ideas on the nature of species differences. After the claims of the early Mendelians that one or a few mutations "made" species had been refuted, we had come to think of the genetic difference between species as a matter of quantity: if enough gene substitutions were piled on top of each other, there would eventually be a different species. Early authors spoke of dozens or scores of gene differences between species, but when it was realized that even individuals within a single population (including the human species) may differ by hundreds of genes, one began to talk in bigger numbers. Haldane (1957) said recently, "Good species, even when closely related, may differ at several thousand loci," and this order of magnitude would probably be supported by most current investigators. Actually such figures tell us relatively little. Indeed, it is becoming increasingly evident that an approach that merely counts the number of gene differences is meaningless, if not misleading.

The genetic analysis of species hybrids has shown that few species differences appear to be controlled by single genes or by a few genes with large phenotypic effects. Where single gene differences do distinguish species, the very same genes may be polymorphic in related species, for instance, black versus yellow wing color in *Papilio*. Most species differences, however, seem to be controlled by a large number of genetic factors with small individual effects. The genetic basis of isolating mechanisms, in particular, seems to consist largely of such genes.

Each isolated gene pool is a different biological system, and the organization that is the result of the coadaptation of the genes may add a new dimension to the difference that cannot be stated as the arithmetic sum of the individual gene differences. It is easy to imagine two conspecific populations that share the same species-specific isolating mechanisms and essential chromosome structure, and yet differ from each other by more individual gene substitutions than some good species. It is evident that a purely quantitative approach may well be misleading. Nor can species difference be expressed in terms of the genetic bits of information, the nucleotide pairs of the DNA. That would be quite as absurd as trying to express the difference between the Bible and Dante's *Divina Commedia* in terms of the difference in the frequency of the letters of the alphabet used in the two works. The meaningful level of integration is well above that of the basic code of information, the nucleotide pairs.

What, then, makes a species different from an intraspecific variant? I believe that Harland hit the nail on the head when he said many years ago that species are characterized by their modifiers. Today we would perhaps use a slightly different terminology for what Harland had in mind. We might say that it is the total system of developmental interactions, the totality of feedbacks and canalizations, that makes a species. Two individuals of *Drosophila melanogaster* that differ in five conspicuous mutations affecting eye color, pigmentation, wing shape, bristle structure, and haltere formation may look strikingly different from each other, yet they still share their "modifiers," their total developmental system, and are thus still *Drosophila melanogaster.* Two wild-type individuals of *D. melanogaster* and *D. simulans,* which are hardly distinct visibly, nevertheless differ from each other by hundreds, if not thousands, of genes and are the possessors of totally different developmental systems. The important point is that *different species are different systems of gene interaction, that is, different epigenetic systems.*

Morphological consequences. That the genetic reconstitution of isolated populations is often rather drastic and affects major homeostatic systems is supported by studies of morphological and physiological characters. Peripheral isolates, no matter how close to the main range of the parental species, almost always exhibit noticeable differences, in contrast to the essential uniformity of the contiguous populations in the main range of the species. Such differences are well illustrated by *Tanysiptera galathea* (Fig. 16.4), *Dicrurus hottentottus* (Fig. 16.2), and other species cited in Chapter 13. An independent evolution of isolates characterizes geographical variation in all regions with insular distribution. The rapidity with which morphological changes take place in such peripheral isolates confirms our conclusion that shifts in the previously existing developmental homeostasis are permitted or induced by the genetic reconstitution of these populations.

The degree of morphological distinctness acquired during the period of isolation is not necessarily an accurate measure of the degree of general genetic difference or, more specifically, of the degree of reproductive isolation. This is true both where, as with the snails of the genus *Cerion* (Fig. 2.1), a high degree of morphological difference is associated with lack of reproductive isolation and where, as is the case in sibling species, the situation is reversed (Chapter 3).

Ecological consequences. The marginal environment of the geographical isolate and the unfavorable properties of the "normal" niche of the species under these marginal conditions often reinforce the genetic revolution. The genetic changes, in turn, may profoundly affect the ecological preferences and adaptations of a population.

The environment in the peripheral isolate is always to some degree unlike the optimal environment of the species in the center of its range. The biotic environment, in particular, is unbalanced at isolated locations. The new isolate will thus be exposed to a considerably changed selection pressure. Even where the physical environment in the peripheral isolate is not very different from the environment in nearby peripheral areas of the species, the response of the isolated population to selection pressure will be quite different from that of a population which is part of a contiguous array of populations held together by gene flow and by all the cohesive devices discussed in Chapter 10. A population that is part of a continuum of populations is forced to compromise between becoming adapted to local conditions and remaining coadapted with the gene pool of the species as a whole. The more distant a population is from the optimal center of the species range, the less suitable its genetic equipment will be to cope with the optimal species-specific niche. The species border represents the stalemate between local ecotypic adaptation and coadaptation with the gene pool of the species as a whole. In contrast to the population that is part of a continuum, the isolated population can respond to its local adaptive needs without having to compromise with the solutions found by other populations.

The best answer to the challenge of an unusual peripheral environment is in many cases a shift into a new niche. Such a shift is greatly facilitated by a genetic revolution and the special properties of isolated populations. In particular, the genetic lability of such populations and the pronounced population fluctuations (in the absence of strong density-dependent factors) facilitate such shifts. In no other situation in evolution is there a greater opportunity for adaptive shifts or evolutionary novelties. That this is not merely a hypothesis is documented by the many ecological shifts in peripherally isolated populations that have been observed (Chapters 11, 13, and 19). Indeed, nearly all aberrant populations of species are peripherally isolated. The ecological shifts on oceanic archipelagos illustrate this phenomenon dramatically.

Requirements for Successful Speciation

A species is an independent genetic system that has the properties of being reproductively isolated from and ecologically compatible with other sympatric species. Speciation means the acquisition of these properties. It may take place almost instantaneously, as in the case of polyploidy, or gradually, as in the case of geographic speciation. The process by which reproductive isolation and ecological compatibility are acquired is sufficiently important to deserve detailed analysis.

The term isolation has been used in evolutionary biology for two very different kinds of phenomena: geographic isolation and reproductive isolation (Mayr 1959a). Some authors have confounded these two phenomena and have thereby been led to erroneous theories of speciation. *Geographic isolation* refers to the division of a single gene pool into two by strictly extrinsic factors. It is a reversible phenomenon that in itself has no effect whatsoever on the two separated gene pools. What it does is to guarantee their independent development and to permit the accumulation of genetic differences. *Reproductive isolation* refers to the devices that guard a harmoniously coadapted gene pool against destruction by genotypes from other gene pools. These protective devices are called isolating mechanisms (Chapter 5). Speciation is characterized by the acquisition of these devices.

Ecological compatibility. An incipient species, in order to complete the process of speciation, must acquire sufficient differences in niche utilization to be able to exist sympatrically with sister species without fatal competition (see Chapter 4). Such differences are due to ecological shifts in the isolate. The longer the populations have been isolated, the more drastic the genetic revolution; and the greater the ecological differences between the areas, the greater is the probability of ecological differences. Ecological compatibility ("exclusion") need be only initiated during geographic isolation. Even if the ecological divergence is only slight when the species begin to overlap, selection can continue to widen the gap. Such selection will be strongly centrifugal (Fig. 17.7), since it will be directed against the individuals in the zone of ecological overlap. That such selection actually takes place is substantiated both by observation (Chapter 4) and by experiment. When *Drosophila melanogaster* and *D. simulans* are put together in the same culture at 25°C, *D. melanogaster* always eliminates the competing species sooner or later. If, after 10 or 15 generations of competition, some of the experimental *simulans* flies are placed in competition with a new (unselected) batch of *D. melanogaster*, they prove to be con-

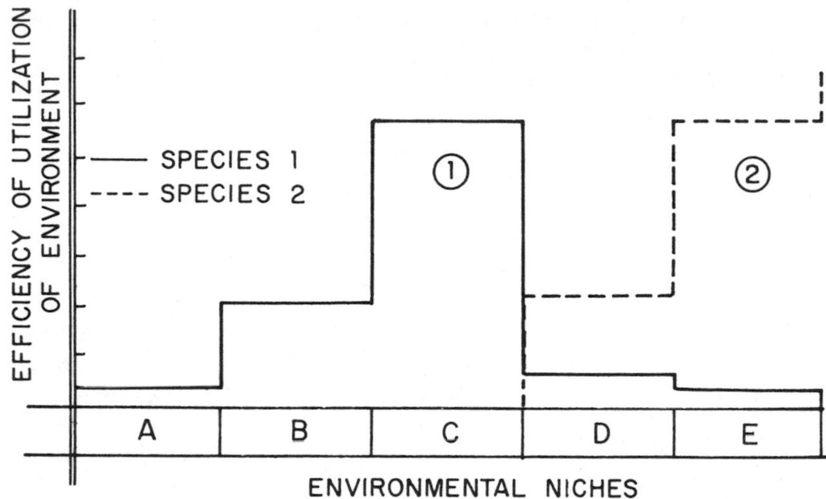

Fig. 17.7. Niche utilization by two different species. Species 1 finds optimal conditions in environmental niche *C*; it utilizes niche *B* inefficiently and niches *A*, *D*, and *E* very poorly. Species 2 cannot utilize niches *A*, *B*, and *C* at all, but finds optimal conditions in niches *E* and *F*. The absence of competition in niche *A* will invite the evolution of a species adapted for this niche. (From Mayr 1949.)

siderably improved as competitors (Moore 1952a,b). This indicates that genes had accumulated in the experimental *D. simulans* population that enhanced their status as competitors.

The Origin of Isolating Mechanisms

The most indispensable step in speciation is the acquisition of isolating mechanisms. Isolating mechanisms have no selective value as such until they are reasonably efficient and can prevent the breaking up of gene complexes. They are ad hoc mechanisms. It is therefore somewhat difficult to comprehend how isolating mechanisms can evolve in isolated populations. This problem has been the subject of considerable discussion during recent years, and scientists are only now approaching agreement.

There are two major theories of the origin of isolating mechanisms. According to the *sympatric theory,* isolating mechanisms are built up through natural selection when two incipient species begin to become sympatric, that is, when their geographic ranges begin to overlap. This hypothesis is based on the observation that hybrids between two species are usually of lowered fitness. It argues that individuals with inefficient isolating mechanisms will be susceptible to hybridization in areas of contact between the parental and the incipient new species. These genotypes will be eliminated from both populations as a consequence of selection against the hybrids that they produce. Genotypes with better-developed isolating mechanisms are not apt to hybridize and will not be eliminated by selection. Their frequency will increase in the population. This process, it is postulated, will in due time lead to an improvement and final perfection of the isolating mechanisms.

Cases in which reproductive isolation between sympatric species was found to be greater than that between allopatric species are usually cited as evidence supporting the theory of the sympatric origin of isolating mechanisms. For instance, Dobzhansky and Koller (1938) found that *Drosophila miranda* was more strongly reproductively isolated from the populations of the two related species *D. persimilis* and *D. pseudoobscura* that occurred in the zone of overlap with *miranda* than from those populations that came from outside the range of *miranda*. Other similar cases have been reported in the literature. Singling out such instances is, however, biased sampling. If isolating mechanisms are simply a by-product of genetic divergence, and if—as is well established—different populations of a species differ in the level of divergence, one might expect, in comparing pairs of closely related species, to find some cases where the degree of reproductive isolation is higher between sympatric populations and other cases where it is higher between allopatric populations. This, indeed, is exactly what Patterson and Stone (1952) have found. Even more frequently, however, no differences exist between sympatric and allopatric populations of two species with respect to the degree of reproductive isolation.

A similar objection may be raised against the citation of evidence that two species are more different in coloration or in call notes in areas of sympatry than where they are allopatric. These instances can easily be matched by cases where the reverse is true. The best evidence for the sympatric theory would be cases where incipient species began to overlap in historical times, hybridized at first quite freely, but at present hybridize much less freely or not at all. Several cases that might qualify are recorded in the literature, but they are susceptible also to a very different interpretation. When the titmouse *Parus cyanus* invaded the range of *P. caeruleus* in western Russia or the woodpecker *Picoides syriacus* that of *P. major* in southeastern Europe, quite a few hybrids were reported during the first invasion but only a few later, except at the front line of expansion. It is quite likely that the first flush of hybridization was due to the fact that the sparse early invaders failed to find conspecific mates and hybridized only after their threshold had lowered. Hybridization decreased or disappeared after the population density of the invading species had risen (Short 1969). Where the Mallard (*Anas platyrhynchos*) and the Black Duck (*A. rubripes*) have been overlapping in the eastern United States, no change in the frequency of hybridization has been noticed in the last 75 years. Hybridization is most frequent (about 4 percent) where both species are almost equally abundant, there being no indication of a reinforcement of isolating mechanisms in the primary zone of contact.

The various objections against the theory of the sympatric origin of isolating mechanisms have been ably stated by Moore (1957). Perhaps the most convincing argument against the power of natural selection is supplied by the "old" hybrid belts (Chapter 13). These have existed in many cases for thousands of years, and the narrowness of the belts proves that the hybrids are indeed being selected against. Yet there is no indication that this selection has led to a strengthening of the isolation in any of the cases. Proponents of the hypothesis that isolating mechanisms originate or are markedly improved by natural selection have not been able to solve the difficulties raised by introgression. If the hybrids are not sterile, some of them will backcross with the parental species, a process that will lead to a further weakening of isolation rather than to its strengthening (but see p. 326).

Another objection to the sympatric theory emerges from a comparison of the strength of isolating mechanisms within zones of overlap with the strength of those outside such zones. The genetic factors responsible for reproductive isolation should, according to the sympatric theory, be restricted to zones of overlap between the related species, since there would be no selective advantage in having these ad hoc mechanisms spread beyond the area where they are favored by selection. However, there is no evidence that isolating mechanisms are geographically thus confined. That selection is not necessary for the perfecting of isolating mechanisms is demonstrated by the numerous cases in which efficiently functioning isolating mechanisms have undoubtedly evolved in geographic isolation without any possibility of their improvement by subsequent selection.

These facts are taken into account in the *allopatric theory of the origin of isolating mechanisms*, according to which they arise as an incidental by-product of genetic divergence in isolated populations. This was the thesis of Darwin (see Mayr 1959a), who could not see how natural selection could produce interspecific sterility. Since Darwin's time many evolutionists have defended this thesis.

According to this theory any drastic genetic reconstitution, such as may take place in isolated populations, particularly if they are subject to a pronounced ecological shift, may simultaneously affect the genetic basis of isolating mechanisms. Since most genes are pleiotropic, selection pressures against one portion of the phenotype very often affect also the genetic basis of an entirely different component of the phenotype. A genetic restructuring of an incipient species, in response to an adaptive shift, may concurrently produce new isolating mechanisms.

This hypothesis is supported by three sets of observations. First, there is much evidence for the geographic variation of isolating mechanisms (Chapter 16), including incipient sterility and ethological isolation. The beginnings of such isolation have been observed even in separate cultures of laboratory stocks, for instance in *Drosophila*. Second, in view of the highly composite and polygenic character of the isolating mechanisms (Chapter 5), it would be unlikely for them not to be affected by genetic reconstitution. Third, many isolating mechanisms have ecological components. The ecological shifts in incipient species are bound to have an effect on their isolating mechanisms. The thesis that reproductive isolation arises as a by-product of the total genetic reconstitution of the speciating population is consistent with all the known facts.

The theory of allopatric origin does not conflict with the fact that natural selection plays a role in the subsequent improvement of subsidiary isolating mechanisms. One primary mechanism, however, or a combination of several, must be fully efficient before contact is first established. Otherwise, a zone of secondary hybridization is inevitable.

Much of the apparent conflict between the opposing theories disappears when the large category "isolating mechanisms" is subdivided. It is quite evident that one primary isolating mechanism, or several, must be acquired in geographical isolation before contact is established. Depending on the group of animals in question, this may be a behavior barrier or cross-sterility (Chapter 5). This single mechanism will prevent hybridization in most cases, and where it does not do so, inferior hybrids will be produced, owing to behavioral, ecological, or cytological incompatibilities. Such hybrids will reproduce poorly, if at all, and there is, thus, no danger of a breakdown of the species barrier. And there will be strong selection in favor of the acquisition of additional isolating mechanisms to prevent such wastage of gametes. Many such cases have been described in the literature (Mayr 1963; Blair 1964). The selection theory is, thus, valid as far as the strengthening of secondary isolating mechanisms is concerned.

It is not certain for most groups of animals whether the behavior barrier or the sterility barrier is the first isolating mechanism to be perfected. In many families of birds, the duck family (Anatidae) for instance, sympatric species may still be quite fertile with each other and yet not hybridize in nature, because of the efficiency of the ethological barriers. Occasional hybrids occur, but at such a low rate that the elimination of the introgressing genes is not too severe a burden on the parental species. Most of the hybrids are in any case excluded from further reproduction owing to behavioral incompatibility. Such "behavioral sterility" of hybrids has also been observed in *Drosophila.* Ever more exceptions are found to the traditional view that the sterility barrier is the first one to be perfected.

Perdeck (1957) shows that reproductive isolation between *Chortippus brunneus* and *C. biguttulus,* two sibling species of grasshoppers, is maintained exclusively by the difference in display song. No other isolating mechanism can be detected. F_1 and backcross hybrids seem to have the same fertility as intraspecific crosses (although the material is limited). The only handicap of the hybrid males is that they are discriminated against by females. The functioning of this ethological isolating mechanism must have been virtually perfect before contact was established, because otherwise the essentially fully viable hybrids would have served as a channel of gene flow between the two species. A number of additional cases among amphibians and insects have been described in which the vocalization of the males seems to be the primary and by far the most important isolating barrier.

Gradual speciation, then, proceeds as follows: an isolated population acquires, during its isolation, the primary isolating mechanisms that guarantee its integrity after establishment of contact with a sister or parent species. It also acquires a minimal amount of that "adaptive property" which permits the two species to be ecologically compatible. Selection pressures, after the establishment of sympatry, will help to improve the isolating mechanisms to such an extent that no more wastage of gametes (at least not of female gametes) occurs, and ecological exclusion will be steadily improved at the same time.

Conclusions

Geographic isolation is a purely extrinsic and completely reversible factor that does not by itself lead to the formation of species. Its role is simply to permit the undisturbed genetic reconstruction of populations that is the prerequisite for the building up of isolating mechanisms.

The need for coadaptation and for the harmonious integration of genes sets severe upper limits to the number of genes that can be accommodated in a gene pool, since many genetic combinations are incompatible. The rapid elimination of disharmonious combinations after hybridization is proof of this conclusion. There is a tendency in the integrated gene complex to establish an ever-greater cohesion, to achieve a steady improvement of developmental and of genetic homeostasis. Numerous feedbacks permit the individual as well as the population to compensate for the unsettling impact of the environment. Heterosis, in particular, tends to diminish the effectiveness of selection for specific effects by raising viability and by decreasing dependence on the environment. A well-integrated genetic system may come into perfect balance with its environment and become so well stabilized that evolutionary change will no longer occur. Such a system will be able to cope with the regular input of mutations and the normal environmental fluctuations without having to undergo any change. Its future is at best *evolutionary inertia* and, more likely, eventual extinction when there is a drastic change of the environment.

Speciation is potentially a process of evolutionary rejuvenation, an escape from too rigid a system of genetic homeostasis. Speciation disrupts the cohesion of the gene pool by temporarily depleting its gene contents and by inevitably forcing the population into a slightly or drastically different environment. If the genetic shake-up is sufficiently severe, it may start a chain reaction, a genetic revolution. The greater the genetic change, the greater the probability that the daughter species can enter a new ecological niche and be successful in it. The genetic chain reaction may thus start an evolutionary chain reaction. This process is most likely to occur in its purest form in peripherally isolated populations.

Speciation is a risky process. The impoverishment of the gene pool and the genetic instability that accompanies it are far more likely to lead to disaster than to success. Even though most incipient species will die out, an occasional one not only completes the process, but also succeeds in entering a new niche or adaptive zone.

The importance of speciation is that it invites evolutionary experimentation. It creates new units of evolution, particularly those that are important for potential macroevolution. Speciation is a progressive, not a retrogressive, process.

18 · The Ecology of Speciation

Geographic speciation means the genetic reconstruction of a population during a period of geographic (spatial) isolation. The genetic factors involved in this reconstruction were discussed in the last chapter. In the present chapter I will attempt to analyze the role of the environmental factors that influence the origin and maintenance of discontinuities between populations. Four sets of factors that have an effect on the rate of speciation will be singled out for special discussion: (1) factors determining the effectiveness of geographical isolation; (2) factors affecting shifts into new ecological niches; (3) factors affecting the frequency with which geographic isolates are established, and (4) factors favoring genetic turnover within isolates.

The Effectiveness of Geographical Isolation

One of the basic properties of species, and of the individuals of which species are composed, is the capacity to spread. Every species has at least one dispersal stage in its life cycle. A study of the geographical barriers that surround every species and every geographical isolate must take this ability for dispersal into consideration. To be sure, "geographical isolation" means the interruption of gene flow by external barriers. But we must realize that the physical nature of these barriers (extrinsic factors) is only one aspect of this isolation. The numerous physiological and psychological characteristics of the individuals that encounter these barriers during their dispersal stage are of crucial importance. Indeed, to a large extent, these intrinsic factors determine the effectiveness of barriers.

The Role of Extrinsic Factors

An understanding of the functioning of the natural barriers that are responsible for the discontinuities between geographical isolates is an indispensable prerequisite for an understanding of speciation. The study of geographical barriers is therefore as important for the evolutionist as it is for the biogeographer and ecologist.

Kinds of barriers. Various authors have attempted to work out a logical classification of distributional barriers (Hesse, Allee, and Schmidt, 1951, have devoted an entire chapter to the topic). Such studies permit the broad generalization that any area which is unsuitable for occupation by a species may serve as a distributional barrier. The action of such barriers in the process of speciation may be illustrated by some examples.

The effectiveness of the sea as a speciation mechanism was fully appreciated by Darwin, Wallace, and other founders of the science of evolution. The birds of the Galapagos Islands (Lack 1947), of the Hawaiian Islands (Amadon 1950), and of the Papuan region (Mayr 1942) are classic examples. The six islands or island groups of the central Solomons (Fig. 18.1) have permitted subspeciation in five of the 53 species of Passerine birds occurring there (Table 18.1). Eleven of the 19 potential barriers (three species are absent from Gizo) have permitted subspeciation. Straits as narrow as 2 kilometers (1.2 miles) have resulted in the evolution of strikingly

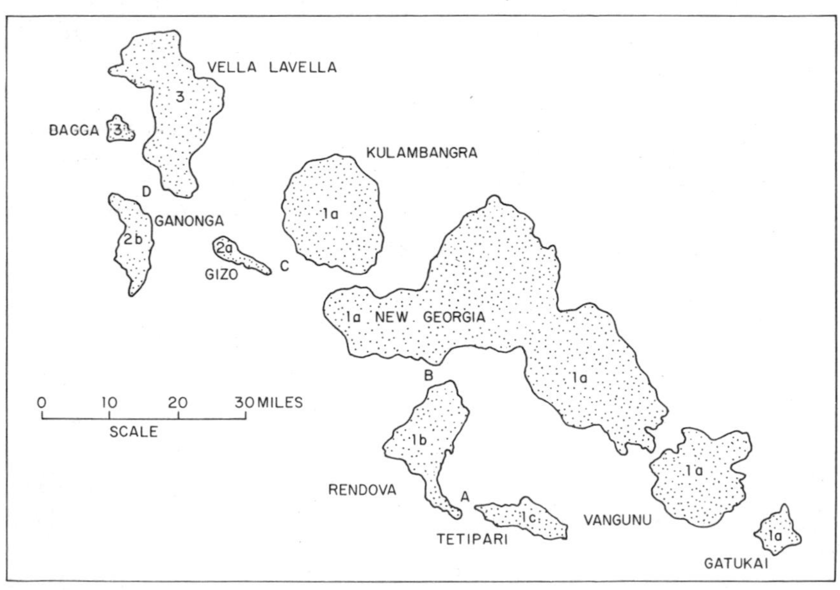

Fig. 18.1. Speciation in the white-eye *Zosterops rendovae* in the central Solomon Islands. 1, *rendovae* group; 2, *luteirostris* group; 3, *vellalavellae.* Groups 1, 2, and 3 are considered full species by some authors, subspecies by others. The shortest distances between the islands are: $A = 1.7$ km; $B = 2$ km; $C = 6$ km; $D = 5$ km. (From Mayr 1942.) See Table 11.1.

Table 18.1. Speciation in birds of the central Solomon Islands.[a]

Species	Islands and water barriers[b]										
	Vella Lavella		Ganonga		Gizo		New Georgia group		Rendova		Tetipari
		I		II		III		IV		V	
Rhipidura cockerelli	a		a				b		b		b
Myzomela eichhorni	a		b		c		c		c		c
Monarcha barbata	a		b				c		d		d
Pachycephala pectoralis	a		b				c		d		d
Zosterops rendovae	a		b		c		d		e		f

[a]The letters a–f refer to distinct subspecies of each of the species listed in the left-hand column; for names see Mayr, *Birds of the Southwest Pacific* (Macmillan, New York, 1945). Many of these subspecies were originally described as full species.

[b]Water barriers indicated by Roman numerals. Width of barriers: I, 5 km; II, 10 km; III, 6 km; IV, 2 km; V, 1.7 km.

different populations in several species. Ocean barriers are even more efficient for lizards and mice. Numerous additional examples of the effectiveness of water barriers (including rivers) are given by Mayr (1942, 1963:557–560).

Mountain ranges may be important barriers, particularly if they simultaneously separate climatic zones, as the Alps separate the Mediterranean from the Central European area, and the Himalayas India from Tibet. Even in such a climatically uniform area as the equatorial island of New Guinea, the central mountain range serves as an important barrier between strikingly different northern and southern populations of some very sedentary tropical species; and the same is true for other tropical mountain ranges (Janzer 1967).

As mountains are efficient barriers for lowland animals, so valleys are for mountain species. Each of the mountain species of birds of paradise is separated from its nearest relatives by valleys or low passes (Fig. 16.3). Snails on tropical islands may be isolated from each other either by valleys or by mountain ridges.

The Pleistocene ice masses of the northern continents were among the most potent barriers in the recent history of the earth. In Europe the Scandinavian ice cap and the Alpine glaciers came within 200 miles of each other, separated merely by a cold steppe that formed an efficient barrier between the unglaciated Atlantic region and the unglaciated areas in the eastern Mediterranean and the Near East. This led to much subspeciation (Mayr 1942), but the evidence for completed speciation among Eurasian birds is equivocal. In America, the speciation process was apparently completed much more frequently. Rand (1948) and Selander (1966) list numerous species pairs that apparently split into a western and an eastern component during one or another phase of the Pleistocene. There were mesophytic refuges in the southeast, presumably two in the west, and one in the north in the Yukon area. Evolutionists agree on the overwhelming importance of Pleistocene barriers in the speciation of temperate zone animals. Detailed analysis of the nature, location, and chronology of such barriers is just beginning.

A study of the barriers which now exist between geographical isolates shows that the conventional barriers (water, mountains) are no more important, at least on continents, than are vegetation zones. The borders of vegetation belts are often exceedingly sharp in the tropical and subtropical zones and form effective geographical barriers for many animals. This has been shown by Moreau (1966) for African birds, by Haffer (1969) for Amazonian birds, and by Keast (1961) for the birds of Australia. In Australia, for instance, the wet sclerophyll forests are broken up into a number of isolated areas along the periphery of the continent, separated by drier areas. Each pocket of forest has served as a local center of differentiation for incipient or fully formed species (Fig. 13.3).

A study of the different types of barriers shows that it is not permissible to make a distinction between geographical and ecological barriers. The Amazonian lowlands are for an Andean mountain species as much of a physical as an ecological barrier. Any terrain that is unsuitable for a species is at once a geographical as well as an ecological barrier.

Barriers in fresh water. The barriers permitting speciation in fresh-water fish and bivalves are self-evident. On the whole, each stream or stream basin is a population unit separated by land from adjacent ones.

Lakes are for water animals what islands are for land animals. Every old fresh-water lake has its own endemic fauna. These faunas are either relatively young, like those of Lake Waccamaw in North Carolina and Lake Nabugabo in East Africa (Greenwood 1965), or very old and rich in peculiar endemics, like those of Lakes Baikal, Nyasa, Tanganyika, and other ancient fresh-water lakes (Brooks 1950; Lowe-McConnell 1969). Furthermore each lake consists of an archipelago of suitable areas, with each habitat island (such as a rocky shore) separated by a barrier (such as a sandy or muddy shore) from other suitable areas.

Climatic barriers. Those borders of the range of species and of geographical isolates that are determined by the climatic tolerance of populations tend to be particularly labile. They are controlled by temporary conditions of temperature and rainfall. Climates are subject to short-term and long-term fluctuations and these cause advances and retreats of such borders. Relict populations are often left behind in suitable situations during periods of range contractions. If conditions are otherwise favorable, these relics may, like other geographical isolates, reach species level.

The ever-present expansions and contractions of species ranges in response to changes of climate are sometimes ignored in the evolutionary literature. A careful study of any climatically determined species border shows that it is in a state of dynamic stability. The general amelioration of the climate in the Northern Hemisphere during the first half of the twentieth century illustrates this quite graphically. Numerous species of birds, mammals, and other animals greatly expanded their breeding range toward the north, while some of the northern species retreated from the south (Kalela 1944). Many existing isolates can be explained only as the result of former range expansions during periods climatically favorable for such expansion.

The efficiency of barriers. Most geographic isolation is relative, few barriers being 100-percent efficient. It would be interesting to know how large the distance between isolated populations must be in order to permit the completion of speciation. The amount of tolerated gene flow probably depends on the properties of the genetic system of the isolated population. No barrier seems to stop gene flow altogether. Even the wide Atlantic is not a complete barrier. An Old World bird, the Cattle Egret (*Bubulcus ibis*), colonized northern South America across the Atlantic around 1930 and has since then expanded its range into the Caribbean and the southern United States. The very isolated Hawaiian Islands have received their entire fauna through transoceanic colonization from Polynesia and America (Chapter 16). The entire mammal life of Madagascar can be interpreted as the result of no more than five colonizations across the ocean separating Madagascar from Africa.

If barriers to gene flow are as important a factor in speciation as is now generally believed, one should find the greatest amount of active speciation in areas richest in geographic barriers. This is indeed the case. Regions which in any sense of the word are insular always show active speciation, whereas continental regions show speciation only where physiographic or climatic barriers produce discontinuities between populations (Table 13.1). The thesis that speciation should be most active and rapid where natural barriers are most frequent and most efficient is richly substantiated by all the known facts (For further examples see Mayr 1942).

The Role of Intrinsic Factors

Geographic barriers are sometimes regarded as purely mechanical devices, like dams that hold water in reservoirs. Such an emphasis on the mechanical aspects of barriers is one-sided. The geographical features of a given area, its mountains, rivers, ocean straits, and treeless plains, affect the population structure of different species in very different ways. The differences can be explained only in part by variations in the physical means of dispersal. To a large extent these differences are the result of *intrinsic factors* (Mayr 1942:238, 1949:288), that is, of physiological and psychological properties, which cause every species to react differently to such barriers. The dispersal ability of individuals of a species, that is, the faculty of moving longer or shorter distances from their birthplace, controls to a large extent the establishment and maintenance of geographical isolates.

Dispersal in plants is normally the task of the seeds. In animals it may occur at almost any stage of the life cycle, but there are a number of regularities. If the adults are sedentary, as in many marine organisms, dispersal will take place during the larval stage, usually through free-swimming larvae. If the adults are mobile, as is the case with most insects, the larvae tend to be sedentary. If the adults are subject to passive dispersal (for instance, aphids), leading to much waste and loss, means for accelerated reproduction (for example, parthenogenesis) are often present.

No reliable method is known for measuring dispersal ability. Individual mobility is definitely one component of it, and this is why so many workers in recent years have attempted to calculate the average amount of dispersal per individual per generation. Yet these investigations give at best an incomplete, and most often a decidedly misleading, answer to our question. Such dispersal studies are based on the arbitrary assumption that the individuals of a population obey in their dispersal the same laws that control the scattering of inanimate objects. Dispersal curves, however, are rarely normal (Bateman 1950); indeed, most animal populations seem to be composed of three classes of individuals: (1) those that scatter slowly and at random like inanimate objects; (2) those that have a definite tendency to remain where they are (philopatry); and (3) those that travel far greater distances than one would expect. Classes 2 and 3, which are responsible for the kurtosis and skewness of dispersal curves, are manifestations of intrinsic factors. Some of these intrinsic factors facilitate the overcoming of barriers, others reinforce them. The capacity of a group of animals to speciate depends to a considerable extent on the relative strength of the two opposing sets of factors.

Factors facilitating dispersal and the crossing of barriers. Dispersal may be active, passive, or a mixture of both. The probability of passive dispersal is increased by numerous factors, some of which are listed by Simpson (1952:168), such as small size, low specific gravity, protective coating, a dormant stage, and so forth. Species that are optimal for all these factors may have world-wide ranges, such as certain tardigrades (Fig. 18.2), rotifers, and fresh-water crustaceans. A successful *cosmopolitan* with an essentially panmictic species population is evidently barred from geographic speciation. A high dispersal ability is a necessity for occupants of temporary habitats, such as most bodies of fresh water. It is likewise characteristic of most marine organisms. More than 70 percent of the bottom-living marine invertebrates have a pelagic larval stage. The ability to overcome barriers (for instance, the deep sea between shallow waters) depends on the length of the larval stage. Marine gastropods with long larval stages, such as *Cypraea*, *Conus*, and *Mitra*, have far wider ranges in the Pacific

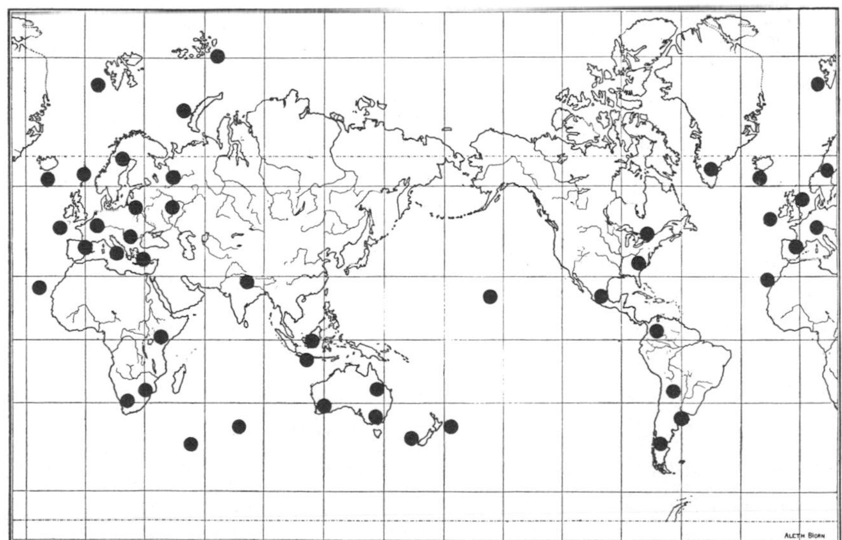

Fig. 18.2. Cosmopolitan distribution of the tardigrade *Macrobiotus hufelandii*.

than bivalves with short larval periods. Most planulae of the reef-building coral *Galaxia aspera* settle within a week, but a few float for more than 60 days. In an ocean current of 2 miles per hour (48 miles per day), such larvae could cover 3000 miles before settling. It is not surprising that species with such dispersal facilities do not speciate even in the vast area of the Pacific and that littoral species are found on such remote oceanic islands as Ascension, St. Helena, and the Hawaiian Islands. Ocean currents are remarkably rich in pelagic larvae.

Differences between species in the degree of active dispersal are even greater than differences in the degree of passive dispersal. Various ill-defined and poorly understood properties of species are responsible for these differences (Mayr 1965a). For instance, the ability to cross barriers, and thus to establish or swamp geographical isolates, is in the higher vertebrates often determined by psychological factors. Around 1850 the small white-eye *Zosterops lateralis* jumped the 2000-kilometer gap between Tasmania and New Zealand and proceeded to become the most common native songbird of New Zealand. In subsequent decades it successfully colonized all the outlying islands of the stormy New Zealand seas. Yet a close relative, the white-eye of the central Solomons, *Zosterops rendovae,* refuses to cross barriers only a few kilometers wide (Fig. 18.1) even though its flying equipment is essentially the same as that of the New Zealand bird. Similar differences in dispersal facility of morphologically equivalent species are known in many families of birds. Fruit- and nectar-feeding birds that have to follow shifting food supplies show greater dispersal and less subspeciation than the more sedentary insect eaters.

The number of instances in which a species has actively crossed an important geographical barrier in historical times and established a new isolate, as the New Zealand white-eye has done, is probably quite large. However, only rarely will a naturalist be present to record the details of the invasion. Such a case is the invasion of Greenland in January 1937 by a flock of the Eurasian thrush *Turdus pilaris,* carefully documented by Salomonsen (1951). By 1949 the bird had become a common breeding bird of southwest Greenland. The histories of many other invasions (most of them due to passive dispersal) have been described by Elton (1958).

Factors reducing dispersal. The probability of passive dispersal is reduced by the reverse of the characters enumerated above as facilitating dispersal. The rate of speciation is higher in wingless genera of carabid and tenebrionid beetles than in flying ones. Retreat into caves sets up new barriers and favors speciation. Reduction of the pelagic larval stage in marine animals makes local barriers more effective.

Again it appears that in many animals, particularly the higher vertebrates, psychological factors are the most important means of reinforcing geographical barriers. The Rendova White-eye (*Zosterops rendovae*) is a typical example of a species showing philopatry, even though a flight of only a few minutes' duration would carry the birds to the next island. Such a preference for staying on the home ground is widespread among animals. It has been documented for mice, lizards, turtles, snakes, fish, snails, butterflies, and, in fact, for nearly all species of animals in which the movements of marked individuals were carefully recorded.

Other intrinsic factors that reduce gene flow include the maintenance of a *territory* and the ability of *homing*. The importance of homing in birds has long been known and explains the extreme localization of the ranges of many subspecies of geese (*Anser, Branta*), in spite of the enormous migrations of these species. Homing is well substantiated not only for birds and mammals, but also for reptiles, amphibians, and fishes. It occurs even among invertebrates.

Any factor that reduces dispersal may facilitate speciation. Parental care, for instance, tends to increase philopatry and reduce dispersal sharply. The extremely late emancipation of the young in the geese is a good illustration. Mouth breeding in the cichlid fishes is another. By the time the young cichlids become independent, they have become thoroughly habituated to the very localized station of their parents. As a consequence, mouth breeding, philopatry, and great efficiency of habitat barriers are closely correlated in the cichlids. The extreme localization of populations in potentially mobile species can be explained only by these intrinsic factors.

Habitat selection. Perhaps the most important of all the intrinsic factors leading to the localization of populations and to a restriction of species to their species-specific ecological niches is a phenomenon usually referred to as *habitat selection*. Members of every species have the ability, as well as the urge, at the end of the dispersal phase in their life cycle, to choose as domicile an area that has a constellation of environmental factors characteristic of the species. Occasionally habitat selection can be studied by direct observation. Often, when an individual of a species is forced out of its proper habitat, it makes every effort to return to it as quickly as possible. Habitat selection is particularly conspicuous in all animals that are cryptically colored and thus agree in color with their substrate. Niethammer (1940:81-82), who experimented with the color preference of South African larks, made the following observations:

> It is very striking in southwest Africa that reddish larks are found only on red soil, and dark ones on dark soil, even where two completely different types of soil meet, as at Waltersdorf, for example, where dark soil, rich in humus, comes in contact with the red Kalahari sand. *Mirafra sabota hoeschi* stayed entirely on the dark soil, in spite of the fact that the area of the red sand began only a few hundred meters from its territories. On the other hand, I met *Mirafra africanoides* on the red sand up to its very edge but never on the dark soil, which was inhabited exclusively by *Mirafra sabota hoeschi.* Similar conditions prevailed at the Farm "Spatzenfeld" with the exception that the red sands here border light lime pans, where *Spizocorys starki* lives. I tried to chase little groups of *Spizocorys starki* to the red sands, but in vain—they turned before the beginning of the red soil and flew back directly, as if they knew, to their accustomed light lime soil. The red *Mirafra fasciolata deserti* which in Spatzenfeld inhabits the red sands did not go astray either in the lime pans. The reverse experiment I made in Lidfontein. I tried to chase the red *Mirafra africanoides gobabisensis* from a red dune to the light lime soil—again in vain. I do not believe that the experiment will ever have a different result because it is obvious that the birds are conscious of the color that corresponds to their own coloration.

This deliberate choosing of the proper habitat and avoidance of unsuitable habitats serves as a most powerful reinforcement of geographical barriers. Habitat selection prevents the wastage of individuals in unsuitable habitats; thus it is the ecological equivalent of isolating mechanisms.

Habitat selection is a conservative factor in speciation because it reduces the probability that new isolates will be established beyond the current species border. The normal habitat of the species usually does not occur beyond the species border. If a species has the ability to change its habitat preference, it can not only expand its range but also change genetically under the pressure of the new environment in the newly established geographical isolate. Such a shift creates conditions that are unusually favorable for rapid speciation. The great importance for speciation of such shifts in habitat preference has long been recognized but has often been erroneously interpreted as evidence for sympatric speciation. All speciation involves a greater or lesser amount of ecological transformation. This is superimposed on the geographical changes; it is not an alternative to geographical speciation.

In sum, dispersal within the range of a species and beyond the species borders is not a purely mechanical phenomenon to be explained solely in terms of the physical means of dispersal. Intrinsic factors, such as philopatry, homing ability, restlessness, length of dispersal stage, parental care, habitat selection, and other psychological and physiological characteristics, strongly influence the amount and the distance of dispersal. Together they control the establishment and maintenance of geographical isolates and are thus one of the determining factors in speciation.

Colonization and the Invasion of New Niches

Dispersal facility alone is not sufficient to make the crossing of a barrier a success. It must be supplemented by an ability to find suitable habitats beyond the barrier and to colonize them. In this respect there appears to be a pronounced difference between plants and higher animals. Plants have high dispersal facilities but also make highly specific demands on their habitat and seem to find it difficult to establish themselves in new areas (weeds excepted). Animals, particularly vertebrates, have more difficulty in crossing barriers, but this is compensated for by a greater ability to cope with new conditions. The result is that the distribution of floras tends to be determined by climate, that of vertebrates by dispersal barriers. In the East Indies, for instance, the entire tropical belt from Malaya to New Guinea is essentially a floristic unit, more or less sharply distinguished from Australia. The higher animals, on the other hand, are clearly separated into a Malayan fauna to the west and an Australo-Papuan fauna to the east of Weber's line (a north-south line between Celebes and the Moluccas).

The Invasion of New Niches

A colonizing organism that has crossed the current species border is often unable to find a habitat or ecological niche equivalent to that which it has left. It will not be able to establish itself unless it has the capacity for a shift in its ecological requirements or its niche. To understand such a shift, one must realize that a species does not necessarily have a typologically fixed "demand" on its environment. There is no such thing as *the* ecology of a given species. As we have seen in Chapter 16, there is often geographical variation in habitat or niche requirements, and peripheral populations deviate especially often from the "norm" of the species. Species consist of local populations, each adapted to its local environment, held together loosely by gene flow. Even within a single population, individuals may differ in their ecological preferences and tolerances. It is, thus, evident that sufficient raw material is available in most species to permit some of the colonizing individuals to shift into new niches. These founders of the new population must have a genetic constitution that will give success in the new environment.

The shift into a new niche must pass, in many cases, through a stage of ecological polymorphism. The new preference or tolerance is simply added to the already existing ecological characteristics of the population (Ludwig 1950). There is, however, a limit to the ecological versatility of a population, and if a new niche preference is superior to a previously existing one, it may displace the older one. When such a shift occurs in a geographical isolate and the isolate acquires reproductive isolation during a lasting period of geographic isolation, it will have acquired all the characteristics of a new species (Fig. 15.3).

The Environmental Constellation

The shift into a new ecological niche, the acquisition of a new ecological trait, is facilitated by some circumstances and definitely impeded by others. Two factors of the biotic environment have recently been singled out by several authors as particularly important for the chance of success of ecological shifts.

Competitors. It is obvious that a population will find it difficult to enter a new niche if this niche is already occupied by another species. Yet there is evidence that there are vacant or partly vacant niches even in well-balanced faunas. The spectacular success of faunal transfers illustrates this point (Elton 1958), as does the sudden breakdown of faunal barriers. Since the opening of the Suez Canal in 1869, numerous Red Sea species of fish and marine invertebrates have colonized the Mediterranean, and some have become very abundant, at least seasonally. Although there is some evidence for competition, the magnitude of the invasion indicates that the invaders filled, at least in part, empty niches. A greater richness of niches—or to put it in different words, a finer partitioning of the natural resources—rather than any difference in the process of speciation would seem to be the explanation of the greater richness of tropical faunas (MacArthur 1965, 1969).

Predators. The effect of predator pressure on the invasion of new niches is still in doubt. Indeed, it may differ from case to case. Sometimes a predator concentrates on a superabundant species and thereby makes room for somewhat less successful species. In other cases, the predator is sufficiently harmful to a species that has a precarious foothold in a habitat to cause its local extinction. The great frequency of successful shifts into radically new niches in the absence of predators that have occurred on oceanic archipelagos such as the Galapagos and the Hawaiian Islands seems to confirm the thesis that the shift into a novel niche makes an individual far more vulnerable to predation than are the members of well-established populations. However, the simultaneous absence of competitors on these oceanic islands precludes ruling out the possibility that the absence of competition is the decisive factor.

Ecological Aspects of Rate of Speciation

There is perhaps no other aspect of speciation about which we know as little as its rate.* Indeed, we shall probably never have very accurate information on this phenomenon. The splitting of one species into two is a short-time event that, as such, is not preserved in the fossil record. For information we rely entirely on inference.

The rate of speciation depends on three sets of factors: (1) the frequency of barriers, that is, of factors producing geographical isolates; (2) the rates at which geographical isolates become genetically transformed and, more specifically, at which they acquire isolating mechanisms; and (3) the degree of ecological diversity offering vacant ecological niches to newly arising species. Among these three sets of factors only the second has a genetic component. Furthermore the rate of genetic transformation itself depends on an array of contributory factors: (a) the size of the population, (b) its genetic system (including its karyotype), (c) the completeness of its isolation, (d) its mutation pressure, and (e) its selection pressure. When all these different and largely independent factors act synergistically, assisting rapid speciation, the rate of speciation can indeed be astonishingly swift (see below). When, on the other hand, all the conditions militate against rapid speciation, an isolate may hardly change at all in millions of years. Tropical archipelagos, like the Hawaiian Islands, and certain fresh-water lakes demonstrate how extraordinarily rapid the rate of speciation can be when population sizes are small and the constellation of ecological factors favorable.

The suggestion, sometimes made, that rate of evolutionary change should be measured in terms of numbers of generations makes little sense. Microorganisms with high numbers of generations per unit of time seem to have low evolutionary rates. And certain types of insects with many generations per unit of time are little changed since their appearance in the Oligocene

*Rate of evolutionary change as such is considered outside the scope of this treatment. For excellent recent discussions see Simpson 1953 and Kurtén 1959.

amber. In contrast, slowly maturing mammals, like proboscideans, evolve very rapidly. Length of generation is usually inversely correlated with population size, and it is thus canceled out as a determinant of evolutionary rate.

Rate of speciation, broadly speaking, is the product of two sets of factors, rate of the *formation (and maintenance) of isolates* and rate of *transformation of isolates.* Neither can be measured directly, although estimates are possible that may perhaps permit a determination of their order of magnitude.

Formation and Maintenance of Isolates

Any factor that facilitates the establishment of isolates and reduces gene flow into the isolate increases the rate of speciation. Accordingly, among birds sedentary species have on the average twice as many subspecies as migratory species. A high number of subspecies indicates a considerable localization of populations, in other words, reduced gene flow. Large-sized species such as herons, storks, and hawks usually have fewer but more widely ranging subspecies and isolates than small songbirds. In a study of desert birds, Hoesch (1953) showed that five species of larks and buntings that were dependent on water had only one subspecies each in southwestern Africa, while six other species that were independent of water had an average of four subspecies in the same area. The species that are independent of water are far more sedentary than the dependent species.

The establishment of isolates is also favored by an ability to discover unoccupied niches or, in a different terminology, previously unutilized resources of the environment. This is well illustrated by the insects, which have a larger number of species than any other group of animals. The reason must be an exceptionally high frequency of successful speciations, since the survival of species is quite evidently not nearly as long as in some other types of organisms, for instance brachiopods or pelecypods. The great potentiality of insects for specialization may be the answer to this puzzling situation. A high percentage of species is either host-specific or at least oligophagous. Indeed, different species may exist in different parts of the same host plant, in the flower buds, in the upper stem, in the lower stem, or in the root stalk. Several species can thus utilize resources that would in different circumstances be occupied by a single polyphagous or otherwise ecologically versatile species. Loss of wings and thus of mobility has resulted in a burst of speciation in flightless beetles and grasshoppers.

Mobility and the ability to select the appropriate niche have given animals the opportunity for high specialization without risk of the loss of colonizing zygotes in unsuitable locations. Insects have utilized this potentiality to a greater extent than any other group of animals, and this is the reason for their high rate of speciation. At the other extreme is the species *Homo sapiens,* which can live in every environment from the pole to the equator, which flourishes on a pure meat diet (Eskimos) or on a virtually pure vegetarian diet, which can live the life of a hunter, a nomad, a farmer, an industrial worker, a minister, or a theoretical physicist. What other niche would there be available, if man were ready to speciate?

If one postulates that the frequency of speciation depends to a considerable extent on the frequency of unoccupied niches, one would expect very little speciation in the pelagic waters of oceans, with their exceedingly stable environment and scarcity of barriers. What little we know about speciation in marine animals seems to verify this assumption. Yet barriers are not absent in the oceans. Benthic faunas, either of shallow waters or of the deep sea, have species populations that are often split into isolates by geographic barriers. Although speciation is relatively slow, it is regular and steady in these faunas.

Transformation of Isolates

Geological evidence. Lacking any other suitable means for measuring the rate by which a geographical isolate is transformed into a separate species, evolutionists have often in the past accepted phylogenetic transformation as a yardstick. Valuable as this evidence is, it has various shortcomings. First of all, interpretation of such evidence rests on somewhat unrealistic assumptions, such as that the more recent forms (in a series of strata) are the direct descendants of the earlier forms and that the amount of morphological change reflects accurately the degree of speciation. However, one must presumably equate a much greater amount of morphological change with completed speciation in groups that are as plastic morphologically as the ammonites than in groups as static morphologically as the bivalves. Second, it is becoming increasingly clear that population size is by far the most important determinant of rate of genetic change. The smaller the population, other things being equal, the greater the probability of rapid genetic change. The fossil record, however, favors widespread species rich in individuals. These, we now suspect, are an exceptional group of species with by far the slowest rates of evolution and speciation. The fossil record, thus, represents a badly skewed sampling of species and the chances are all against the fossil preservation of remnants of a rapidly speciating, small geographical isolate. At best, the fossil record can give us data on the length of time during which certain species seem to have undergone no appreciable morphological change. For instance, the fairy shrimp *Triops cancriformis* is known from the upper Triassic to the present, a span of about 180 million years. In strata about 2000 million years old, certain microorganisms have been found that cannot with certainty be distinguished from living forms. However, such rate of unchanging survival is only tenuously correlated with rate of speciation.

The opposite of such evolutionary inertia is the occurrence of veritable bursts of speciation, such as are found by paleontologists and by the students of fresh-water lakes. The foraminiferan family Fusulinidae originated in the latest Mississippian, flourished in the Pennsylvanian, and died out in the Permian, not much longer than 50 million years after its origin. In this short period it developed 6 subfamilies, 48 genera, and more than 1000 species. Fusulinids are often more abundant in fossil faunas of this period than all other contemporary invertebrates combined.

Fossil mammals show perhaps the fastest known transformation of species (Simpson 1953; Kurtén 1959). Birds apparently change far more slowly: most Miocene Nonpasseres belong to modern genera and many Pleistocene birds cannot even be separated specifically.

Geographical evidence. The rate of transformation of isolates can also be determined by comparing allopatric populations that have been separated for a known period. Here one runs into the same three difficulties one had to face in the evaluation of the fossil evidence: (1) the date of separation can usually be arrived at only by inference; (2) only phenotypes can be studied, and the genetic basis of the differences is left undecided; and (3) the degree of correlation between the acquisition of morphological differences and of genetic isolating mechanisms is unknown. The age of subspecies, thus, does not give us by extrapolation the age of species. Let me illustrate these difficulties on the basis of a concrete case. The British Red Deer (*Cervus elaphus scoticus*) evolved during the 8000 years since the submergence of the English Channel. Yet when this form was introduced into New Zealand, it assumed within a generation or two the phenotype of the Carpathian Deer, thus becoming far more different from its parental stock than is the British Red Deer from the Continental race.

Equally frustrating is the fact that the length of the isolation is always uncertain. Much subspeciation and even speciation in Australia is due to a previous isolation of populations, during arid periods, in more humid pockets ("drought refuges"). Yet the timing of these arid periods is still uncertain, and there are even some specialists who deny their existence. The time of origin of eastern and western races and species of European birds is likewise highly controversial (Mayr 1963:578). The evidence for

Africa has been presented by Moreau (1966) and Hall and Moreau (1970). Haffer (1969) has attempted to determine the chronology of speciational events in South America and Selander (1964) the chronology in North America. The evolutionist can work out the relative aspects of speciation, but he depends entirely on the Quaternary geologist and geographer to give him an absolute chronology that can be used as a framework and yardstick for speciational events.

Even where absolute times of geological events are available, it is found that rates of speciation are different in different genera and species groups. Cameron (1958) points out that the island of Newfoundland at the mouth of the St. Lawrence River became habitable for mammals less than 12,000 years ago. Of the 14 species of native mammals, 10 have evolved well-defined subspecies in that period, some, like the Newfoundland beaver (*Castor canadensis caecator*), almost distinct enough to be considered a separate species. Mayr (1963:576-580) gives an extensive review of the literature. When, as in the case of Newfoundland, several species become isolated simultaneously, it is invariably found that some species diverge more rapidly than others. Weismann (1902) pointed this out long ago, citing butterflies in the Alps that are evident Pleistocene relicts. They show every degree of divergence from their close arctic relatives.

Rates of subspeciation tell us little about rates of speciation. Morphological differentiation, leading to the recognition of subspecies, is not a halfway point toward the acquisition of isolating mechanisms. Even in a species where it takes only 10,000 years to develop a well-defined island subspecies, it might well take 100,000 or perhaps 1,000,000 years for the completion of the speciation process.

Rate of acquisition of isolating mechanisms. Since speciation ultimately means the acquisition of isolating mechanisms, one way of determining rate of speciation is to determine the rate at which effective isolating mechanisms are acquired in geographically isolated populations. This, again, varies from case to case. The frequency of zones of allopatric hybridization (Chapter 13) indicates that speciation is normally completed slowly. The other extreme is shown by species in fresh-water lakes and springs, where speciation may occur exceedingly fast. It is becoming increasingly evident that the population size of the incipient species is the crucial factor.

Population size and rate of speciation. The closing of the last Central American portal, toward the end of the Pliocene, divided each marine species of this region into a Pacific and a Caribbean population. The time since this event, perhaps 5 million years, should have been more than sufficient to complete the separation of each parental species into a Pacific and a Caribbean daughter species. However, this did not happen. Among the crabs, for instance, 11 species remained identical, while 13 other species split into allopatric pairs of subspecies or species. No up-to-date information on fish is available, though it seems that most coastal species have evolved into species pairs, while among the more pelagic tropical species many are still considered identical in the two oceans. Unfortunately, this conclusion is based on inference from the morphological evidence rather than on the actual testing of the isolating mechanisms. However, Rubinoff is now conducting such tests on coastal species pairs.

There is good evidence that speciation in isolated small bodies of fresh water might proceed 1000 times as fast as in the ocean. This has long been inferred for the fish fauna in the creeks and springs of the western North American deserts, where full species seem to have originated since the last pluvial (Miller 1961). The situation in Lake Lanao (Philippines) is not quite as simple as it was first thought, but there is no doubt that speciation must have been very rapid (Myers 1960; Kosswig and Villwock 1964). Fairly rapid rates of speciation must be assumed to account for the swarms of *Tilapia, Haplochromis,* and other cichlid fishes in the large East African lakes (Lowe-McConnell 1969). In one well analyzed case more precise information can be given. In a lagoon (Lake Nabugabo) off Lake Victoria but completely separated from it, live six species of the cichlid fish genus *Haplochromis.* One is indistinguishable from the approximately 170 species of *Haplochromis* in Lake Victoria, the other five are endemic, even though each species is more or less obviously related to a sister species in Lake Victoria. It was possible through carbon dating and other methods to determine the approximate age of the sand bar that separates the two lakes. The answer is astonishingly low: less than 5,000 years (Greenwood 1965). Equally rapid rates of speciation have been suggested for terrestrial vertebrates (for example, lizards) similarly isolated on small ecological "islands." The island of Hawaii is believed to be no older than 800,000 years, and yet it is the home of a rich endemic fauna with some secondary speciation having taken place on the island itself.

What is or what are the factors responsible for these enormously different rates of speciation? Obviously it is not fresh water vs. salt water, or land vs. water. The factor that seems to make by far the greatest contribution to rate of speciation is population size. A species with millions of individuals has a gene pool of such enormous size that the replacement of a gene by another allele is a very slow process, and the replacement of an entire well-balanced epistatic system by another one is almost impossible. Species with large populations are, therefore, from the evolutionary point of view highly inert. This conclusion is confirmed by the study of evolutionary rates in phyletic lines. Even in relatively "rapidly evolving lines like the dinosaurs and the ancestors of the horses [both probably consisting of partially isolated small herds!], measurable lengths, such as those between homologous points on homologous teeth or body length in general, changed by quantities of the order of [only] 1-10 percent per million years" (Haldane 1954b). The rate of evolutionary change is even slower, sometimes much slower, in most populous, widespread species.

The situation is quite different in the case of small populations. Here are great opportunities for a genetic revolution and for the acquisition of new translocations, chromosomal fusions (or dislocations), pericentric inversions, and other structural chromosomal changes, following a period of severe inbreeding. The acquisition of isolating mechanisms is presumably greatly accelerated under these conditions (Chapter 17). It would not surprise me if under these circumstances new species could arise in a period measured only in thousands or even hundreds of years. The constellation of factors that would have to be just right to permit such rapid speciation is, however, sufficiently improbable (in the statistical sense) that such a rapid rate will be exceedingly rare. This much, however, is certain: there is no "standard" rate of speciation. Each case is different and the range between the possible extremes is enormous.

The question is sometimes raised whether there has really been enough time available in the geological history of the earth to produce the millions of species of animals and plants that are known to live and to have lived on the earth. A few simple calculations show, however, that this doubt has no foundation. If a species were to produce only four new species every 3 million years, half of which became extinct without further speciation, there would be 65,000 species after 50 million years and this sum would double every 3 million years. This is by no means a particularly rapid rate of speciation since species seem to arise quite often in less than 1 million years in normally speciating groups, and species may be budded off at the periphery of a parental species at many different places. The large number of geographic isolates (Chapter 16) that has been revealed by systematic analysis substantiates this point.

Several broad generalizations emerge from the analysis of the pattern of speciation in the various groups of animals. The rapidity with which an isolate is converted into a separate species depends on the size of the isolated area (number of contained demes), the ability of the population to shift into a new niche, the selection pressure to which the isolated population is exposed, and the effectiveness of the isolation. This in turn depends on the distance from potential sources of immigration and on the dispersal efficiency of the species. To these factors must be added the genetic and cytological factors that have been discussed in Chapter 17. Although each of these factors is important, it must be stressed that their relative importance differs from case to case.

19 · Species and Transpecific Evolution

The nature and cause of transpecific evolution has been a highly controversial subject during the first half of this century. The proponents of the synthetic theory maintain that all evolution is due to the accumulation of small genetic changes, guided by natural selection, and that transpecific evolution is nothing but an extrapolation and magnification of the events that take place within populations and species. A well-informed minority, however, including such outstanding authorities as the geneticist Goldschmidt, the paleontologist Schindewolf, and the zoologists Jeannel, Cuénot, and Cannon, maintained until the 1950's that neither evolution within species nor geographic speciation could explain the phenomena of "macroevolution," or, as it is better called, *transpecific evolution.* These authors contended that the origin of new "types" and of new organs could not be explained by the known facts of genetics and systematics. As alternatives they advanced two explanations, both in conflict with the synthetic theory: saltations (the sudden origin of new types) and intrinsic (orthogenetic) trends.

It is not the purpose of this volume, which concentrates on the evolutionary problems of populations and species, to refute these two explanations and to discuss transpecific evolution in detail. This task has been superbly performed by Simpson (1953) with emphasis on the paleontological evidence and by Rensch (1960) with emphasis on the general zoological evidence. They agree that essentially the same genetic and selective factors are responsible for evolutionary changes on the specific and on the transpecific levels and that it is misleading to make a distinction between causation in microevolution and macroevolution. If used at all, these terms should be considered purely descriptive. The manifestations of transpecific evolution are, of course, in many respects different from those of infraspecific evolution, even though the underlying mechanisms are the same. In this chapter I will discuss primarily those macroevolutionary phenomena to the elucidation of which the study of species can contribute.

A consideration by the systematist (of Recent species) of the problems connected with the origin of new higher taxa supplements in many respects the approach of the paleontologist and the comparative anatomist. Stating the problems of macroevolution in terms of species and populations as "units of evolution" reveals previously neglected problems and leads to a new conceptualization of macroevolution.

The Species as a Potential Evolutionary Pioneer

The key role played by the species in the evolutionary process is based on the following facts. Every species (1) is a different aggregate of genes that controls a unique epigenetic system, (2) occupies a unique niche, having found its own specific answer to the demands of the environment, (3) is to some extent polymorphic and polytypic, thus able to adjust to changes and variations in its total environment, and (4) is ever ready to bud off populations that experiment with new niches. Occasionally a population of a species may, (a) acquire a new combination of genes, a new epigenetic system, that constitutes a novel, more successful adaptation to the environment, and (b) shift to a new ecological niche which is so favorable that it becomes an entirely novel adaptive zone. Every population that makes such a shift is an evolutionary pioneer and may become the founder of a new type, a new higher taxon. The first honey creeper on Hawaii, the first weevil, the first bird, the first amphibian, all were successful in finding a new adaptive zone, and their occupation of that zone set the stage for an avalanche of successful speciations that continued until the zone was completely filled by adaptive radiation.

It cannot be emphasized too strongly that the population is ultimately the key to every evolutionary problem and that any evolutionary theory which overlooks the importance of populations is doomed to failure. It is the fatal weakness of the saltationist theories that they operate with mutated individuals. In actuality it is not the individual but the local population in which selection leads to the increase or decrease of certain genes and gene combinations; it is the local population which becomes conditioned to new habits and which may acquire new ecological preferences; it is the local population in which all intraspecific competition takes place and which in turn is in competition with its sympatric competitors. In short, to repeat, it is the local population which is the key to the solution of all evolutionary problems. Every higher taxon ultimately originated as the local population of a species. The task before us, then, is to analyze the steps by which a population may become a new higher taxon—a new type.

The Origin of a New Type

Different forms of life were referred to as *types* by comparative anatomists in the last century and even earlier. Bats, whales, birds, penguins, snails, sea urchins, and all the other well-known kinds of animals and plants are such types. Various kinds of types have been recognized in the phylogenetic literature. Purely metaphysical constructs like the archetypes of the idealistic morphologists, which are based on Plato's concept of the eidos, have led to such confusion that systematists now use the term "type"* (if they use it at all) only in the strictly descriptive sense of the totality of characteristics of a given taxon. There is thus essential coincidence between types, the systematist's taxa that they represent, and the structures that characterize them. It is evident that the origin of new types, the origin of new morphological and other biological characteristics, and the origin of the higher taxa are merely three different aspects of the same problem.

When trying to explain the origin of a new, different type, we are at once stymied by an inability to define unambiguously the term "different type." A bird is surely a different type from a terrestrial reptile, and a penguin a different type from an albatross. But is not the Emperor Penguin (*Aptenodytes forsteri*), with its superb adaptation to the rigorous life on the antarctic ice, also a type quite different from the small, hole-nesting penguins of the southern temperate zone, and even from its subantarctic cousin, the King Penguin (*A. patagonicus*)? Is not every good species a separate type, considering its numerous morphological, physiological, and other biological adaptations for the unique niche it occupies? Indeed, does not all the available evidence suggest that the differences between minor and major types are merely matters of degree? Every species is an incipient new genus, every genus an incipient new family, and so forth. And the same is true on the ecological side: there is an insensible gradation from the new niche to the new adaptive zone. And it can probably never be predicted which new niche is a cul-de-sac and which the entrance to a new adaptive zone. Simpson (1953) says quite rightly, "The event that leads, forthwith or later, to the development of a higher category is the occupation of a new adaptive zone," and, he continues, "the broader the zone the higher the category *when fully developed*; yet the new occupant when first entering the zone will hardly, if at all, differ from the parental population.

The higher taxa with which the student of phylogeny deals are ranked as orders, classes, and phyla, but these are far too "fully developed" to tell us much about the first origin of a higher category. For this we have

*For the use of the term in zoological nomenclature see Mayr (1969).

to study taxa in lower categories such as families, genera, and species, which are incipient higher taxa. Species have been shown (Chapter 16) to originate from geographical isolates by acquiring not only isolating mechanisms but also sufficient niche differentiation to be able to coexist with the parental or with sister species. And some of these species are indeed different enough to qualify as potential higher categories. When we map the distribution of monotypic genera we find occasionally that they are nothing but highly differentiated geographical isolates of sister species. Mayr (1942, 1963) has cited many cases of the geographic variation of generic characters.

One of the most spectacular cases of the geographic variation of a basic character was described by Amadon (1947). In the Hawaiian genus *Hemignathus* (honey creepers) the bill is long and strongly curved (Fig. 19.1). In the species *H. obscurus* the lower mandible is about as long as the upper; in *H. lucidus* it is shortened and thickened. When creeping along the trunks of trees in search of insects, *H. lucidus* used its heavier lower mandible to pry or chip off bits of bark. The development of the mandible is carried one step further in *H.* (*lucidus*) *wilsoni*, the geographical representative of *lucidus* on the island of Hawaii. Here the lower mandible has become straight and heavy and is used in woodpecker fashion like a chisel; as a result an entirely new niche was opened to the bird. This population was far more numerous than any *lucidus* population on the other islands and has all the earmarks of a new "type."

The objection may be raised that taxa which are merely aberrant geo-

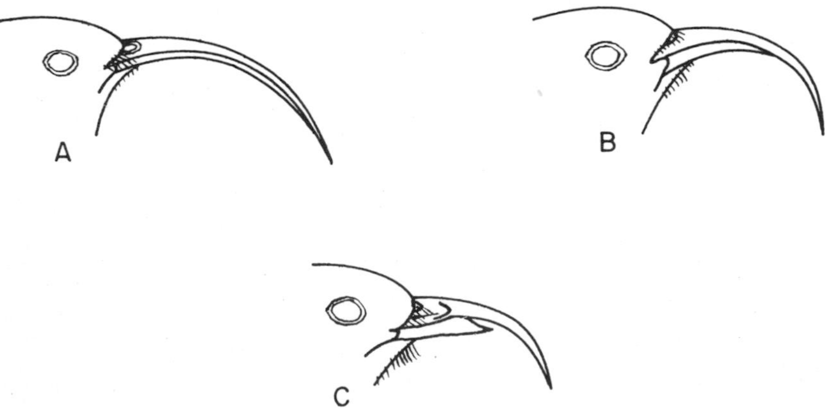

Fig. 19.1. Geographic variation of bill function in the Hawaiian honey creeper *Hemignathus lucidus*. A = *H. o. obscurus*; B = *H. lucidus hanapepe*; C = *H.* (*lucidus*) *wilsoni*. (From Amadon 1947.)

graphical isolates are not true genera but merely artifacts of the taxonomist. This objection is invalid because these taxa are indistinguishable from other genera in their morphological differentiation and ecological role. Indeed, they are like other genera, except for their geographical location. The geographic origin of genera is far more common than one would infer from the literature. The concealment of this interesting fact is largely the fault of the taxonomist. If an allopatric population is strikingly different, it is usually recorded in the literature as a separate genus without stress on its obvious origin by geographic speciation. If it is not very different, it is simply listed as a subspecies without it being emphasized that geographic variation has produced what is normally considered a generic character.

Adaptiveness of Taxonomic Characters

The question of the adaptiveness, the "selective value," of taxonomic characters has been a source of bitter argument since Darwin's days. It is now possible to make a far more objective analysis than heretofore. A close analysis has revealed again and again that "useless," "deleterious," and "neutral" taxonomic characters may give cryptic selective advantages. On the other hand, it is equally true that not all differences between species and genera or between certain diagnostic characters of higher categories are necessarily the result of an ad hoc selection for that particular component of the phenotype.

> To use *Drosophila* as an illustration: the more than 600 known species of this genus all have three orbital bristles on either side of their heads, and the anterior of these bristles is always proclinate (bent forward), while the other two are reclinate (bent backward). Now, why should this character be retained so tenaciously in so many species? Is it really important for the flies of this genus to have one proclinate and two reclinate orbital bristles? (Dobzhanský 1956).

There are usually numerous pathways open to achieve a certain biological end; which pathway is chosen depends on the particular genetic constitution of the incipient species. Nothing can be incorporated into the phenotype that definitely lowers fitness, but, to cite one example, the choice of prey in a parasitic wasp, the way it is handled, the building and provisioning of the brood cells, and many other aspects of the insect's life cycle and behavior have components that were "permitted" by natural selection rather than dictated by it. These are "correlated responses" of selection for other components of the phenotype. Only the adherents of the one-gene–one-character hypothesis insist that every aspect of the phenotype is the result of ad hoc selection.

Preadaptation

In order to be able to enter a new niche or adaptive zone successfully a species must be preadapted for it. The term *preadaptation* was coined by Cuénot during the heyday of mutationism. All evolutionary change was believed by the early Mendelians to be due to major saltations, with the new mutant being either preadapted for a new niche or doomed to immediate extinction. This saltational concept of preadaptation has been abandoned by proponents of the theory of gradual evolution. A redefined concept of *preadaptation* is, however, useful in evolutionary biology: an organism is said to be preadapted if it is able to shift into a new habitat; a structure is preadapted if it can assume a new function without interference with the original function. (See Bock 1959 for a critical discussion of the preadapted aspects of structures.) Either type of shift may lead to the occupation of a new adaptive zone.

The invasion of new adaptive zones. Habitat shifts of a minor nature, such as those of geographical isolates discussed in Chapter 16, require little special preadaptation and have little evolutionary potential. At the other extreme are shifts of fundamental significance such as those from aquatic to terrestrial life, or from terrestrial to aerial life. Such shifts are possible only for the possessor of a highly unlikely combination of characteristics, and this is the reason for the infrequency of such shifts. Let us look, for instance, at the shift from water to land. Although land is such a favorable and diverse environment that 85 to 90 percent of all species of animals live on dry land, they all belong to only 3 of the 25 known phyla of animals, the arthropods, the mollusks, and the higher vertebrates. The reasons for the infrequency of the ecological shift from water to land are obvious when one considers the following facts (Rensch 1960).

(1) *Weight.* Water is 800 times as heavy as air. Only animals with a strong skeleton or armor can become terrestrial. A jellyfish needs no support in water, but collapses completely as soon as it is brought on land. Ciliary locomotion, so common in water, is useless in air. Locomotion on land requires strong muscles.

(2) *Protection against the environment.* A land animal must be protected against the danger of drying out and against strong fluctuations of temperature. There is a high selective premium in favor of a tough skin, armor, or scales.

(3) *The excretory system.* Permanent life on land, in contrast to amphibious existence, requires the excretion of metabolic end products in such a manner as to reduce water loss to a minimum.

(4) *The respiratory system.* It must be possible for the land animal to take in oxygen directly from the air.

(5) *Sense organs.* On land, there is much greater need than in water for sense organs that are effective at long range, particularly among the rapidly moving land animals; hence land animals must develop long-distance vision and hearing.

It is true that as soon as the amphibious mode of life is adopted all of these preadaptations will continue to be perfected owing to greatly increased selection pressure, but the very beginning of an amphibious life is impossible without considerable preadaptation. Individual species, genera, and families of many other phyla have gone on land, such as certain turbellarians (land planarians), nemerteans, rotifers, ostracods, nematodes, oligochaetes (earthworms), and land leeches. However, all of these are forced to live in a moist environment; none has succeeded in becoming truly terrestrial. In spite of all preadaptations the shift of the tetrapods from water to land was a slow and painful process. The first truly terrestrial reptiles did not appear until some 60–75 million years after the origin of the tetrapods.

Most invasions of new adaptive zones do not require nearly so formidable a set of preadaptations as the shift from water to land. To become a member of the rich fauna that lives in the interstitial spaces of sea-bottom sand and silt, small size, an ability to tolerate temporary oxygen deficiencies, and certain types of locomotion seem to be the major prerequisites. This extraordinary habitat has been colonized by some coelenterates, mollusks, holothurians, a mobile solitary bryozooan, various crustaceans, and nemerteans, and it is one of the major areas of radiation for turbellarians, gnathostomulids, gastrotrichs, archiannelids, copepods, nematodes, and ciliates. The colonization of fresh water by salt-water forms and vice versa involves primarily preadaptations of water and salt metabolism.

Sometimes no active invasion of a new habitat is involved, only an ability to take advantage of a climatic or vegetational shift. This was the case when some line of browsing horses began to feed on grasses (grasslands had then already existed for millions of years) and gave rise to the prosperous branch of modern horses.

The availability of a new resource in the environment is invariably a stimulus for the origin of types adopting it as niche. The evolution of land plants preceded the evolution of terrestrial faunas. The rise of the angiosperms was followed by the explosive evolution of insects and insect eaters. And the origin of any kind of higher organisms is almost invariably followed by the origin of some new species of parasitic arthropod, helminth, or protozoan. Among the 40,000 species of animals that occur in Germany no less than 10,000 are parasites on the remaining 30,000.

Each major shift of habitat is an evolutionary experiment. Each of the successful branches of the animal kingdom, for instance the insects, the tetrapods, or the birds, is the product of such a shift, with the decisive step quite likely taken by a single species. "However, not all such shifts are equally successful. No spectacular adaptive radiation has followed the invasion of the sand niche by a coelenterate. The shift of a carnivore (Giant Panda) to a herbivorous diet has not led to a new phylogenetic breakthrough . . . The tree kangaroos, the return of a specialized line of terrestrial marsupials to arboreal life, likewise seem to have reached an evolutionary dead end" (Mayr 1960:369). Not every evolutionary experiment is a success; in fact most of them are failures.

Key Characters and Mosaic Evolution

According to the saltationist, a new type comes suddenly into existence and evolves harmoniously from that point on. The original ancestor, the archetype, lacks all the specializations of its descendants, and subsequent evolution affects all structures at about the same rate. Missing links are in every respect halfway between the types which they connect.

Paleontologists and comparative anatomists have shown that none of these postulates is correct. The rates of evolution of different organs are often drastically different. Some may rush far ahead while others stagnate. *Archaeopteryx*, the "missing link" between reptiles and birds, is a typical pseudosuchian reptile in nearly all of its characters, but in its feathers it is like a modern bird (Table 19.1). One of the oldest known amphibians,

Table 19.1. Mixture of reptilian and avian characters in *Archaeopteryx*.

Reptilian	Intermediate	Avian
Teeth Free tail vertebrae Unkeeled sternum Free metacarpals Ilia and ischia separated Cranial kinesis Reptilian snout	Metatarsals (partly fused) Brain Reversed, but small hallux Incompletely fused synsacrum	Feathers Furcula Pelvis with backward pubes Large eyes Gliding wing

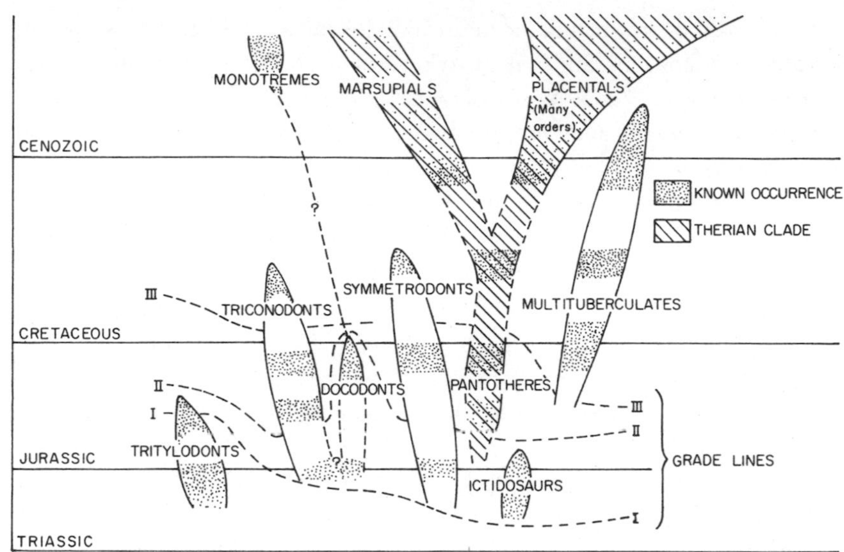

Fig. 19.2. Repeated and independent acquisition of mammalian characters (grades I, II, III) by various lines of mammal-like reptiles (therapsids). The triconodonts, symmetrodonts and multituberculates became extinct. The docodonts probably gave rise to the monotremes, the pantotheres to the genuine eutherian mammals. (From Simpson 1959a; for details see Simpson 1959b.)

the stegocephalian *Ichthyostega* from the upper Devonian of Greenland, had as many (or more) fish characters as amphibian characters. The earliest mammals were so much like their ancestors, the therapsid reptiles, except for the jaw region and certain tooth characters, that the experts still argue where to draw the line between the two classes and how many separate phyletic lines are involved (Fig. 19.2).

During the shift into a new adaptive zone, one structure or structural complex is usually under particularly strong selection pressure, like the wing in the avian ancestor. This structure then evolves very rapidly while all others lag behind. As a result there is not a steady and harmonious change of all parts of the "type," as envisioned by the school of idealistic morphology, but rather a *mosaic evolution.* Every evolutionary type is a mosaic of primitive and advanced characters, of general and specialized features. A realization of this unequal evolution of the components of the type has been implicit in the work of outstanding paleontologists for nearly 100 years; indeed, it was commented on by Lamarck.

Even missing links may be odd mixtures of advanced and primitive characters. When the South African ape-man *Australopithecus africanus* was discovered, an outstanding anthropologist assured me that it was surely an aberrant side line and not anywhere near the direct human ancestry because it was a "disharmonious type," a composite of hominid and anthropoid characters, and consequently a failure and doomed to extinction. *Australopithecus* is, indeed, very much of a mixture of advanced and ancestral characters, yet it is now quite evident that man's ancestors must have passed through an *Australopithecus*-like stage (Chapter 20).

Which key structure it is that is responsible for an evolutionary breakthrough is sometimes quite obvious. For birds it is the development of the feather, and for the artiodactyls the reconstruction of the tarsal joints. In other cases there are several structures involved, as lobed fins, lungs, and internal nares in the crossopterygians that gave rise to the earliest tetrapods. Sometimes it is a new behavior pattern, rather than a structure, which is the key invention that opens up the new adaptive zone. The adoption of internal fertilization in some branches of the amphibians and early reptiles, as well as in all terrestrial invertebrates, was an absolute prerequisite for a truly terrestrial existence. This, in turn, permitted the development of the hard-shelled amniote egg which, so Romer (1957) suggests, was developed by the early reptiles before they had become truly terrestrial. The invention of mastication by some therapsid reptiles was perhaps the behavior invention that gave rise to the mammals. Adaptive shifts occur probably most commonly in organisms that show—at first—no structural change at all. Many tree-trunk–climbing birds, for instance, lack the stiffened tail feathers of woodpeckers and true tree creepers. And the more primitive woodpeckers get along perfectly well with a normal perching foot; only the more advanced genera have evolved specialized foot structures.

The Origin of Higher Taxa

The hierarchy of categories that the classifying taxonomist recognizes is an attempt to express degrees of similarity ("characters in common") and recency of common descent. The methods employed and the pitfalls encountered in this endeavor are discussed by Simpson (1961) and Mayr (1969). The most closely related species are combined into genera, groups of related genera into subfamilies and families, these into orders, classes, and phyla. Our delimitation of a higher taxon and its ranking in the hierarchy have a large arbitrary component. We would never consider *Archaeopteryx* the representative of a separate class of vertebrates if this branch of the reptiles had not given rise to the flourishing multitude of birds, outnumbering by far all the "other" living reptiles. "A higher category is higher because it *became* distinctive, varied, or both to a higher degree and not directly because of characteristics it had when it was arising" (Simpson 1953).

The earliest representatives of a new higher taxon would certainly be included in the taxon from which they originated if it were not for their known descendants. The fossil genera of placental mammals, for instance, that are known from the early and middle Paleocene are far less different from each other than all the known marsupials, which have always been classed as a single order. "If we knew no placentals after the middle Paleocene we would certainly place them in a single order. *As of then* their proper comparative categorical rank was in fact that of an order. They are placed in six different orders because we recognize in them ancestors and allies of what *later* became six orders" (Simpson 1953). A new taxon does not arise as an order, class, or phylum. It arises as a new species and eventually becomes a new genus that we assign to a new order only because its subsequent descendants show the degree of distinctness and of discontinuity (after much extinction) that by convention is considered to signify ordinal rank. There is an early Paleocene genus, *Protogonodon,* some species of which on the basis of certain details of structure could be classified, according to Simpson, as carnivores and some as ungulates. The experienced specialist can see here the beginnings of two orders within a single genus. Similar cases have been described for fossil taxa in various groups of animals.

Where a new group is the result of a broad and varied adaptation, it may arise cryptically and its distinguishing characters may not become apparent until some time after the group has branched off. The origin of the mammals is a prime example. In other cases, the origin of a new group is very closely correlated with, and even sometimes preceded by, a structural invention, the origin of an evolutionary novelty.

The Emergence of Evolutionary Novelties

The origin of a new type is sometimes correlated with the "invention" of a new structure. How can such a new structure come into being, except by the saltational origin of the new type that has it? Furthermore, how can an entirely new structure originate without a complete reconstruction of the entire type? Can it be gradually acquired when the incipient structure has no selective advantage until it has reached considerable size and complexity?

I have treated these questions in detail elsewhere (Mayr 1960) and will here only summarize my findings. I will begin by defining an *evolutionary novelty as any newly acquired structure or property that permits the performance of a new function, which, in turn, will open a new adaptive zone.* Many evolutionary novelties, such as new habits and behavior patterns, are not primarily morphological, although they may have morphological consequences. The study of metabolic pathways has revealed the occurrence of molecular evolutionary novelties that are as important as such new structures as lungs, extremities, or brood pouches. Chemical inventions on the cellular level are the prerequisite of some of the most important adaptive shifts. Alas, our knowledge of comparative biochemistry is still far too rudimentary to tell us whether or not it was a biochemical invention that gave the mollusks, crustaceans, and other now dominant groups of marine invertebrates their ascendancy over eurypterids, trilobites, graptolites, and brachiopods, once the rulers of the seas.

A more difficult problem is posed by new structures, like the bird feather, the mammalian middle ear, the swim bladder of fishes, the wings of insects, and the sting of aculeate hymenopterans. Two major modes of origin of new structures have been suggested.

(1) A preexisting structure may be modified owing to an *intensification of function.* Sewertzoff (1931) has made a special study of this process. An intensification of the running function has led to the conversion of the five-toed foot of primitive ungulates into the two-toed foot of the artiodactyls or the one-toed foot of the perissodactyls. Looking at it somewhat differently, one could also call this an "intensification of selection pressure." Since this pressure is directed against an existing structure, "intensification" does not lead to the emergence of anything that is basically new, and yet it may result in so drastic a reorganization of the phenotype that one's first impression is that an entirely new organ has emerged. During this process there is never a stage in which the incipient structure is "not yet of selective value," as is sometimes claimed by antiselectionists.

The striking differences between birds or mammals and their reptilian ancestors are not due to the origin of any truly new structures. Most differences are merely shifts in proportions, fusions, losses, secondary duplications, or other modifications that do not materially affect what the morphologist calls the "plan" of the particular type. The decisive role in most of these modifications is an intensification of function of one component.

(2) The most important cause of the origin of new structures is a *change of function.* This factor was recognized by Darwin (1859:454). The origin of a new structure through change of function depends on two prerequisites:

the capacity of a structure to perform simultaneously two functions (and more specifically its preadaptation for a second function) and the duplication of one of these functions by another structure. Many primitive fishes, for instance, had two independent organs of respiration, gills and primitive lungs. In the terrestrial tetrapods the simple baglike lungs were converted into the highly complex respiratory organ of mammals and birds, and the gill arches and pouches into endocrine glands and accessory organs of the digestive system. In the later fishes the gills were elaborated and the lungs often converted into a swim bladder or accessory sense organ. Cases of drastic changes of function of an organ are found in every phyletic line of animals. To give a complete catalogue would mean listing a good portion of all animal structures. The change of the ovipositor of bees into a sting, of scales into teeth, of the *Daphnia* antennae into paddles, and of arthropod legs into mouth parts or copulatory organs are a few examples. In all these cases the structure evolved under the selection pressure of the primary function until it was large enough to take on the additional secondary function. The finding that not every function requires its own executive structure invalidates one of the favorite arguments of the antiselectionists. A new function can be performed by a structure that evolved for a different reason.

The Role of Behavior in Evolutionary Shifts

A shift into a new niche or adaptive zone is, almost without exception, initiated by a change in behavior. The other adaptations to the new niche, particularly the structural ones, are acquired secondarily (Mayr 1958, 1960). Habitat and food selection are also behavioral phenomena, both playing a major role in the shift into new adaptive zones. It is no exaggeration to say that, as far as animals are concerned, behavior is the most important evolutionary determinant, particularly in the initiation of new evolutionary trends. In the case of sibling species, which often show remarkable behavioral differences, it is particularly obvious how relatively unimportant the morphological changes are. Most recent shifts into new ecological niches are, at first, unaccompanied by structural modifications. Where a new habit develops, structural reinforcements follow sooner or later. Miller (1950) has shown how the characteristics of the thrasher genus *Toxostoma* can be gradually traced from a mockingbird-like arboreal ancestor. As the various species of thrashers became more and more terrestrial and more and more adapted to scratch on the ground and to use their bills for digging and removal of leaf litter, a selective premium was placed on strengthening of tarsus and toes and on increased curvature of the bill.

The same is true of special structures that facilitate specific behavior elements. Many birds raise their nape feathers, but only some of these have developed elongated crests in this area. However, no birds have developed a crest without the ability to raise it. The precedence of behavior over structure is most easily demonstrated with respect to courtship patterns in birds and the accumulation of morphological characteristics that make the courtship more conspicuous and stimulating.

New habits and behavior elements always start in a concrete local population. If the new behavior adds to fitness, it will be favored by selection and so will be all genes that contribute to its efficiency. That new habits occur all the time in natural populations is abundantly documented in the natural history literature. A particularly striking example is the recently developed habit of British titmice, mostly *Parus major*, of opening milk bottles and drinking the cream. If the milk bottles had been an unoccupied natural niche, it is evident that a selection pressure would have been set up, on one hand, for the titmice to develop a more efficient milk-bottle opener and, on the other, for the milk bottles to become less easily opened, assuming the milk bottles to be organic material that could be modified by selection. The enormous role played by behavior in initiating transpecific evolution is being increasingly appreciated by evolutionists.

Baldwin Effect

Prior to the rediscovery of particulate inheritance, Baldwin (1896) proposed a "new" evolutionary principle in an effort to reconcile Lamarckism with Weismann's Neo-Darwinism. The Baldwin effect (later so called) designates the situation where, owing to a suitable modification of the phenotype, an organism can stay in a favorable environment until selection has achieved the genetic fixation of this phenotype. Baldwin described his "organic selection" (as he called it) as a strict alternative to natural selection, which he discounted as an evolutionary force owing to, he said, its purely negative effects.

The conceptual assumptions underlying the hypothesis of the Baldwin effect make it desirable to discard this concept altogether. Mayr (1963: 610–612) has fully analyzed the flaws in these assumptions. It has been known since Darwin's time that changes in habit often precede structural adaptations. Chance, as well as inherent genetic and phenotypic variability, permits shifts in niche occupation that set up new selection pressures. Such shifts, contrary to Baldwin's contention, are completely consistent with natural selection. If the phenotype is highly plastic, the selection pressure may actually be reduced because there is no selective advantage in changing the genotype when an individual can adjust itself phenotypically to a current condition. Invoking a "Baldwin effect" in no way helps to clarify the evolutionary process.

Evolutionary Potential and Predisposition

The reevaluation of mutation as the source of variability rather than as a direct evolutionary force and of the character as a product of the whole genotype rather than of a single gene has changed the interpretation of many evolutionary phenomena. The new concepts make it easier to understand why there are such definite limits to variation in any given group of animals and why a change in one character often produces such serious "correlated effects" (Chapter 10), phenomena that had puzzled Darwin greatly. Every group of animals is predisposed to vary in certain of its structures and to be amazingly stable in others. Mayr and Vaurie (1948) pointed out that whenever the plumage of drongos (Dicruridae) varies it is the frontal crest or the tail or both that are affected. Certain groups of mammals have a predisposition to develop horns on the forehead, others to develop them on the top of the head, others not to have horns at all. Taxonomy would be difficult if it were not for these inherent differences between groups. Only part of these differences can be explained by the differences in selection pressures to which the organisms are exposed; the remainder are due to the developmental and evolutionary limitation set by the organisms' genotype and its epigenetic system.

The felicitous term *grade* was introduced into the evolutionary literature by Huxley to designate "a step of anagenetic advance, or unit of biological improvement." Several related lines may reach the same adaptive or structural grade independently. It is, of course, quite unreasonable to demand that major levels in phyletic evolution should always coincide with the branching of lineages. Often they do, but often they do not.

True *parallelism* is due to response of a common heritage to similar demands of the environment (similar selection pressures). Where no common heritage exists, evolutionary parallelism is more correctly called *convergence*. The animal world is full of convergences (and so is the plant world!) where similar demands by the environment have evoked similar phenotypic responses in unrelated or at least not closely related organisms. If there is only one efficient solution for a certain functional demand, very different gene complexes will come up with the same solution, no matter how different the pathway by which it is achieved. The saying "Many roads lead to Rome" is as true in evolution as in daily affairs.

The Genetics of Transpecific Evolution

In the past generations the idea was widespread that transpecific evolution—the origin of higher taxa and of evolutionary novelties—was controlled by a genetic system different from that controlling populations. All that we now know about the nature of genetic systems and about the genetic basis of characters refutes any such dualism. To be sure, the analysis of genetic differences between higher taxa through crossbreeding is impossible, but extrapolation from intraspecific genetic analysis permits conclusions that are entirely consistent with the observed phenomena. The comparative study of molecular differences in enzyme systems has given us an additional powerful analytical tool.

It is most important to clear up first some misconceptions still held by a few people not familiar with modern genetics:

(1) Evolution is *not* primarily a genetic event. Mutation merely supplies the gene pool with genetic variation; it is selection that induces evolutionary change.

(2) A character is *not* (normally) the product of a single gene, and a change in a character is therefore not an indication that a single gene has mutated. Virtually all characters are highly polygenic and, since most genes are pleiotropic, the change of a character indicates a minor or major reconstruction of the genotype.

(3) The assumption that genes can be classified into those that are superior and will automatically be incorporated into the gene pool and those that are inferior and will inexorably be eliminated is quite misleading. The selective value of a gene is not absolute: in any individual case, it is to a large extent determined by the external environment and the epigenetic system in which the gene operates. Natural selection consequently is not a problem of simple arithmetic. It must, furthermore, be assumed that the genes of a gene pool, having been selected for their coadaptational capacity, will make on the average a greater contribution to fitness than randomly added new genes.

Transpecific evolution is not a matter of isolated genes and mutations, but of whole coadapted gene complexes. As soon as this is clearly understood, the interpretation of macroevolutionary phenomena becomes much less difficult. The genotype is not a beanbag full of unconnected genes.

Most actions of genes are interactions during development, and the genotype is therefore an epigenetic system, or, as Waddington called it, an epigenotype. This includes the entire system of regulatory genes.

The "wholeness" of the genotype, owing to its epigenetic system, explains many other phenomena that are difficult to interpret in terms of beanbag genetics, for instance, "tendencies" in certain families and orders that are absent in others. So-called orthogenetic trends are due to the fact that the evolutionary changes of the phenotype owing to natural selection are limited by the possible amplitude of response of the epigenotype.

The evolution of the genotype, as a whole, may explain also the well-known phenomenon of *evolutionary inertia.* The larger the number of genes that contribute to the shaping of a phenotypic trait, a "character," the less likely it is that such a character will be modified through natural selection. Because many of these genes will be pleiotropic, they will simultaneously affect also components of fitness. This is the explanation of the morphological similarity of sibling species and of the existence of generic, family, and ordinal characters. Other factors that contribute to stability, on the level of the population, are allelic balance (heterosis), epistatic (internal) balance, and the entire system of regulatory genes, discussed in Chapters 9 and 10. Almost any change, but particularly a major change, of the phenotype in such a well-balanced system will be deleterious. Although minor gene substitutions may be frequent, the well-buffered system of developmental canalizations shields the phenotype from major changes. There is opportunity for speciation, but a major alteration of the morphotype is impossible as long as the epigenotype is intact. A gene pool rich in genetic variation and with a well-buffered epigenetic system has great fitness, but there is much evidence to indicate that its evolutionary potential is not as great as is generally believed. We find again and again in the fossil record abundant and apparently highly successful genera that remained essentially unchanged for millions of years while far less common types simultaneously underwent a process of rapid evolution. One of the reasons for the absence in the fossil record of some of the links between major groups is that they belonged to such rare lineages.

Evolutionary stagnation has always been a puzzle. To mention a few stable types, how could *Limulus* (horseshoe crab) and *Triops* (fairy shrimp) have stayed so nearly the same for more than 200 million years? About 5 species of insects and 1 myriapod in Oligocene amber (35 million years old) are indistinguishable from still living species. More amazing, some 2 billion year old microorganisms are indistinguishable from living types. It is probable that an exceptionally well-balanced epigenotype is responsible for such phenotypic and genotypic stability. Indeed, it appears to me that to store genetic variability is perhaps not nearly so great an evolutionary problem as to escape the constraints of the well-balanced genotype. Since new characters are not produced by mutations (as the typologists thought) but by a reorganization of the genotype, it may require a "genetic revolution" (Chapter 17) to break up the perfectly buffered genotype.

The loosening of a tightly knit, coadapted gene complex can presumably be achieved in many ways. One way is a rapid change of population size accompanied by a temporary depletion of the gene pool, resulting in a drastic change of selective values of the included genes (Chapter 17). In plants, hybridization might achieve the same end. A biochemical invention might affect a sufficient number of different metabolic pathways to result in a genetic shake-up. Finally, the emergence of any new structure, for the reasons given above, may set up a selection pressure sufficiently strong and varied to break down the genetic homeostasis. Since we were unable to study these processes when they occurred, and only see their consequences, we will probably never get definitive answers to these questions.

It is evident that there are several kinds of genetic systems, some favoring specialization, some broad, general adaptation, and some evolutionary pioneering. A lineage must be able to switch back and forth between these systems to achieve the greatest evolutionary success. A pioneering line, after having made a breakthrough, must develop broadly adapted species to have a future (Chapter 14).

The Life History of a Higher Taxon

Some evolutionists have compared the evolution of a higher taxon to the life cycle of an individual. They contend that a higher taxon goes through a series of stages that can be designated by the terms birth, growth, maturity, senescence, and death. Furthermore they claim that such life cycles are sometimes paralleled in various groups to such an extent that an entire fauna simultaneously goes through such a cycle. The more detailed analysis of fossil faunas shows, however, that such contentions are greatly exaggerated and that the stated regularities are largely absent. To be sure, every group of animals has originated at some time and has eventually died out unless it has persisted into the Recent fauna. But this is about all the regularity one can find. Most evolutionary lines with an adequate fossil record have had one or several periods of flowering ("bursts"), while a few lines seem to have persisted from beginning to end, for periods of sometimes more than 100 million years, without marked ups and downs in generic and familial diversification. The great period of proliferation in the lower categories often comes immediately after the origin of the new type, as with the trilobites, nautiloids, and some of the brachiopods and corals. The reptiles, however, although well diversified when first recorded in the Early Pennsylvanian, did not have their great period of dominance and diversity until Triassic and Jurassic times. Mammals have existed since the Triassic, but their flowering did not take place until more than 100 million years later, in the Paleocene and Eocene.

The concept of a life cycle is also contradicted by the fact that many groups have had several periods of eruptive evolution, such as the crinoids in the Silurian, Mississippian, and Permian, and various groups of mollusks (Fig. 19.3). The period of proliferation may be short, such as that of the mammals in the early Tertiary and that of the land pulmonates, or it may be long—that of the marine prosobranchs extended for 150 million years. Periods of strong eruptive evolution in marine animals occur in different geological periods from those of land animals: in the oceans in the Ordovician, Triassic, and Jurassic, on land in the Carboniferous and Permian.

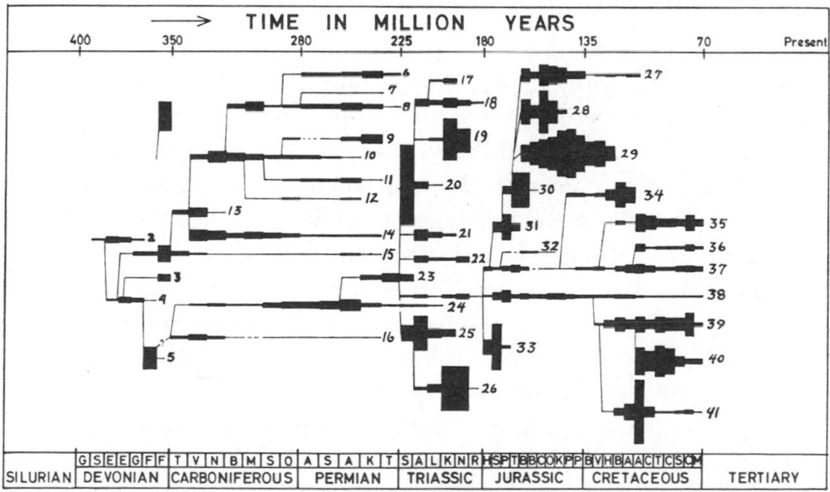

Fig. 19.3. History of the 41 superfamilies of ammonoids showing mass extinctions at the close of the Devonian, Permian, Triassic, and Cretaceous periods. Height of the blocks is proportional to the number of genera known from each stage. Number 26 represents 48 genera. Letters in boxes above the names of the periods are the initial letters of the standard stages. (From Newell 1967.)

All these facts prove that the resemblance of the life cycle of a higher taxon to that of an individual is quite spurious. Indeed, the rise and fall of higher taxa is merely the total effect of the superposition of a number of largely independent evolutionary phenomena. Five of these are clearly distinguishable and the existence of others is indicated by the facts. These five phenomena are: (1) the origin of the new type, (2) a great multiplication of species of this type, (3) their divergent evolution and adaptive radiation, (4) increases or decreases in genetic homeostasis, and (5) a gradual or sudden decrease and extinction. Each of these phenomena can be analyzed individually, as has been done by Simpson, Rensch, Stebbins, and others.

Recent research on the evolution of phyletic lines leaves many questions unanswered. Why did the dinosaurs on land and the ammonites in the sea suddenly become extinct in the Cretaceous? Why did some classes and phyla of invertebrates have their great burst in the Ordovician, others in the Silurian, or in the Mississippian? Why were such dominant groups as the trilobites, graptolites, brachiopods, fusulinid foraminifera, and nautiloids displaced by later groups? Many of these events took place at such a remote time that we may never get an answer. Yet it has become clear that there is nothing in the past history of the earth that cannot be interpreted in terms of the processes that are known to occur in the Recent fauna. There is no need to invoke unknown vital forces, mutational avalanches, or cosmic catastrophes.

Geographic speciation, adaptation to the available niches (guided by selection), and competition are largely responsible for the observed phenomena (see Chapters 16 and 18). In other words, the interaction between organisms and environment is the most important single determinant in the rise and fall of evolutionary types. For instance, when the continental shelves are flooded and covered with shallow seas, there are numerous partially isolated embayments that provide abundant opportunity for speciation. A change in sea level may result in an intermingling of the previously isolated faunas of such embayments and manifest itself in the fossil record as a very drastic change of fauna.

The occurrence of speciation, even of a very active multiplication of species, does not always signify great evolutionary activity. This is well documented by speciation in stable types, such as *Drosophila*, *Culex*, or the weevils. Yet, with few exceptions, a breakthrough into a new adaptive zone is followed by two events: a colossal speed-up in the rate of evolutionary change and a period of adaptive radiation. Simpson in particular (1953, 1960) has pointed out how rapidly a new type may reach a new phylogenetic "grade," but once this grade is reached the type remains essentially stable. The bats (Chiroptera), for instance, presumably evolved from insectivore-like ancestors sometime in the Paleocene, but the first known fossil (middle Eocene) is essentially a modern bat. About twice as much phylogenetic change occurred in the lungfishes (Dipnoi) during a period of about 30 million years in the middle and late Devonian as in the 250 million years since (Westoll 1949; Fig. 19.4). A similarly rapid early evolution is typical for most better-known phyletic lines. In view of these exceedingly uneven rates of evolution, it is quite misleading to extrapolate from the rate of evolution along one part of a lineage to others. Once an ancestral type has been "loosened up," by the unbalancing of its structural,

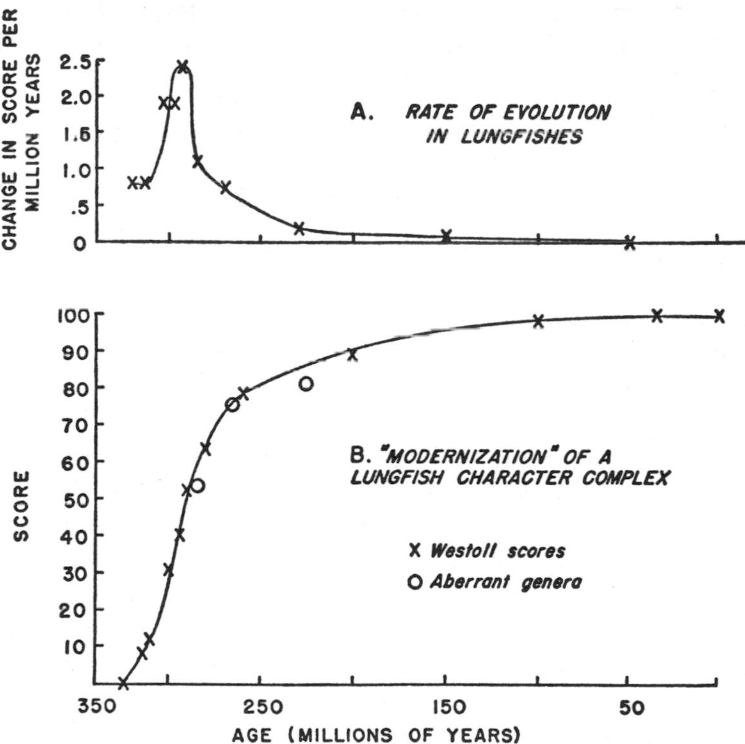

Fig. 19.4. Rate of evolution in lungfishes from (reconstructed) Silurian condition to the present. "Score" is the number of advanced structural characters or character grades. (0 is the inferred most primitive condition.) A = rate of change per million years, B = level of "modernization" reached at a given geological period. (From Simpson 1953, after Westoll 1949.)

epigenetic, and genetic homeostasis, a rapid shift seems to be possible under the greatly increased selection pressure in the new adaptive zone. This will proceed until a new balance is achieved, and a new grade has been integrated and stabilized.

The second aspect of every major breakthrough, and in fact even of most minor ones, is a great development of minor types in the new adaptive zone, a phenomenon usually referred to as *adaptive radiation.* Every kind of animal is adapted for a certain mode of living, but there are numerous niches possible within this mode. For instance, birds are adapted for life in the air, but a warm-blooded, air-inhabiting, feathered vertebrate has a choice of thousands of niches, ranging from that of a swift to that of a woodpecker, a duck, or a penguin. On the adaptive plateau "bird" there are numerous minor peaks that can be occupied. The fossil history indicates that during periods of proliferation numerous experiments are made which have no lasting success. Aberrant types produced during such periods disappear as rapidly as they appear.

Extinction. Considering that the number of living species presumably adds up to a good deal less than 1 percent of those that have ever existed, it is evident that extinction is one of the most conspicuous evolutionary phenomena. Although a certain number of species become extinct at all times, some geological periods, for instance the end of the Permian, have witnessed far more extinction than others. Cosmic events, like the passing of the earth through a radioactive cloud, are sometimes invoked as explanations, but the fact that the great periods of faunal turnover on land and in the seas do not coincide deprives the cosmic theories of all probability. Climatic events, orogeny, and other geophysical processes that produce fluctuations of the sea level on the continental shelves are far more likely causes. Individual species may become extinct owing to new or newly invading diseases or changes in the biotic environment (particularly a loss of habitat or the arrival of a more successful competitor), yet ultimately their extinction is due to an inability of their genotype to respond to new selection pressures. The smaller the species population the more vulnerable to extinction it is (Mayr 1965a). Large-scale extinction often occurs when two whole faunas mingle owing to geological events (fusing of two continents or two ocean basins). The reaching of certain threshold values in the changing chemistry or nutrient content of ocean water may also be a factor in large-scale extinction. The actual cause of the extinction of any fossil species will presumably always remain uncertain. Indeed, we often have trouble determining the cause of the extinction of Recent species.

It is certain, however, that any major epidemic of extinction is always correlated with a major environmental upheaval.

The Evolutionary Role of Species

This consideration of macroevolution and of the higher taxa has led to new interpretations of the evolutionary significance of species. Closely related species are variations on a theme. Even though each species in a related group occupies a different niche, these niches are, on the whole, quite similar. Such niches are only subdivisions of a single "adaptive zone," as Simpson would call it. The adaptive zones, particularly the major ones, are occupied by distinct types, such as rodents or bats, birds or snakes, sharks or eels. Each of these adaptive types is characterized by a unique set of attributes. It requires a singular combination of genetic, physiological, and morphological properties and a unique constellation of environmental conditions to enable an animal to invade a unique major adaptive zone. Only one species in 10,000 or 100,000 will have the improbable combination of characteristics that preadapts it to undertake a major ecological shift.

The real difficulty is that ecological "space" is not continuous. The terrestrial insectivores and the aerial bats are separated by an adaptive discontinuity, and so are the diving petrels and the penguins. Adaptive zones are separated from each other by zones of adaptive disequilibrium. To cross them is hazardous and the chances of success are slight. Every new species is an ecological experiment, an attempt to occupy a new niche. The number of successful shifts into new adaptive zones will be directly proportional to the total number of new species that come into existence.

For this reason I do not agree with Huxley (1942) when he says, "Species formation constitutes one aspect of evolution; but a large fraction of it is, in a sense, accident, a biological luxury, without bearing upon the major and continuing trends of evolutionary process." On the contrary, I am convinced that it is the very process of creating so many species that leads to evolutionary progress. Species, in the sense of evolution, are quite comparable to mutations. Like species, mutations are a necessity for evolutionary progress, even though only one out of many mutations leads to a significant improvement of the genotype. Since each coadapted gene complex has different properties and since these properties are not predictable, it requires the creation of a large number of such gene complexes before one appears that will lead to real evolutionary advance. Clearly, then, a prodigious multiplication of species is a prerequisite for evolutionary diversification.

Each species is a biological experiment. The probability is very high that the new niche into which it shifts is an evolutionary dead-end street. There is no way to predict, as far as the incipient species is concerned, whether the new niche it enters is a dead end or the entrance into a large new adaptive zone. The tetrapods go back to a single ancestral species, as do the insects and presumably the angiosperms. If this ancestor had not experimented with a niche shift, the new adaptive empire would not have been discovered.

The evolutionary significance of species is now quite clear. Although the evolutionist may speak of broad phenomena, such as trends, adaptations, specializations, and regressions, they are really not separable from the progression of the entities that display these trends, the species. The species are the real units of evolution, as the temporary incarnation of harmonious, well-integrated gene complexes. And speciation, the production of new gene complexes capable of ecological shifts, is the method by which evolution advances. Without speciation there would be no diversification of the organic world, no adaptive radiation, and very little evolutionary progress. The species, then, is the keystone of evolution.

20 · Man as a Biological Species

Man is a product of evolution. Much that is puzzling about man can be understood only when man is considered as evolved and evolving. A thorough knowledge of the principles and mechanisms of evolution is therefore a prerequisite for the understanding of man.

Man is a species of animal, as is self-evident as soon as one applies the concept of evolution to man. He shares many characteristics with other animals, and to study man from the viewpoint of the biologists reveals aspects of man's nature that are omitted from the traditional picture presented by the humanities. But no more tragic mistake could be made than to consider man "merely an animal." Man is unique; he differs from all other animals in many properties, such as speech, tradition, culture, and an enormously extended period of growth and parental care. This has been pointed out perceptively by Huxley, Haldane, Simpson, Dobzhansky, and other recent writers. In my own discussion the emphasis will be on the biological aspects of man's evolution and on those questions that must be asked as a consequence of recent advances in our understanding of biological evolution.

The Phylogeny of Man

Man is as much a product of evolution as is any other organism. Indeed, man is the historical creature par excellence. He has not only a biological heritage but also a cultural one, making him history-bound in two separate ways. Man's gradual shift from the status of an animal to that of "not merely an animal," and the forces that brought about this evolution, are by no means fully understood and are a source of much controversy, in part because the reconstruction of man's phylogeny is still largely a matter of guesswork. Yet considerable progress has been made in recent years owing not only to the discovery of many additional fossil hominids in southern Asia and in Africa but also, and perhaps more important, to the revision of some basic concepts of evolution and phylogeny. Our thinking on missing links, the evolution of "types," irreversibility, the role of mutations and of the environment in evolution, the variability of population samples, and the meaning of species has changed so much in recent years that even the previously known fossil hominids are now regarded in a very different light than they were only 25 years ago.

Man is so strikingly similar to certain other mammals that no biologist can question the closest relationship. Even long before evolution was seriously considered, Linnaeus placed man in the order Primates together with apes and monkeys. Although not conceived in terms of evolutionary biology, this arrangement was generally accepted by post-Linnaean authors. Comparative anatomical studies fully confirmed the great similarity between man and the anthropoid apes (Pongidae).

All phylogenetic investigations involve two aspects of evolution—the branching off from the original line and the development of the new line—and these must be carefully distinguished in order to avoid confusion. In the case of man, one is the branching of the hominid from the pongid line (speciation) and the other is the reaching of the human level within the hominid line (assuming that the earlier representatives of the hominid line were still anthropoid apes). Consequently we must answer two questions:

(1) When and where did the hominid line branch off from the anthropoid line that gave rise to the Pongidae and what did the missing link look like?

(2) Through what stages did the hominid line pass, after its separation from the pongid line, before the truly human level was reached?

The Search for the Missing Link

Man's closest relatives among the living primates are the so-called anthropoid apes. They consist of three groups, perhaps best regarded as three genera. These are the chimpanzee and the gorilla (genus *Pan*) in Africa, the orang (*Pongo*) in the East Indies (Borneo, Sumatra), and the gibbon group (*Hylobates*) in southeastern Asia and the East Indies. Some of the differences between man and the living anthropoids are listed in Table 20.1.

The living apes, in spite of all their differences, share several so-called "anthropoid characters," such as powerful canines, large incisors, a sectorial form of the first lower premolar, a simian shelf of the mandible, specialized feet, and powerful brachiating arms. These characteristics are striking and delineate a pronounced gap between man and the living anthropoids. Yet

Table 20.1. Some of the differences between man and the anthropoids.

Characteristic	Man	Anthropoids
Long bones of lower extremities	Longer than those of upper extremity	Shorter than those of upper extremity
Tarsal bones	Rather long, toes rather short	Rather short, toes rather long
Trunk	Short compared to lower extremities	Long compared to lower extremities
Vertebral column	Alternately curved backward and forward	Straight or curved uniformly backward
Leg	In upright posture, straight in knee and hip joint	Curved, the knees turned outward
Joint between skull and vertebral column	Almost in center of base of skull	At back of skull
Canines	No larger than premolars	Large fangs
Crown of first lower premolar	Unspecialized (bicuspid)	Has blade-like cutting edge
Dental arch	Rounded without sharp angles	Laterally compressed with side rows of teeth almost parallel
Jaws	Short	Long and large
Face	Short, steep, under brain	Long, in front of brain, protruding
Brain	Large	Average size $\frac{1}{3}$ that of human

it is obvious that man and these anthropoids must have descended from a common ancestor.

For a long time, the study of fossil man was essentially a search for a connecting form, the "missing link." At first no one knew quite what to look for. The earliest reconstructions, for instance that by Haeckel, pictured a creature which, character by character, was intermediate between man and chimpanzee. This implied that man had the chimpanzee as his direct ancestor, that the chimpanzee stopped evolving as soon as it had given rise to the human line, and that all organs evolved at the same rate! All three of these assumptions are wrong. Equally unfounded is the additional assumption that the living anthropoids are primitive and that man had to go through a stage represented by these anthropoids. Numerous recent fossil discoveries have made it clear that the anthropoids have evolved as much since branching off from the common anthropoid-hominid stem as has the hominid line. In fact, in many morphological respects the anthropoid lines are apparently less similar to the common ancestor than is modern man.

To find the common ancestor of the Hominidae and the Pongidae, we must look for a creature that lacks the various specializations of the living Pongidae and the complete bipedalism and brain development of the recent Hominidae, but that possesses some of the characteristics by which the anthropoids (including *Homo*) differ from the cercopithecoid monkeys. Fossil forms close to this postulated creature have been found in the Oligocene and Miocene of Africa.

Fossil Anthropoids

Although some fossil primates from the Eocene of Burma (*Pondaungia* and *Amphipithecus*) and the early Oligocene of Texas (*Rooneyia*) show anthropoid features, the first fossil that is rather clearly an anthropoid ape is *Aegyptopithecus zeuxis* of the late Oligocene (about 28 million years ago) of Fayum, Egypt. He could well be ancestral to the genus *Dryopithecus* (including *Proconsul*), which in several species was widespread in the Miocene and Pliocene (18–8 million years ago) from Africa and Europe to India and China (Simons and Pilbeam 1965), and which gave rise to the African apes (*Pan*) and possibly other anthropoids. The first fossil that is clearly a hominid is *Ramapithecus punjabicus,* which is known from East Africa, India, China, and possibly Europe from the late Miocene (14 million years ago) to the middle Pliocene (7–10 million years ago). It is likely that *Ramapithecus* split off somewhere from the diversified genus *Dryopithecus;* if so the branching took place at least 15 million and possibly more than 20 million years ago.

The relative size of the incisors, canines, and premolars (as compared to the molars) is considerably reduced in *Ramapithecus,* which means that this first hominid had less of a snout than other anthropoids and perhaps a less exclusively vegetarian diet. It has even been suggested that tool use might have favored this reduction of the anterior dentition, but there is no solid evidence for this. All characters by which *Ramapithecus* differs from *Dryopithecus* are on the line of development toward later hominid characters, and it seems quite legitimate to assign this fossil to the hominids. What needs special emphasis is the wide geographic range of this early hominid species, now known only from some jaw and mandible fragments.

What can we say about the relation of this hominid line to the three kinds of living anthropoid apes? The gibbons are an ancient group that branched off before the *Dryopithecus* stage and surely have nothing to do with the hominid ancestry. Still unanswered is the question of whether the hominid line branched off from the pongids (*Pongo* in Asia and *Pan* in Africa) before the Pongidae split into an Asiatic and African line or afterward. Unfortunately, there is no fossil record of the orang (*Pongo*) and our decision must be based on inferences from a comparison of the living anthropoids. In view of the superficial similarity of the anthropoids to each other and the unique attributes of man, it was heretofore rather generally assumed that the hominid line had branched off long before the anthropoids split into orang, gorilla, and chimpanzee. A study of proteins (hemoglobin and serum proteins) and chromosomes, however, indicates rather decisively that the African apes are more similar to man than they are to the orang. Their external and internal parasites confirm this conclusion. It is reasonably certain therefore that the orang line branched off before the split between the remaining pongids and the hominid line (Fig. 20.1).

The Australopithecines

First discovered in 1924, the Australopithecines, late Pliocene to middle Pleistocene, are now the best-known fossil hominids. Although they have been described under a bewildering variety of names, they are now usually arranged into two species, *Australopithecus africanus* and *A.* (*Paran-*

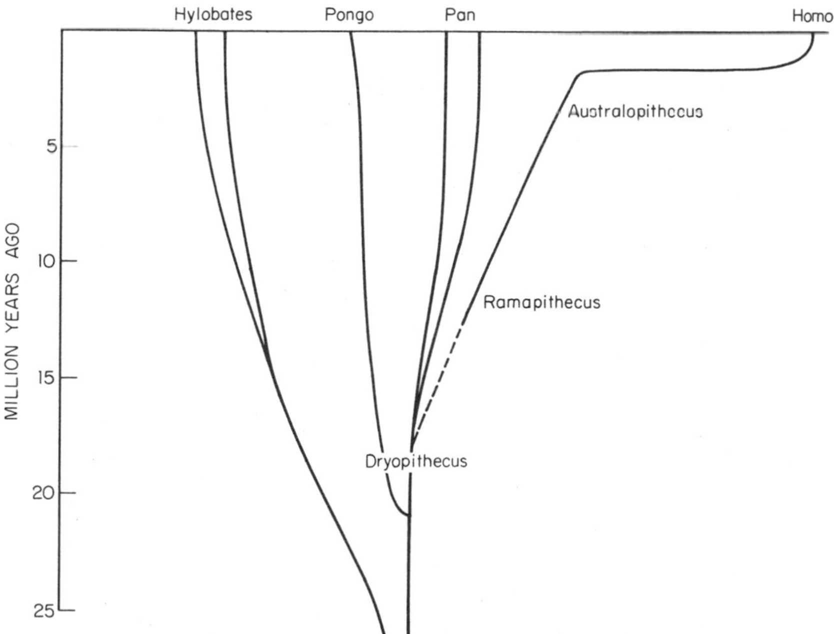

Fig. 20.1. Hominoid phylogeny. The branching off of the hominid line may have taken place almost 20 million years ago, but its major divergence (brain evolution!) from the anthropoids took place less than 3 million years ago.

thropus) *robustus*. As its name indicates, *robustus* is a more robust species, somewhat more ape-like, believed to have specialized in a diet of grass seed. The earliest known specimen of *robustus* (found near Lake Rudolph) is 2.8 million years old; *robustus* has also been found in Olduvai and Lake Natron (E. Africa) and in South Africa (Swartkrans, Kromdraai). An incomplete jaw fragment from Java ("*Meganthropus*") is believed to belong to the same species. The most characteristic features of *robustus* are its large grinding cheek teeth (molars) and its small front teeth (canines and incisors).

Ever since their discovery there has been a wide divergence of opinion concerning the relationship of *africanus* and *robustus*. According to one opinion, *robustus* is so different as to deserve separation in the genus *Paranthropus*, differing in food niche (grass seeds), locomotion (less erect), and possibly habitat. It succumbed to competition (or predation) from his fellow hominid (*africanus-habilis*), when the latter evolved into *Homo erectus*. According to others some of the differences, at least among the South African Australopithecines, merely reflect sexual dimorphism. Further finds and a more careful analysis of the fossil material will resolve these differences of opinion. As the dating of the various exposures gains in precision, relationships often become clearer. For instance, "*Homo*" *habilis*, an East African Australopithecine found in Olduvai Bed I (1.8 million years old), is now believed sufficiently younger than South African *africanus* to be considered its descendant. Likewise, A. *africanus*, formerly believed to be only 0.5 million years old, is now known to have lived much earlier (2.2–5.5 million years ago [Table 20.2]).

Table 20.2. Brain evolution.

	Years ago	Estimated brain size (cc)
Homo sapiens		
Modern Man	40,000	1450
Neanderthal	100,000	1500
Steinheim, etc.	200,000	±1200
Homo erectus		
Peking Man	400,000	900–1100
Java Man	600,000	700–900
Australopithecus		
habilis	1.75 million	660
africanus	2.2–5.5 million	430–485
Chimpanzee, Gorilla		325–650

Past generations of human paleontologists liked to stress the ape-like characteristics of our ancestors. It is therefore important to point out how much like *Homo* the Australopithecines were. The small brain is the only drastic difference. *Australopithecus* is characterized by an essentially hominid dentition consisting of small canines, typically bicuspid premolars, and a molar cusp pattern similar to that of *Homo erectus*. The teeth are arranged in an even arcade of elliptical form with no gap (diastema) between the canines and adjacent teeth. Pelvis and limb bones are constructed along hominid lines and indicate a posture that was upright, though not as highly perfected as that of modern man. Upright posture is also indicated by the location of the occipital condyle (the articulation of the skull with the spinal column), which faces essentially downward, as in modern man, rather than backward, as in the anthropoids. Other hominid features become apparent in a close study of the skull. However, the cranial capacity is hardly larger (relative to body size) than that of the modern large apes. The jaws are exceedingly large compared to the cranium, and the skull bones, particularly the jaw bones, are tremendously thick and heavy.

A word must be said at this point about the general size of these early man-apes. If one compares only their massive skulls and jaws and their huge teeth with those of modern man, one may conclude that these forms must have been giants. Now that material of the body skeleton has become available, it is evident that *Australopithecus* probably did not exceed 5 feet in height, being probably somewhat shorter than a modern Bushman. This evidence flatly contradicts the popular notion that man descended from a line of giants.

The discovery of *Australopithecus* was a shock to those orthodox typologists who believed in the "harmonious evolution of the type." Pelvis, extremities, the shape of the tooth row, and the cusp pattern of the molars are very similar to the corresponding structures in modern man, but the huge jaws, the prognathous face, and the small brain are very ape-like. The evolution of the hominids provides a classical demonstration of mosaic evolution of any type that shifts into a new adaptive zone. Yet, evolution can be quite "harmonious," even though one structure evolves much faster than the others. It is selection pressure which determines how different the rates of evolution of different portions of the body will be.

Considering the essential morphological similarity between *Australopithecus* and *Homo* and the fact that quite clearly the *africanus-habilis* series gave rise to *Homo erectus*, it has been suggested that *africanus* be included in the genus *Homo*. Yet such a taxonomic treatment conflicts with the prevailing concept of the taxonomic categories. Generic separation does not indicate a degree of difference so much as a difference in the utilization of the environment. *Australopithecus africanus* with a mean brain size of 460 cc filled such an entirely different niche from *sapiens* with a brain of ±1500 cc that generic separation is abundantly justified.

Homo erectus

The most famous fossil hominid before the discovery of *Australopithecus* was *Pithecanthropus*. Haeckel had coined this name for a postulated "missing link" between man and the apes and thus greatly stirred the imagination of many a young man to find it. A young Dutch anatomist, E. Dubois, obtained an appointment as army doctor in the East Indies so that he could search there for it, and much to everyone's surprise (and probably his own) he actually succeeded. In 1891 he found a skull cap at Trinil in eastern Java, and the next year, some 40 or 50 feet away in the same fluvial deposits, a thigh bone (femur) as well as other skeletal remains. Dubois's report on *Pithecanthropus erectus*, published in 1894, precipitated one of the most heated controversies in the history of anthropology (rich as it is in hot controversies!). Some authors regarded the remains as human, others as anthropoid. As far as the femur was concerned, the majority opinion was that it could not belong with the skull cap since it was "of a different type." *Pithecanthropus* remained highly controversial until von Koenigswald in the 1930's systematically and with enormous energy explored the fossil beds of Java, bringing to light many additional and much better specimens.

Java Man (*erectus*) was of great importance in the history of the discovery of fossil hominids, being the first hominid known outside the range of variation of polytypic *Homo sapiens* (*sensu lato*). Consequently all differences from *H. sapiens* were magnified, such as a low forehead, the small brain, the high line of attachment of the occipital bones, and the heaviness of the skull. Yet, in its body skeleton *erectus* does not differ from *sapiens* in any essential point. The differences between *sapiens* and *erectus* are valid species differences, but they do not justify recognition of the genus *Pithecanthropus*. Java Man is now classified as *Homo erectus*.

This species (or species group) ranged all the way from China ("*Sinanthropus pekinensis*") to Europe (*Homo heidelbergensis*), North Africa (Ternifine), East Africa (Olduvai Bed II), and possibly South Africa (*Telanthropus* at Swartkrans). Nothing illustrates the closeness of *Homo erectus* and *Australopithecus africanus* better than the fact that there is still controversy whether to classify *"Homo habilis,"* the Olduvai Bed II hominid, and *Telanthropus* as *Homo erectus* or *A. africanus*. The existence of somewhat intermediate populations is, of course, only natural, considering that the former evolved from the latter. Furthermore, *H. erectus* itself is intermediate between *africanus* and *sapiens*.

Selection Pressures and Evolutionary Trends in the Hominid Line

How did man, now so utterly different from all animals, acquire his unique human characteristics? Even the earliest known hominid (*Ramapithecus*) shows by his short face and reduced anterior dentition that he might have differed from contemporary anthropoids in the preparation of food and tool use. The origin of man surely was not a sudden event but a response to continuing selection pressures. It is the nature of the selective factors which now concerns us. What are the major factors that shifted human evolution into channels so totally different from those of the anthropoids?

There are two classes of such factors: changes in behavior (see below) and changes in the environment.

Changes in the environment. The second half of the Tertiary (Miocene-Pliocene) was characterized by an increasing desiccation of the portion of Africa and Asia inhabited by hominoids. This resulted in opening up an ever larger habitat, ranging from wooded savannas to areas that were very arid, almost desert-like. The occupation of this newly available habitat by the hominoids favored not only bipedal locomotion but also a shift in diet toward a greater proportion of meat. Various behavioral changes were correlated with the ecological changes.

Bipedal Locomotion

It is often stated that man's ancestors adopted upright posture and walking when shifting from an arboreal to a terrestrial mode of living. There is, however, no necessary correlation: none of the other largely terrestrial primates has adopted bipedalism. Gorillas and chimpanzees practice knuckle walking and baboons have reverted to strictly quadrupedal locomotion. Since no postcranial skeletons have yet been found in the fossil record of early hominids (*Ramapithecus* and ancestors), all one can suggest at this point is that some peculiarity of their arboreal ancestors (? semi-brachiation) as well as some anatomical preadaptations favored bipedalism.

Bipedal locomotion, particularly in its early stages, must have been a rather inefficient form of locomotion for a four-legged mammal. Its greatest selective advantage was presumably that it freed the forelimbs for new behavioral tasks. It permitted the use of the hands for the efficient manipulation of tools, for the handling of weapons (sticks and rocks), and for the carrying of food. It is possible that the beginnings of bipedalism go back to the beginnings of the hominid line, but its perfection presumably occupied most of the ensuing time. The differences in the pelvis and posterior extremity between *Australopithecus* and *Homo* indicate how long it took for bipedalism to become perfected.

Tool Use and Tool Making

Our ideas on the evolutionary role of tool use have undergone drastic changes in recent decades. It was believed at one time that tool use and tool making were quite recent in hominid history and that the dramatic increase in brain size in the *erectus* stage was the result of this behavioral change. It is now realized that tool use and even the making of tools is widespread in the animal kingdom. Chimpanzees in particular are skilled tool users and fully capable of adapting natural implements for their purposes. It is therefore not surprising that *Australopithecus,* with a brain no bigger than that of an anthropoid ape, manufactured his own stone tools. The making of simple tools apparently did not set up a strong selection pressure for increased brain size nor did it require a major reconstruction of the anterior extremity. Arm and hand changed remarkably little from the time the hand was used largely for grasping a branch to the time it was first used for piano playing or the repair of a fine watch.

It is only at a later period that tools became crucial for human evolution and survival depended increasingly on the skillful use of tools and on the invention of better and new kinds of tools. It is presumably during the Neolithic period that improvements in tool use and tool making were of the greatest selective significance.

Increase in Brain Size

The most important character by which man differs from the apes is surely the enormously enlarged brain (together with the faculties that this new brain allows). Considering the definite hominid trends displayed by *Ramapithecus,* it is distinctly surprising that *Australopithecus africanus,* appearing 9 million years later than *Ramapithecus,* still had a pongid brain size with a mean value of 460 cc. Evidently, then, the use and even the making of simple stone tools was not the decisive factor that led to the spectacular increase in brain size in the 1 million years between 1.3 and 0.3 million years ago. The late J. B. S. Haldane liked to emphasize that this dramatic increase in brain size was the most rapid evolutionary change known to him. It is even more astonishing because it does not involve a superficial structure but an organ as elaborate as the central nervous system. What then is responsible for this dramatic evolutionary event?

All answers to this question are largely conjectural. It is, however, almost certain that a combination of selection pressures was responsible and that the development of large brains coincided with the shift of man into an entirely new adaptive zone. Far more detailed comparative studies than those now available will be required to determine what the components of this shift were and which were cause and which effect. Among the factors that were probably causal, three seem to be particularly important.

(1) *The hunting of big game.* Man's closest relatives, chimpanzees and gorillas, are not pure vegetarians; occasionally they catch a hare or small gazelle. Other terrestrial primates, such as baboons, are also hunters to a limited extent. What distinguished the hunting of primitive man from the desultory hunting of his relatives was that for man hunting had become the major source of food and that man's method of hunting was quite different from that of any known animal (Washburn and Lancaster 1968). Early man's reliance on hunting, particularly the killing of big mammals, had several important consequences:

The need for the invention and manufacture of improved weapons and tools was increased, especially so when the quarry was the large ungulates on the African plains. Man had to learn to cut up large prey into pieces small enough for distribution and transport back to base camps.

Cooperation among several males was required to achieve success in the hunting of large game. Eventually this resulted in a considerable division of labor and responsibility between leaders, scouts, caretakers of the base camps, and perhaps specialists in various weapons.

The hunting of big game often required lengthy hunting expeditions involving strenuous traveling. The consequent prolonged absences of the men necessitated the establishment of base camps where nursing mothers, pregnant females, and children could be left under the care of guards.

Successful hunting of big game required a great deal of planning, a knowledge of the movement of herds and the location of water holes, a memory for seasonal events and for the habits and peculiarities of the various prey species, and, finally, careful surveillance of competitors. There can be little doubt that the rewards of successful hunting set up a strong selection pressure for an improved brain—for increased planning abilities, increased information storage (memory), and, most important, refined communication techniques.

(2) *Speech.* Planning, cooperation, a division of labor, and memory would not be particularly useful without a far more efficient system of communi-

cation than is available to the anthropoids. The capacity for speech is the most distinctive human characteristic, and it is quite likely that speech is the key invention which triggered the step from hominid to man. It permitted community structure and it permitted man to become a truly social organism. As such man developed a need for mechanisms promoting social homeostasis, communal rights, myths and beliefs, and finally primitive religions. All these developments exerted a high pressure on improving speech, enlarging the vocabulary, and increasing the storage capacity of memory. This chain of developments included a number of positive feedback mechanisms; thus each advance in turn exerted a selection pressure in favor of still greater brain development.

(3) *Structure of the breeding group.* By now we may well have identified the main factors that favored an increase in brain size. One question, however, remains unanswered: Why did the increase occur at such a rapid rate? It seems probable that the family structure of these early hominids favored rapid evolution. Anthropologists and sociologists have a tendency to assume that monogamy was the original reproductive system in mankind. And indeed most instances of polygamy now found in human societies, as for instance among Moslems, are clearly recent institutions, derived from monogamy. Recent evidence, however, indicates that monogamy itself might have been a derived condition. Among Alaskan Eskimos, among New Guinea mountain Papuans, and among relatively untouched South American Indians polygamy is widespread, and it is the individual with leadership qualities who has the greatest chance to have several wives. In their study of the Xavante Indians from South America, Neel et al. (1964:127) concluded:

> The evidence suggests that fertility differentials have far more genetic significance in the Xavantes than for civilized man today. The position of chief or head of clan is not inherited but won on the basis of a combination of attributes (prowess in hunting and war, oratory, skill in wrestling, etc.) The greater fertility of these leaders must have genetic implications. Indeed, it may be that the single most dysgenic event in the history of mankind was departure from a pattern of polygamy based on leadership, ability, and initiative.

As astounding as the precipitous increase in brain size is the fact that it came to a sudden halt some 100 thousand to 200 thousand years ago. There has been no increase in brain size since the time of Neanderthal. And there is no evidence whatsoever indicating internal improvement of the brain without enlargement of the cranial capacity, though we cannot decisively refute such a possibility. As evolutionists, we must ask ourselves why all the factors favoring an increase in brain size suddenly lost their power after the *Homo sapiens* level had been reached. Of all the possible answers that can be given to this question, the one that seems most convincing is that the breeding structure of *Homo sapiens* changed.

There was apparently a stage in human evolution where the most successful groups grew so large that the fertility advantage of the leaders became minimal. The larger a population group is, the relatively less will the genes of its leader contribute to the gene pool of the next generation, and the more protected (biologically) will be the average or below-average individual of the group. Reproductive success in such large groups will not be closely correlated with genetic superiority. In recent times there has been an even greater reduction of the selective premium on the characteristics that had previously been favored during human evolution. Add to this the dysgenic effects of urbanization and of density-dependent diseases, and it becomes apparent why the trend that created man has not continued to produce superman. The social structure of contemporary society no longer awards superiority with reproductive success. The development of cultural tradition and the steady improvement in means of communication greatly reduces selection pressures. All the members of the community benefit equally from the technological and other achievements of the superior individuals. Thus the below-average individual, provided he is not too far below average, can make a living and reproduce as successfully as the above-average individual. These, then, are the factors most probably responsible for the stabilization of brain size—an unusually abrupt cessation of an exceedingly steep evolutionary advance.

The reconstruction of the skull. The increase in brain size was the main factor responsible for a complete reconstruction of the skull. Two additional selection pressures also favored this reconstruction. One was the forward shift of the skull support resulting from the assumption of upright posture. The other was a lessening of selection pressure for strong jaws and big teeth, no longer needed with the prepared food (tools, cooking, shift of diet) and the new means of attack and defense (weapons) made possible by the enlarged brain. All this resulted in a reduction of the jaws, the teeth, and the facial part of the skull and in a simultaneous enlargement of the cerebral part. It resulted also in reduction of the facial muscles and all bony crests and ridges to which these muscles are attached. It is these developments which cause the skull of *Homo sapiens* to look so strikingly different from that of *Australopithecus africanus*.

The Role of Behavior

Behavior is perhaps the strongest selection pressure operating in the animal kingdom. Hominid evolution has been rich in behavioral shifts:

arboreal ⟶ terrestrial
forest ⟶ savanna
vegetarian diet ⟶ increasing meat diet
small game ⟶ large game
tool use ⟶ tool making
polygamy ⟶ monogamy
hunting ⟶ farming
farming ⟶ urbanization and industrialization

Each of these shifts initiated new selection pressures, some facilitating and speeding up hominization, some of the later ones reversing or at least halting previous trends. One of the most significant aspects of hominid behavior is parental care.

Parental care. Much of the mortality in lower animals (for example, that in the pelagic larvae of marine organisms) is accidental and haphazard. The institution of parental care permits a marked decrease in such random mortality. Survival then depends increasingly on the goodness of the care bestowed upon the child by the parents. The increase in brain size induced by the practice of parental care requires a lengthening of the period of development of the human infant and consequently a lengthening of the period during which parental care is needed. These developments then reinforce the selective value of parental care and exert more intense selection pressure in favor of increased brain size in the parents.

Highly developed parental care permits a fundamental change in the genetics of the behavioral program. Most short-lived lower animals are born with a ready answer to most of the environmental situations with which they are confronted. They react to them "instinctively." They have very little capacity for learning, that is, the storage of new, useful information. We may call an inherited system of ready responses to environmental challenges a "closed program." Higher organisms, especially the higher primates and man, have a far greater capacity for "learning," that is, the storing of new information which permits them to react adequately to environmental situations. They have an "open program" (Mayr 1965a). Not having a ready set of instinctive answers to the challenges of the environment makes the young of such species highly vulnerable. Parental care protects the young while they are acquiring all the learned information that they need to cope successfully with the challenges of the environment. The open program permits a far more fine-grained response to outside stimuli and thus greatly improves the chances of survival once adequate information has been accumulated. Such a program also requires a far larger central nervous system than a closed program, and thus exerts additional selection pressure in favor of an increase in brain size.

Parental care is a typical manifestation of kinship selection (Chapter 8). The primitive hominids faced their adverse environment not as individuals but as family groups or small bands. In an animal with the social population structure of the primitive hominids, natural selection will favor inventiveness, foresight, leadership, and, in many cases, cooperation far more than it will favor crude force or egotism, as Darwin pointed out long ago. He who makes the greatest contribution to the harmony and well-being of the group may thereby become the ancestor of the greatest number of surviving descendants. There is no conflict between natural selection and human ethics. Ethical qualities in a social organism are important components of fitness.

An ability to learn is not enough for the acquisition of certain types of information. There has to be a definite desire or drive or readiness to store the appropriate information. This information is of a very special kind when it consists of ethical principles, myths, or religious dogmas. In order to be able to acquire these, as Waddington (1960) has rightly stressed, there has to be a readiness to accept authority: "The newborn infant has to be ready to believe what it is told." The capacity to accept concepts, dogmas, and codes of behavior is one of the many forms of imprinting. The greater the amount of parental care and education and the more highly developed the means of communication, the more important becomes conceptual imprinting. Man has a great capacity for such imprinting, as is documented not only by his acceptance of ethical systems and religions but also by his vulnerability to propaganda and the mass media. It is one of many traits that had high selective value in the small hominid communities where they evolved but that may well be liabilities in modern mass society.

The Gradualness of Man's Emergence

This discussion of the multiplicity of trends in human evolution makes apparent the futility of trying to designate a particular moment at which our hominid ancestor changed into man. Throughout the history of anthropology there have been attempts to pinpoint this moment. Was it the first use of fire or the first hunting of a large ungulate? Was it the first manufacturing of tools or the first communication by speech? Was it an increase of brain capacity beyond 1000 cc or beyond 1200 cc? Considering the continuity of all these trends, any decision would be arbitrary, any cut-off point unjustified. Furthermore, no two individuals would have been the same in regard to the stated criteria nor would have been any two local populations of the widespread polytypic species which evolved from *Australopithecus africanus* through *Homo erectus* to *Homo sapiens*. Indeed, is man as he now exists the end point of an evolutionary development? One might claim, or at least hope, that we are still in the midst of an evolution toward a more nearly perfect stage of that which is now represented by *Homo sapiens* with all of his glaring imperfections. It is obvious that we must consider the emergence of man a continuous process.

Polytypic and Sympatric Species in Hominids

The study of man's evolution benefited enormously from the application of the new insights of evolutionary systematics. It had been hopeless to try making sense of hominid phylogeny as long as every fossil specimen was named (far more than 100 species names were given!) and considered a separate type. This all changed as soon as these fossils were regarded as samples of populations with a definable distribution in space and time. Because Recent Man is a polytypic species and because most species of mammals are polytypic, it can be assumed that the species of fossil hominids likewise were polytypic. Furthermore we are justified in assuming that throughout hominid history some of the geographic isolates (Chapter 13) reached species rank (reproductive isolation) and were then able to overlap or exterminate sister species. Finally we can assume that rates of phyletic change were different in the various isolates of a polytypic species, as they are in the isolates of polytypic species of living animals, and that consequently some advanced and retarded races were contemporaries. Applying all these principles to the known fossil hominids, what do we learn about speciation in the hominid phyletic line?

The basis of all scientific interpretation is the rule of parsimony, which demands in each case the simplest explanation consistent with the facts. As far as fossil hominids are concerned, the simplest assumption would be that only a single polytypic species of hominid existed at any given time, and that the variety of observed types is merely a manifestation of individual and geographic variation (Table 20.3). Let us now see what difficulties this simplifying model entails.

Table 20.3. The chronology of hominid evolution.

	Millions of years ago	First date of
Pleistocene	3	300,000 *Homo sapiens* 700,000 *Homo*
Pliocene	7	5 *Australopithecus*
Miocene	25	14 *Ramapithecus*
Oligocene	40	30 *Aegyptopithecus*
Eocene	60	
Paleocene	70	
Cretaceous		75 First primates

Ramapithecus punjabicus

The remains of this Miocene-Pliocene hominid are too scanty to determine whether or not the African and Asiatic populations differed only racially or whether they differed specifically. They must be considered members of a single species until clear-cut specific differences are demonstrated.

Australopithecus

It is now well established that two types of hominids existed side by side throughout the early Pleistocene, *Australopithecus africanus* and *A. robustus*. These have been universally regarded as two distinct species or even genera. Unquestionably *robustus* has much more robust molars and correlated with this heavier musculature and corresponding skull features.

It has been suggested that they occupied different niches, with *robustus* being a vegetarian or granivore.

There are good reasons for believing that the widespread species *Australopithecus africanus* showed considerable geographic variation. However, no careful comparison has yet been made between the South African and the East African samples. A considerable time span (2.2–5.5 million years ago) is involved, so that a visible amount of evolutionary change might well have occurred between the earliest and the latest recorded populations of *A. africanus*. The situation was greatly clarified by the discovery and correct dating of *"Homo" habilis* (1.8 million years ago) at Olduvai. The considerable increase in brain size (660 cc), against 430–485 (revised estimates) in *africanus*, shows that *habilis* is quite a distinct taxon, although apparently closer to *africanus* than to *Homo erectus* (see Table 20.2). *A. habilis* is perhaps best recognized as a distinct species of *Australopithecus*, although it might also be treated as a chronological subspecies of *A. africanus*.

Homo erectus

No one doubts any longer that *H. erectus* is a direct descendant of *A. africanus*. Which population or group of populations of *africanus* gave rise to *erectus* is still unknown. No one doubts any longer that *H. erectus* is a direct descendant of the *A. africanus-habilis* line. But which population or populations were the spearhead in this evolution is still entirely obscure. The problem is the same as with the evolution of *H. sapiens* from *H. erectus*.

Modern man approaches more and more the status of a panmictic species. In the early hominids, however, gene flow was comparatively slight. Fossil man consisted of many localized family groups and small bands, isolated by numerous geographical and ecological barriers. There was probably very little mixing between neighboring bands, and local differentiation must have been high. One can assume that the allopatric populations of the earlier hominids were more distinct from each other than are the races of modern man. This does not necessarily mean that they had reached species level.

The *Homo erectus* stage (1 million to 350,000 years ago) illustrates these geographical distinctions well. Java Man, Peking Man, Heidelberg Man, and the specimens found in Hungary and in north, east, and south Africa all have certain peculiarities superimposed on the general *Homo erectus* features. We do not know whether any of these populations had reached the level of a separate species. The differences between mid-Pleistocene artifacts found in Africa, Europe, and western Asia and those found in eastern Asia indicate lack of gene flow but do not prove specific distinction. Some isolated populations with *Homo erectus* characters (including a small brain) seem to have survived until 30,000 years ago, long after *H. sapiens* had evolved. In general, the picture presented by *H. erectus* is that of a widespread polytypic species with numerous local races or subspecies.

Homo sapiens

According to Coon's (1962) thesis, *Homo sapiens* is not the product of a narrow isolate, but a more evolved stage of *Homo erectus*. Coon's argument is as follows: The cultural shifts that had occurred during the *H. erectus* stage exerted a uniform selection pressure on most *Homo* populations (excluding perhaps a few peripheral ones in Africa, Asia, and Australia), activating in them parallel evolutionary trends toward larger brains and a reduction of the facial skull. Even though the gene exchange among them was limited (see below under races), they all crossed sooner or later the arbitrary dividing line between *erectus* and *sapiens*, just as various lines of therapsid reptiles had independently and at different times crossed the line from reptiles to mammals. Coon's thesis rests on very fragmentary documentation and may well be refuted through further discoveries of fossil man. At the present time, however, it is consistent with the known facts.

The first fossils that seem closer to *H. sapiens* than to *H. erectus* were found in England (Swanscombe), France (Fontechevade) and Germany (Steinheim) in deposits of the Second Interglacial. Their calculated brain capacity averages smaller than that of modern *sapiens*, but in some respects (small supraorbitals, general outline of the skull) they are more similar to Recent Man than is Neanderthal.

Neanderthal Man, the ice-age man of Europe, was the first fossil man known, and has played a far greater role in the literature of human evolution than his importance for the ancestry of man justifies. The exact taxonomic assignment of Neanderthal is still controversial, but the conclusion now most widely supported is to consider him a geographic race of *H. sapiens* that became extinct around the time of the first Würm Interstadial (Howell 1957).

Although Neanderthal was distinguished from *sapiens sapiens* by various skull characters, such as a flattened brain case, bun-shaped protuberant occiput, marked projection of jaws, virtual absence of chin, high attachment of muscles on occipital bone, immense supraorbital ridges, very large orbits, and a powerful mandible, he had a cranial capacity at least as large as that of Recent Man and a highly developed paleolithic culture. More important for the decision whether Neanderthal is a separate species is the distributional pattern. In not a single paleolithic site has Neanderthal been found unquestionably associated with Recent Man. Neanderthal, on the whole, is a western type, with his center of distribution in Europe, although some finds in North Africa, Palestine (Tabun), Iraq (Shanidar), and Turkestan (Teshik-Tash) indicate a vastly enlarged area of distribution. When only the European sites of classical Neanderthal were known, it was tempting to consider him an arctic ecotype, the "Eskimo" of the first stage of the Würm glaciation. However, the finds south and east of the Mediterranean disprove such an assumption. In view of the absence of established sympatry, it seems justified to consider Neanderthal a more northerly and westerly geographical representative of *Homo sapiens sapiens*. The time progression from Steinheim through Fontechevade, Ehringsdorf, and Sacco-Pastore to classical Neanderthal shows that the Neanderthals are not primitive. They seem to owe their characteristic features to the selection pressures prevalent in western Eurasia during the latter portion of the Pleistocene.

Whether or not this climatically adapted geographic race of the western Palearctic ever reached species status is one of the many unsolved Neanderthal problems. The ultimate fate of this type is equally puzzling. Wherever it is found, it is associated with artifacts of a flake culture (Mousterian). It was suddenly replaced by typical modern *Homo sapiens sapiens*, associated with a blade culture (Perigordian). Whether Neanderthal had become extinct before Cro-Magnon Man arrived or whether the latter exterminated Neanderthal is unknown. Remnants of Neanderthal might have been absorbed by Cro-Magnon without leaving demonstrable traces.

The best evidence of mixture between Neanderthal and modern *sapiens* comes from the two caves of Mount Carmel in Palestine. Both caves were inhabited early in the Würm glaciation. The older cave (Tabun) was inhabited by Neanderthals with a slight admixture of modern characters, the younger cave (Skhul) by an essentially modern population with distinct Neanderthaloid characters. The date is too late for these populations to have belonged to the ancestral stock that gave rise to both Neanderthal and Recent Man. The differences between Tabun and Skhul are too great for both caves to have been inhabited by a single population coming from the area of geographical intergradation between Neanderthal and Recent Man, although this could be true of the Skhul population. Hybridization between invading Cro-Magnon Man and Neanderthal remnants is perhaps still the most plausible interpretation for the Skhul population, while there is no good reason for not considering that of Tabun a regular Neanderthal population, particularly in view of its similarity to the Shanidar specimens (Stewart 1960).

Homo sapiens sapiens, represented by Cro-Magnon Man, suddenly appeared in Europe some 35,000 years ago and has been dominant ever since. Where did this invader come from? This is one of the greatest unsolved puzzles of human evolution. There is much to favor an Asian or south-Asian origin. Very early (about 40,000 years ago) *Homo sapiens sapiens* was found in Niah Cave, Borneo, and also in China (date uncertain). Unfortunately no fossil man is known from India from the crucial period. Since Neanderthal occurred as far east as western Iran (Shanidar) and Turkestan, the home of *H. s. sapiens* must have been still further east or south, even Africa being a possibility.

The Polytypic Species Homo Sapiens

All the different kinds of living man on the face of the earth belong to a single species. They form a single set of intercommunicating gene pools. As a matter of fact, the various races of man are less different from each other than are the subspecies of many polytypic species of animals. Yet a few misguided individuals have applied a typological species definition to man and have divided him into five or six separate species by using such artificial criteria as white, yellow, red, or black skin color. Such a division not only fails to account for a considerable portion of mankind—intermediates and relic primitives—but also is completely contrary to the biological species concept (Chapters 2 and 12). There are no genetic isolating mechanisms separating any of the races of mankind, and even the social barriers function inefficiently where different races come into contact.

It is often asked whether man is in the process of speciating and whether the races of man should be considered incipient species. In attempting to answer this question one must recall that the hominids occupy one of the most spectacularly distinctive adaptive zones on earth. In the animal kingdom the invasion of a new adaptive zone usually results in a burst of adaptive radiation into various subniches. This has not happened in the history of the family Hominidae. Mayr (1950) has pointed out that the failure of man to speciate is due to two causes. "It seems to me that [one] reason is man's great ecological diversity. Man has, so to speak, specialized in despecialization. Man occupies more different ecological niches than any known animal. If the single species man occupies successfully all the niches that are open for *Homo*-like creatures, it is obvious that he cannot speciate." The second reason is that isolating mechanisms in hominids apparently develop only slowly. There have been many isolates in the polytypic species *Homo sapiens* and in the species ancestral to it, but isolation never lasted sufficiently long for isolating mechanisms to become perfected. Man's great mobility and independence of the environment have made effective geographic isolation impossible. As a consequence, all parts of the globe, including all climatic zones, are now occupied by a single species. What other species of animal includes populations adapted to the arctic as well as to the tropics, and ranging from almost pure vegetarians to almost pure carnivores? The probability of man's breaking up into several species has become smaller and smaller with the steady improvement of communication and means of transport. The internal cohesion of the genetic system of man is being constantly strengthened.

The Races of Man

There is no agreement yet on the formal classification of the subdivisions of *Homo sapiens*. Of two highly competent recent treatises on the races of man, one recognized 6 races, the other recognized 30. Both classifications are legitimate. Yet even the division into 30 races is by no means exhaustive. Lumped together in one of these races are numerous relic populations ranging from the Congo in Africa to southeast Asia, the Philippines, and New Guinea, populations that (if they are related at all) are less closely related than are the four usually recognized European races. All races are collective groupings of more or less differentiated local populations (Coon 1965). When we go back in history we find chronological subdivisions of *Homo sapiens* such as, let us say, Cro-Magnon Man or more differentiated Neanderthal and finally the Steinheim-Swanscombe Man, the earliest *Homo* that cannot clearly be separated from the polytypic species *Homo sapiens*. Biologically, it is immaterial how many subspecies and races of man one wants to recognize. The essential point is to recognize the genetic and biological continuity of all these gene pools, localized in space and time, and to recognize the biological meaning of their adaptations and specializations.

Adaptive Aspects of Human Races

A large component of the geographic variation of animals is adaptive, that is, each local race is to a lesser or greater extent adapted to the climatic and other environmental conditions of the given area (Chapter 11). There is no reason why man should be immune to this type of natural selection, and yet there have been many authors who ascribe the differences between human races to "accidents of variation." That this is not the case can be demonstrated, and Coon (1965) has gathered a great deal of evidence for trends of adaptive variation in human races. Pigmentation is almost without exception more intense in the humid tropics than in the more arid and cooler regions (Gloger's rule). Prominent body parts, and the heat-radiating body surface as a whole, are reduced in races exposed to the full brunt of cold winters (central Asia, arctic); the opposite trend can be observed in tribes that live in subtropical and tropical savannahs—they have comparatively small bodies with elongated extremities. The effect of selection is particularly obvious where a single racial group has invaded different climatic zones, as have the American Indians, who live from arctic Canada through the South American tropics to Tierra del Fuego. Another case is that of the Lapps in northern Scandinavia. By their blood groups, facial height, and other characters they can be diagnosed rather decisively as a European race. And yet, owing to living in an arctic climate, they have by convergent evolution acquired many of the features of Mongoloid races living in similar climates. Information on the physiological adaptations of human races is now also beginning to accumulate (Barnicot 1959).

A special problem is that of the origin of the white race, particularly in its extreme blue-eyed, blond form. This combination of characteristics has a distinctly negative selective value in tropical areas. To have become the dominant type in certain regions the white race must, one assumes, have had a positive selective value in those areas. The suggestion has been made that the cloudy, foggy climate of western Europe with cold rainy summers that prevailed during the last Interstadial and after the retreat of the ice might have favored the origin of this racial type. Nothing further can be said on this subject until we know more about the differential susceptibility to diseases of blond, blue-eyed individuals under the stated environmental conditions. Reduced susceptibility to, let us say, colds or arthritis might be a factor favoring such individuals.

The Amount of Difference among Human Races

It is relatively easy to describe the differences among human races in terms of dimensions, proportions, pigmentation, hair shape, and other morphological characters. Yet the question is raised again and again: just how meaningful are such morphological data? First of all, they have only a partial genetic basis since, as is well known, pigmentation may depend on exposure to the sun, and size on nutrition. More serious is the objection that neither size nor pigmentation is a critical human characteristic. The critical characteristics are intelligence, inventiveness, imagination, compassion, and other traits that are difficult to measure and to compare. The extreme viewpoint has been to deny that such differences among human races exist. A more conservative view would be to assume that as a consequence of the manifold genetic differences among human populations there will be average differences for any kind of trait that has at least in part a genetic basis. The comparison of identical with dizygotic twins has produced much evidence that certain mental traits, including intelligence, have—at least in part—a genetic basis (Shields 1962; Gottesman 1963). Traits that show genetically controlled differences among individuals of a single population are the ones that are most apt to show also racial differences. The study of such traits has been greatly handicapped in the past by a typological approach which, by necessity, led to racist interpretations. Human racial differences cannot be understood except on the basis of populational thinking.

The Population Concept in Man

The outstanding conceptual revolution in physical anthropology, as in biology, was the replacement of typological thinking by population thinking. This shift has affected every concept in anthropology, although none as strongly as the race concept. The typological race concept of the racists is thoroughly odious; the statistically defined race of the botanist and zoologist is a fact of nature. The basis for race formation is the same for all sexually reproducing organisms and consists in the fact that no two individuals are identical nor are any two local populations. No individual can therefore be "typical" of a race. Indeed, in polymorphic races different individuals may be strikingly different. To speak of "pure races" is sheer nonsense. Variability is inherent in any natural population and is favored by natural selection on account of the frequent superiority of heterozygotes and the diversity of the environment (Chapters 9 and 10). What differs from race to race is the degree of variability, and this depends on the size of the population, the amount of gene exchange with adjacent populations, the variability of the habitat, and other factors discussed in previous chapters. Much of the phenotypic variability of mankind is presumably due to the occasional production of homozygotes by heterozygous parents. An example is constitutional extremes—individuals that are exceptionally large, small, obese, or thin.

There are other reasons for man's phenotypic variability. Man is a restless creature and since prehistoric times he has embarked on large-scale migrations. The repeated colonizations of America by Asiatic tribes, the conquest of the South Seas by the Polynesians, the great Bantu migration, and the massive movements of Slavic and Germanic tribes in the dying days of the Roman Empire are only a few spectacular examples. Conquerors almost invariably absorb part of the defeated tribe or nation or are absorbed by it. On this basis one might expect man to have high individual variability, and this is indeed the case. Yet some of the anthropoids and many other animals far exceed man in amount of variability. Some human populations that are clearly the product of race mixture do not seem to have significantly higher variability than so-called unmixed races.

Marriage in primitive man usually takes place between members of the same tribe or group and leads to moderate inbreeding, even though there are nearly always taboos against extreme inbreeding as well as provisions for intertribal marriages. The amount of outbreeding has dramatically increased in modern man, and this increase has led to a great enlargement of the effective breeding population and has substantially increased the genetic variability of the individual deme. This, quite rightly, is generally considered a favorable development, since it reduces the danger of deleterious genes becoming homozygous. It is, unfortunately, not an unmixed blessing, since such outbreeding merely postpones the day of reckoning. "Genetic death" is controlled by rate of mutation regardless of population size.

It is often ignored in discussions of outbreeding that the beneficial effects may not materialize if the mixed gene complexes are too different. Hybridization between species leads almost invariably to unbalance through deleterious combinations of genes. I have presented evidence in Chapter 10, based on the work of Stone, Brncic, Vetukhiv, and Wallace, that in *Drosophila* even the hybridization of races may lead to destruction of well-integrated gene combinations. To what extent the findings of F_2 incompatibility can be applied to man is still uncertain. Medawar (1960) has cautiously stated the case as follows: "Hybridization between people of different races need not be expected to lead to an improvement, because both races will probably have adopted the well-balanced genetic constitution that matches their own environment." All investigations of race mixtures in man have failed to produce any evidence of decreased fitness. Physically, these mixed populations seem to be exceptionally vigorous and well adapted. It must be admitted, however, that the data are not precise enough (mortality and fertility data for F_2, F_3, and so on, are scarce or absent) to permit far-reaching conclusions; nor is there much evidence on nonphysical traits.

Polymorphism in Man

In addition to ordinary quantitative variation, which has a polygenic inheritance, we also find genuine polymorphism in man (Chapter 7). The human blood groups are the best-known examples of variable characters with a simple genetic basis. Another interesting case of human polymorphism is that of a deficient type of hemoglobin which produces the so-called sickle-cell anemia in regions with subtertian malaria (Chapter 9). Literally scores of gene loci are now known in man (mostly of enzyme-controlling genes) that are polymorphic in human populations. Additional ones are described every year.

The assumption that superiority of the heterozygotes is responsible for most cases of human polymorphism (Ford 1945) is based mainly on analogy with the better-analyzed cases of polymorphism in lower organisms. The average type, the average individual, seems very often best adapted. Extreme constitutional types are quite obviously less well adapted and often visibly more susceptible to illness. Still they are a normal component of every human population, since their production (through recombination of parental gene combinations) is the inevitable by-product of the turnover of the gene pool in every generation. Other cases of human polymorphism may be selectively favored by the diversity of the environment (Chapter 9).

Identity versus Equality

That no two individuals are alike is as true of the human population as of all other sexually reproducing organisms. Every individual is unique and differs in a large number of morphological, physiological, and psychological characteristics from all other individuals. Each individual is a different combination of characters and of the genetic factors on which these characters are based. Much evidence for this high individuality of man has been gathered by Williams (1956). The humanitarian answer to the challenge of genetic variability within and between human populations is the principle of equality. Simply stated, equality means equal status before the law and equal status in human social relations in spite of genetic difference. Equality, as Dobzhansky has stressed, is a social and ethical concept, not a biological one. Equality means equal opportunity to make the best of one's genetic endowment.

> That society enjoys the greatest amount of liberty in which the greatest number of human genotypes can develop their peculiar abilities. It is generally admitted that liberty demands equality of opportunity. It is not equally realized that it also demands a variety of opportunities and a tolerance of those who fail to conform to standards which may be culturally desirable but are not essential for the functioning of society . . . If a nation were a pure line there would be little scope for liberty . . . There would be no freedom, no deviants, and no progress (Haldane 1949a:410).

Equality in spite of evident nonidentity is a somewhat sophisticated concept and requires a moral stature of which many individuals seem to be incapable. They deny human variability and equate equality with identity. Or they claim that the human species is exceptional in the organic world, in that only morphological characteristics are controlled by genes and that all other traits of the mind or character are due entirely to "conditioning" or other nongenetic factors. Such authors conveniently ignore the results of twin studies and of the genetic analysis of nonmorphological traits in animals. An ideology built on such obviously wrong premises can lead only to disaster. Its championship of human equality is based on a claim of human identity. As soon as it is proved that the latter does not exist, the support for equality is likewise lost.

The denial of genetic difference among human beings with respect to intellectual and character traits is based on fallacy. This ideology is particularly pernicious when applied to education. The widely preached principle of "the same education for everybody" denies equal opportunities because differently endowed pupils would undoubtedly obtain different kinds, rates, and degrees of education if truly given "equal opportunities." Educational identicism is antidemocratic. According to the concepts of true democracy, as developed in the period of the Enlightenment, the free spirits of young men and women should not be shackled by the leveling restrictions of a false identicism. Every school class is a variable population and true equality (of opportunity) for the pupils can be achieved only by making allowance for these differences, not by suppressing them. There is still much to be learned about the relative contribution of inheritance and environment to individual human traits. This much, however, is already known: inheritance *does* play a considerable role and it can only do harm to ignore this role.

Claims of human identity are the outcome of typological thinking, of a belief that within the human type there is "no essential variation." Political theorizers have invariably applied such typological formulas when trying to resolve the difficulties posed by man's variability. The racism of the Nazis, for instance, was an outcome of such thinking. They rigidly defined each race by absolute characteristics: the *X* race "is lazy," the *Y* race is "of great intelligence," the *Z* race is "musical," and, worst of all, the *A* race is "superior." This practice allowed neither for the fact that many of the characteristics mentioned have only a partial (often a very small) genetic component, nor for the fact that many members of the various races do not have these characteristics at all. Another fallacy of typological racism is that it claims perfect correlation between the various characteristics ascribed to each race. Accordingly it claims an association between a particular color of the eyes or the hair and certain traits of the mind or the character. All available evidence negates the existence of such absolute correlations.

Typological thinking has been rampant in recent controversies concerning differences between human races with respect to intelligence. Suppose that calculations showed the mean I.Q. value of race A to be 10 points below that of race *B*; even if valid, this conclusion would not in the least justify the claim that every individual of *A* is 10 I.Q. points below every individual of race *B*. Indeed about a fifth of the individuals of *A* would have an I.Q. higher than half the individuals of *B*, assuming a normal distribution. This example shows how inexcusable it is to assign to individuals characteristics that are the mean values of the races to which they belong. Even if mean differences between human races should be established for mental traits, which seems likely, most individuals of a given race would fall in the area of overlap with other races. Every individual must, therefore, be treated on the basis of his own characteristics, never those of his race.

Every politician, clergyman, educator, or physician—in short, anyone dealing with human individuals—is bound to make grave mistakes if he ignores these two great truths of population zoology: (1) no two individuals are alike, and (2) both environment and genetic endowment make a contribution to nearly every trait.

Biological Problems of Modern Man

Students of man have long been aware that no biological phenomenon can be fully understood unless its evolutionary aspects are also understood. As a consequence they have given much consideration to man's evolution, including his evolutionary future (for instance, Medawar 1960; Simpson 1964; and Tax 1960). Evolutionary thinking is no longer considered the source of all evil in economic and sociological theory, as it was only a few decades ago. We have advanced far beyond the antievolutionary prejudices of the early decades of this century, but there is still confusion in the contemporary literature. Applying the findings of population genetics and population systematics to man might help to clarify some of our thinking.

The evolution of an organism has two aspects that, although they always go hand in hand in nature, should not be confused. One is evolution toward ecotypic adaptation and the other is progressive or phyletic evolution. The often posed question "Is man going downhill biologically?" cannot be answered until it is phrased more precisely. Which evolutionary aspect is meant: whether man as an organism is becoming less well adapted to his environment, or whether man is losing some of his most characteristically human attributes? It seems to me that in the current literature these two questions are frequently confused.

Is Man Becoming Less Well-Adapted to His Environment?

Modern man is an almost supreme master of his environment. Clothing, housing, heating, travel, transport, food production, and food storage have made him independent of the fluctuations of the environment to a degree unparalleled elsewhere in the animal kingdom. The physical environment as such has become a far less severe selective factor in modern man than in any wild animal, and even than in the ancestral hominids. The conquest of disease and the mitigation of the effects of aging have achieved spectacular results. And these developments will continue to make steady progress regardless of minor changes in the genetic composition of mankind. The genetic results will be the survival of genotypes, one might say a normalization of genotypes, that formerly were highly deleterious, such as those of diabetics, sufferers from Addison's disease, or poor antibody formers. It seems to me that in discussions of man's future far too much space is given to the role of genes controlling such diseases or, let us say, to the various metabolic disturbances that characterize most genetic diseases. A rise in the frequency of such genes will have no drastic effect on the future of mankind as long as adequate medical facilities are available. The relaxation of normalizing selection (Chapter 8) indicated by the increase in frequency of such genes in the human species has exceedingly little to do with man's phyletic evolution. Such genes will become a real problem for mankind only if sufferers from the diseases they control should occur in a large number of families and thereby interfere with the proper functioning of family life. This indeed could conceivably happen as a result of greatly increased atomic-radiation damage. A breakdown of civilization owing to overpopulation or wartime destruction, resulting in a widespread unavailability of modern heating, housing, food, and medicines, might likewise convert many genes now "normal" into lethal ones.

Natural Selection in Man

In view of his general emancipation from the environment, it is legitimate to ask whether natural selection still operates in man. Let us remember that the tremendous ecological evolution of man from a semistarved, hunting and hunted caveman to the machine man of the atomic age has taken place without visible biological evolution. Cro-Magnon Man, who entered history about 30,000 years ago, differs physically from modern man no more than do various modern races of man from each other. The great humanization of man during recent millennia is primarily a result of his ability to transmit nongenetic components of culture, including all sorts of scientific and technological information. This has been interpreted by some authors to mean that man has emancipated himself from natural selection. It has been asserted that within his own species man has almost abolished the evolutionary significance of heredity and natural selection. This assertion is based on one or both of two unspoken assumptions: that there are no genetic differences among human individuals in any but physical features; and that whatever differences do exist have no selective significance. Neither of these assumptions is tenable. The considerable mortality of human zygotes before the age of reproduction and the highly differential fertility among human individuals indicate clearly how active natural selection still is today. The importance of natural selection for modern man has recently been reaffirmed by several authors (for instance, Dobzhansky 1962). Of course the nature of selection pressures changes from generation to generation and the intensity of selection is relaxed in times of rapid expansion of the human population and of great medical discoveries. Yet wherever there is an acute shortage of food or poor hygienic conditions—and these conditions apply to more than 50 percent of mankind—natural selection still operates in its crudest form. And where prereproductive mortality has been largely eliminated as a selective factor, it has been replaced by unequal rates and ages of reproduction, both of which make an enormous potential contribution to Darwinian fitness (Crow 1961). Opportunity for selection owing to mortality before the end of the reproductive age has dropped in the United States by about 90 percent between the late nineteenth century and 1950. The importance of selection by unequal reproduction has increased correspondingly. Let me illustrate this by one example. The living Ramah Navaho Indians go back to 29 founders. Fourteen of these produced 84.48 percent of the descendants, 15 others a total of only 15.52 percent of the descendants. This certainly represents an enormous selection differential for whatever genetic differences existed between the two groups of founders. Many comparisons of rates of reproduction of different social classes and of sympatric human races indicate that *differential reproduction* nowadays makes a far greater contribution to selection than prereproductive mortality. Together, these two components add up to such a high value that claims that natural selection is unimportant for modern man appear quite absurd.

We are now prepared to answer the question whether modern man is less well adapted than was man in former generations. I think we can say confidently that, allowing for the change in environment, there is no appreciable difference. Perhaps we are losing the ecotypic adaptations that permitted human races to flourish in special environments. We may be losing the extremes of pigmentation, body proportions, and resistance to cold or heat that make the Eskimos successful in the arctic and the Watusi in tropical savannahs. The number of genes involved in ecotypic adaptation must be very large, considering how many generations of counterselection are required to modify such genotypes after migration to a different climatic zone. Yet, with the increasing emancipation of civilized man from the selective forces of the physical environment, the importance of these ecotypic adaptations is being steadily reduced. Their gradual loss will not make modern man less well adapted in the new environment he occupies. Nor can relaxation of the pressure of normalizing selection under the conditions of civilization be considered a serious loss of adaptation.

The danger that man will become extinct is negligible, unless he exterminates himself through atomic war or some other folly. No other organism can live successfully in so many climatic zones and in so many habitats. Man is sufficiently polymorphic that even the most devastating diseases should leave survivors. The semi-isolated life of many primitive societies of man increases this probability of survival. On purely biological grounds, then, there is not much reason to worry about the genetic continuity of mankind. But this still leaves our second question unanswered.

Is Man Losing His Most Typically Human Characteristics?

Man's phyletic evolution was characterized by the acquisition of upright posture, the adjustment of the hand for all the skills for which the hand is now used, and, most important, the development of speech, the ability of abstract thinking, and the many other uniquely human characteristics associated with man's large brain and cultural evolution. Considering the magnificent achievements of human art, literature, science, and technology, man has good reason to be proud of past progress. Looking at these positive aspects of man's evolutionary history, one can understand the arrogant assertion sometimes made that man is good enough as he is. Yet, when one considers with what futility the human brain has tackled the major problems of our world, one is forced to display more humility. And when one observes with what speed human individuals and whole societies can degrade themselves, one cannot avoid feeling that man could still go a long way on the road toward improvement. Nevertheless, there is no evidence of any biological improvement in at least the last 30,000 years. On the contrary, as Huxley (1953:172–173) has correctly pointed out, it is probable that

> man's genetic nature has degenerated and is still doing so . . . There is also the fact that modern industrial civilization favors the differential decrease of the genes concerned with intelligence. It seems now to be established that, both in communist Russia and in most capitalist countries, people with higher intelligence have, on the average, a lower reproductive rate than the less intelligent; and that some of this difference [in intelligence] is genetically determined. The genetic differences are slight, but . . . such slight differences speedily multiply to produce large effects. If this process were to continue, the results would be extremely grave.

There is hardly a more controversial question in the field of human biology than the question of the genetic contribution to intelligence and its correlation with fertility. Typologists are utterly incapable of discussing this subject because they fail to realize that these are statistical phenomena with highly incomplete correlations. Everyone now recognizes the inadequacy of I.Q. tests, particularly the verbal-facility tests. Everyone also realizes the considerable contribution made by home background, motivation, and general education to performance in such tests. Yet, when all these factors are duly considered, there is no doubt that there is some correlation between performance in I.Q. tests and genuine intelligence (Jensen 1969). Nor can there be any question that intelligence has in part a genetic basis, and it is quite immaterial for our argument whether the heritability of intelligence is 25 percent or 75 percent. Finally, there is abundant statistical evidence that in most communities those people whose professions require high intelligence produce on the average smaller families and produce them at a later age than do people like unskilled laborers, whose professions do not make such requirements. Though Huxley's argument is still being heatedly denied by identicists, the weight of the available evidence fully supports his conclusion that those who are intellectually best endowed contribute less to the gene pool of the next generation than do the average and, indeed, most of the less than average. Here we have another illustration of the principle discussed in Chapter 8, that natural selection is unable to discriminate between reproductive success as such and reproductive success owing to the possession of characteristics benefiting the adaptation of the species as a whole.

It has been claimed that the higher intellects are nothing but a homozygous "fringe" due to the segregation of previously balanced heterozygous gene complexes and that their lower reproductivity is a result of homozygous inferiority. I have no doubt that this is true in some cases, but it seems to me that there is much evidence to discredit it as a broad generalization. First of all, this claim is based on the assumption that the low fertility of those with higher intelligence is a biological characteristic. All special investigations, however, indicate that the low fertility of professionals is largely due to family planning, and not to neurosis or physiological deficiencies. A study of exceptionally high I.Q.'s made by Terman and Oden (1959) in California revealed that they surpassed the average of the population in mental and physical health, indeed, that they were highly superior. Further refutation of the claim that most of the low fertility of the superior intellects is biological consists in the fact that in the same circle of intellectuals in which it was fashionable in the 1920's and 1930's to have two children it became in the 1950's and 1960's fashionable to have four or five children. There is no evidence whatsoever to suggest that the lowered fertility of the higher intellects is comparable, let us say, to the decreased fertility of *Drosophila* strains selected for high bristle number. Intelligence, at least in the human species, is in itself a strong viability characteristic and will therefore be different in its response to selection from such quantitative aspects of the phenotype as size or bristle number, which, to a considerable extent, are pleiotropic by-products of genes forming the general genetic background of the viability genes. This consideration is another objection to the balanced-polymorphism theory of intelligence and fertility.

The assumption that all directional selection leads to a loss of fitness is a fallacy widely accepted among students of selection. It is based on experiments, discussed in Chapter 10, in which closed populations were exposed to extreme selection pressures. Such completely artificial conditions should not be used as the basis for broad generalizations. In natural populations, directional selection pressure, no matter how strong, is always accompanied by extremely high selection in favor of general fitness. In these open populations, new gene combinations are constantly being formed and tested for their ability to produce the desired phenotype. The rapid increase in brain size between *Australopithecus* and *Homo sapiens* proves how drastic a change of phenotype is possible without any loss of fitness. The spectacular achievement of selection in domestic animals and cultivated plants is further proof of the possibility of combining progressive selection with retention of fitness. There is thus no genetic reason why the increase in brain size could not be continued if there were a selective premium on such a process. The possibility exists that the plateau in brain size reached in the curve of hominid brain evolution is indicative of an upper ceiling. The head of the human baby has to pass at birth through the pelvis of the mother, and too large a brain unquestionably adds to the perils of the birth act and would be selected against. Yet natural selection would have alternative pathways if the selection for large brain size were strong enough: (1) increase in the size of the female pelvis, (2) shortening of the length of pregnancy, (3) shift of a greater portion of the growth period of the brain to the postnatal stage. These and other considerations support the conclusion that increase of brain size stopped not because it ran into a road block of direct counterselection, but rather because it was no longer rewarded by a reproductive premium.

Eugenic Measures

Those who are convinced that counterselection is operating in modern man and that the most desirable genes and gene combinations are not maintained at current frequencies in the total gene pool of the human species are at present a very small minority. They may even be a minority among the geneticists and students of human populations. But let us assume, for the sake of the argument, that they are right. It would then be our duty to propose countermeasures. Many such measures have indeed been proposed. Unhappily, most of them are unpalatable to liberal-minded, freedom-loving modern man. Worse than that, many of them are merely negative eugenics and will not materially contribute to the desired end. However, the situation is not altogether hopeless.

Animal breeding has long abandoned all attempts to discover superior genes individually. In fact, such desirable economic features as high egg production in chickens or high milk output in dairy cows are exceedingly difficult to analyze, not only genetically but even physiologically. All sorts of generalized factors, such as resistance to disease, superior utilization of food, and so forth, contribute substantially to the goal of selection. In man also one should favor the selection of generalized desirable characteristics rather than selecting for highly specific traits. Perhaps it is not unreasonable to assume that a person with a good record of achievement in certain areas of human endeavor has on the average a more desirable gene combination than a person whose achievements are less spectacular. In our present society, the superior person is punished by the government in numerous ways, by taxes and otherwise, which make it more difficult for him to raise a large family. Why, for instance, should tax exemption for children be a fixed sum rather than a percentage of earned income? Why should tuition in school be based, in large part, on the ability of the father to pay rather than inversely on the achievement of the student? Innumerable administrative rules and laws of the government discriminate inadvertently against the most gifted members of the community. Changing these laws so as to place a premium on performance (the "opportunity" of true democracy, rather than identicism) is entirely different from distributing privileges according to the artificial, arbitrary criteria of the racists, such as blond hair and blue eyes, or on the basis of descent, as in feudal society. I firmly believe that such positive measures would do far more to increase desirable genes in the human gene pool than all the negative measures proposed by eugenicists of former generations.

Overpopulation

Alas, the progressive loss of valuable genes is not the only danger facing the human species. Indeed, overpopulation is now a far more serious problem. I am not speaking of the material aspects of the problem, such as the exhaustion of mineral and soil resources and the increasing difficulty of supplying food for 6, 8, or 10 billion people. Human technology may find answers to all these difficulties. Yet I cannot see how all the best things in man can prosper—his spiritual life, his enjoyment of the beauty of nature, and whatever else distinguishes him from the animals—if there is "standing room only" (Darwin 1859:64). It seems to me that long before that point has been reached man's struggle and preoccupation with social, economic, and engineering problems would become so great, and the undesirable by-products of crowded cities so deleterious, that little opportunity would be left for the cultivation of man's highest and most specifically human attributes. Nor do I see where natural selection could enter the picture to halt this trend. Man may continue to prosper physically under these circumstances, but will he still be anywhere near what we consider to be the ideal of man?

Most primitive societies had a far clearer understanding of the carrying capacity of their environment and the need for limiting population size than western man. A true unbalance between birth rate and mortality rate has become a problem only in the last 100 years. It seems that we are not conceptually prepared to cope with this dilemma. "Thou shall not kill" is accepted by us without question as a legitimate restriction of the individual's freedom. But we are not yet ready to accept as equally important the maxim "Thou shall not produce more than 2–3 children." And yet we can no longer include unrestricted procreation among the natural "freedoms" of man, if we define as freedom the right to do what we want, *provided it does not hurt others.* Now that we realize how many of the evils of modern society are directly or indirectly the result of overpopulation (Ehrlich 1968), we know that the right of unlimited procreation should be removed from man's freedoms (Hardin 1968). If we do not grasp this in the very near future, it will be too late. It is most regrettable that the adoption of a healthier way of thinking about the perils of overpopulation is impeded by medieval, and in their effects extremely vicious, church dogmas. What good is it if those with enough brains to have a sense of responsibility limit themselves to 2 children, while those who lack it have 8 or 10? And what does this portend for the future of mankind? With governments, legislatures and church leaders displaying a callousness toward these problems that borders on the criminal, it becomes the duty of every single individual to fight for sense and sanity in population policy. The only alternative is disaster for mankind.

Glossary

Bibliography

Index

Glossary

Acrocentric chromosomes. Chromosomes with the centromere at or near one of the ends; rod-shaped chromosomes.

Adaptation. The condition of showing fitness for a particular environment, as applied to characteristics of a structure, function, or entire organism; also the process by which such fitness is acquired.

Adaptive radiation. Evolutionary divergence of members of a single phyletic line into a series of different niches or adaptive zones.

Additive variance. Variance due to the average value of the different genes.

Allele. Any of the alternative expressions (states) of a gene (locus).

Allen's rule. Protruding body parts of warm-blooded animals are shorter in cooler than in warmer climates.

Allochronic speciation. The splitting of a species into two owing to a separation of their breeding seasons.

Allometric growth. Growth in which the growth rate of one part of an organism is different from that of another part or of the body as a whole.

Allopatric. Of populations or species, occupying mutually exclusive (but usually adjacent) geographical areas.

Allopatric hybridization. The hybridization of two previously isolated populations or species in a zone of contact.

Allopatric speciation. Geographic speciation.

Allopolyploid. A polyploid produced by the chromosome doubling of a species hybrid, that is, of an individual with two different chromosome sets.

Amphiploid. Allopolyploid.

Anadromous. Wandering upstream (from the sea) in order to spawn, like the salmon.

Anagenesis. Progressive ("upward") evolution.

Antigen. A substance capable of inducing the formation of antibodies when introduced into the blood stream of an animal.

Apostatic. Widely departing from the norm; of a phenotype that differs strikingly from the search image of a predator (B. Clarke 1962).

Arrhenotoky. The haplodiploid parthenogenesis in which males arise from unfertilized, hence haploid, egg cells.

Autogamy. Formation of a diploid nucleus in certain protozoa by the union of male and female gamete nuclei in the same individual; a process of self-fertilization.

Autopolyploid. A polyploid originating by the doubling of a diploid chromosome set.

Autosome. A chromosome other than a sex chromosome.

Balanced load. The amount by which the overall fitness of a population is depressed owing to the segregation of inferior genotypes the component genes of which are maintained in the population because they add to fitness in different combinations (for instance, as heterozygotes).

Balanced polymorphism. A polymorphism maintained by a selective superiority of the heterozygotes over either type of homozygotes.

Baldwin effect. The condition in which an organism can stay in a favorable environment, with the help of a suitable modification of the phenotype, until selection has achieved the genetic fixation of this phenotype.

Batesian mimicry. The mimicking (similarity) of a species distasteful or poisonous to a predator by unrelated edible species.

Bergmann's rule. Body size in geographically variable species of warm-blooded animals is larger in the cooler parts of the range of the species.

Biological races. Noninterbreeding sympatric populations that differ in biological characteristics but not, or scarcely, in morphology. Supposedly prevented from interbreeding by preference for different food plants or other hosts.

Biological species concept. A concept of the species category stressing reproductive isolation and the possession of a genetic program effecting such isolation.

Biomass. The total mass (or weight) of a population or other specified group of individuals.

Biota. Fauna and flora of an area or region.

Biotype. A group of genetically identical individuals.

Blending inheritance. The complete fusion of the genetic factors of father and mother in the offspring. Now known not to occur. See Particulate inheritance.

Canalization. The property of developmental pathways of achieving a standard phenotype in spite of genetic or environmental disturbances.

Canalizing selection. The selection of genes that would stabilize the developmental pathways so as to make the phenotype less susceptible to the effect of environmental or genetic disturbances.

Catastrophism. Cuvier's explanation of the existence of fossil faunas as the result of geological catastrophies.

Category (taxonomic). One of a hierarchy of levels to which taxa are assigned, such as subspecies, species, genus, and so forth.

Centers of diversification. Geographic areas with the greatest number of different cultivated strains.

Centric fusion. The fusion of two acrocentric (rod) chromosomes into a single metacentric (V) chromosome through translocation and loss of a centromere.

Centromere. A special region on the chromosome where it becomes attached to the spindle.

Character. An attribute of a member of a population or taxon by which it differs from a member of a different group or taxon.

Character displacement. A divergence of equivalent characters in sympatric species resulting from the selective effects of competition.

Character divergence. The name given by Darwin to the differences developing in two (or more) related species in their area of sympatry owing to the selective effects of competition.

Chiasma (pl. chiasmata). An X-shaped chromosome configuration caused by the breakage, exchange, and reciprocal fusion of equivalent segments of homologous chromatids (in meiotic cell division).

Chromosome. A deeply staining DNA-containing body in the nucleus of the cell best seen during certain phases of cell division; the carriers of the (nuclear) genes.

Chronocline. A character gradient in the time dimension.

Circular overlap. The phenomenon in which a chain of contiguous and intergrading populations curves back until the terminal links overlap with each other and behave like good (noninterbreeding) species.

Cistron. The functional gene; the totality of sites on a gene locus that jointly control a unitary function (for example, the formation of an enzyme) (as shown by noncomplementarity and a cis-trans effect of a set of recessive mutants).

Cladogenesis. Branching evolution.

Climatic rules. Rules describing regularities in geographic variation correlated with climatic gradients.

Cline. A gradual and essentially continuous change of a character in a series of contiguous populations; a character gradient.

Clone. All the individuals derived by asexual reproduction from a single sexually produced individual.

Closed population. A population with no genetic input other than that by mutation.

Coadaption. The harmonious epistatic interactions of genes brought together by natural selection.

Codon. A triplet of base pairs (nucleotides) specifying one of the 20 common amino acids.

Competition. The simultaneous seeking of an essential resource of the environment that is in limited supply.

Competitive exclusion. The principle that no two species can coexist at the same locality if their ecological requirements are identical.

Conditioning. The process of acquisition by an animal of the capacity to respond to a given stimulus with the reaction proper to another stimulus when the two stimuli are applied concurrently for a number of times.

Conservative characters. Characters that change only slowly during evolution.

Conspecific. Individuals or populations of the same species.

Controlling factor. With reference to competition, any factor the effect of which becomes more severe as the density of the population increases.

Correlated response. Change of the phenotype occurring as an incidental consequence of selection for a seemingly independent character, such as sterility resulting from selection for high bristle number in *Drosophila.*

Crossing over. The exchange, during meiosis, of corresponding segments between homologous chromosomes.

Cryptic species. A species the diagnostic characters of which are not easily perceived; a sibling species.

Cyclomorphosis. A cyclic, seasonal change of form in a series of genetically identical populations, as in cladocerans and rotifers.

Cytogenetics. The comparative study of chromosomal mechanisms and behavior in populations and taxa, and their effect on inheritance and evolution.

Cytoplasmic factor. A genetic factor in the cytoplasm.

Deme. A local population, as defined in Chapter 7, p. 82.

Density-dependent factors. Causes of mortality and fecundity that become more effective as the density of a population increases.

Developmental homeostasis. The capacity of the developmental pathways to produce a normal phenotype in spite of developmental or environmental disturbances.

De Vriesianism. The hypothesis that evolution in general and speciation in particular are the results of drastic mutations (saltations).

Diagnostic. Referring to a character that unambiguously differentiates a taxon from others.

Diapause. A temporary interruption of growth in the embryos or larvae of insects, usually during hibernation or aestivation.

Dimorphism. Occurrence of two distinct morphological types (morphs, phena) in a single population.

Dioecious. Having the male and the female reproductive organs segregated into different individuals.

Diploid. Having a double set (2n) of chromosomes; the normal chromosome number of the cells (except of mature germ cells) in any individual derived from a fertilized egg.

Disruptive (diversifying) selection. Selection for phenotypic extremes in a population (until a discontinuity is achieved).

DNA. Deoxyribonucleic acid.

Dominant. An allele that determines the phenotype of a heterozygote.

Dosage compensation. The effect produced by modifying genes that compensate for the difference between the dosage of major sex-linked genes present in male and female.

Ecogeographical rules. The formulation of regularities in geographic variation (of size, pigmentation, and so forth) correlated with environmental conditions.

Ecological race. A local race that owes its most conspicuous attributes to the selective effect of a specific environment (see Ecotype).

Ecophenotype. A nongenetic modification of the phenotype in response to an environmental condition.

Ecotype. A local race that owes its most conspicuous characters to the selective effects of local environments (see Ecological race).

Ectothermal. Having the body temperature determined by the temperature of the environment; poikilothermal.

Edaphic race. A race that is affected by the properties of the substrate (soil) rather than by other environmental factors.

Eidos. Any of the fixed types (ideas) that Plato conceived to underlie the apparent variability of phenomena.

Electrophoresis. A technique that separates mixtures of molecules (particularly proteins and polypeptides) by their different rates of travel in an electric field.

Epigamic. Serving to attract or stimulate individuals of the opposite sex during courtship.

Epigenetic. Developmental—referring to the interaction of genetic factors during the developmental process.

Epigenotype. The total developmental system; the totality of interactions among genes resulting in the phenotype.

Epistatic interaction. An interaction of genes at different loci.

Ethological. Behavioral, particularly with reference to species-specific components of behavior the phenotypic expression of which is largely determined genetically.

Ethological barriers. Isolating mechanisms caused by behavioral incompatibilities of potential mates.

Evolutionary novelty. A newly acquired structure or other property that permits the performance of a new function.

Exclusion principle. The principle stating that two species cannot coexist at the same locality if they have identical ecological requirements.

Expressivity. The degree to which a gene is expressed in the phenotype.

F_1. First filial generation.

Fecundity. Reproductive potential as measured by the quantity of gametes, particularly eggs, produced.

Fertility. Reproductive potential as measured by the quantity or percentage of developing eggs or of fertile matings.

Fission. The splitting of a single (metacentric) chromosome into two (usually acrocentric) chromosomes.

Founder principle. The principle that the founders of a new colony (or population) contain only a small fraction of the total genetic variation of the parental population (or species).

Fusion (Robertsonian). The union of two (usually acrocentric) chromosomes to form a single (metacentric) chromosome. See also Fission.

Gametes. Functional germ cells (= eggs and spermatozoa).

Gause principle. Exclusion principle.

Gene. A unit of inheritance, carried in a chromosome, transmitted from generation to generation by the gametes, and controlling the development and characteristics of an individual.

Gene arrangements. Alternative gene sequences on chromosomes, owing to inversion, translocation, or other chromosomal changes.

Gene flow. The exchange of genetic factors between populations owing to the dispersal of gametes or zygotes.

Gene pool. The totality of the genes of a given population existing at a given time.

Genetic drift. Genetic changes in populations caused by random phenomena rather than by selection.

Genetic homeostasis. The property of the population of equilibrating its genetic composition and of resisting sudden changes.

Genetic load. The depression of fitness (from a theoretical optimum) caused by deleterious genes (for example, not yet eliminated recessives).

Genome. The genetic contents of the chromosomes, that is, the genotype.

Genotype. The totality of genetic factors that make up the genetic constitution of an individual.

Geoffroyism. The belief in an adaptive response of the genotype to the demands of the environment; environmental induction of appropriate genetic changes. Usually (though not strictly correctly) included with Lamarckism.

Geographic barrier. Any terrain that prevents gene flow between populations.

Geographic isolate. A population or group of populations prevented by an extrinsic barrier from free gene exchange with other populations of the species.

Geographic race. A geographically delimited race, usually a subspecies.

Geographic speciation. The acquisition in a population—while it is geographically isolated from other populations of its parental species—of characters that promote or guarantee reproductive isolation after the external barriers break down.

Geographic variation. The differences between spatially segregated populations of a species; population differences in the space dimension.
Geographic vicariance. Geographic replacement of populations or species by each other.
Gloger's rule. Races in warm and humid areas are more heavily pigmented than those in cool and dry areas.
Gonochorism. The possession of gonads of only one sex (either male or female) in an individual.
Gonochoristic. Of individuals, having functional gonads of only one sex; of breeding populations, composed of male and female individuals.
Grade. A group of animals similar in level of organization; a level of anagenetic advance.
Group selection. A postulated process of selection in which a whole group of individuals rather than a single one is the target of selection, a process by which characters can be selected for that benefit a group but not an individual.
Gynogenesis. Pseudogamy.
Habitat selection. The capacity of a dispersing individual to select an appropriate (the species-specific) habitat.
Haploid. Having only a single set of chromosomes—gametes are usually haploid.
Hardy-Weinberg law. The fact that, owing to particulate inheritance, the frequency of genes in a population remains constant in the absence of selection, of nonrandom mating, and of accidents of sampling.
Hemizygous. Of unpaired genes in sex chromosomes in the heterogametic sex (sex with unequal sex chromosomes, such as XY).
Heritability. The genetic component of phenotypic variability.
Hermaphroditism. The occurrence of gonads of both sexes in a single individual.
Heterochromatic. Of chromosomal sections, staining differently from the major portions of the chromosomes.
Heterogamy. The mating of unlike individuals; the preference of an individual to mate with an individual of unlike phenotype or genotype (opposed to Homogamy).
Heterosis. Selective superiority of heterozygotes.
Heterostyly. A polymorphism of flowers in which styles and stamens are of two (or more) unequal lengths, a system that insures cross fertilization.
Heterozygote. An individual with different genetic factors (alleles) at the homologous (corresponding) loci of the two parental chromosomes.
Higher category. A taxonomic category of higher rank than the species: genus, family, order, and so forth.
Holometabolic. Of an insect, undergoing a complete metamorphosis between larval and adult stage.
Homeostatic mechanism. Self-regulatory device that tends to restore conditions existing prior to a disturbance or shift.
Homeotic mutant. In insects, the mutational change of one in a series of structures to the form of another structure in the series, as wing into haltere, arista into leg, wing into leg, and the like.
Homing ability. Orientation in unfamiliar terrain that permits return to the original home.
Hominid. A man-like organism belonging to the phyletic line leading to man.

Homogamy. The preference of a mating individual for another with similar phenotype or genotype (opposed to Heterogamy).
Homologous. Of a feature in two or more taxa that can be traced back to the same (or an equivalent) feature in the common ancestor of these taxa.
Homosequential. Of species, with the same linear sequence of genes on their chromosomes.
Homostyly. A system of flowers in which styles and stamens are of equal length.
Homozygous. Having identical alleles at the two homologous loci of a diploid chromosome set.
Hybridization. The crossing of individuals belonging to two unlike natural populations that have secondarily come into contact.
Imprinting. A process of rapid learning of highly specific information (like the parent image) during a critical period in the life cycle.
Inbreeding. Crossing with genetically similar individuals, particularly with close relatives.
Inbreeding depression. A reduction of fitness owing to severe inbreeding; often manifested by loss of fertility, growth anomalies, and metabolic disturbances.
Industrial melanism. The evolution of a darkened population owing to selection of melanistic individuals that better blend with their substrate in the sooty surroundings of an industrial area.
Input load. The load of inferior alleles in a gene pool, caused by mutation and immigration.
Instantaneous speciation. The production of a single individual that is reproductively isolated from the species to which the parents belong and is reproductively and ecologically capable of establishing a new species population.
Intergrading. Of populations, that are intermediate in characters between the populations adjacent on either side.
Internal balance. The harmonious epistatic interactions of genes at different loci.
Introgression. The incorporation of genes of one species into the gene pool of another species.
Introgressive hybridization. An hybridization leading to introgression.
Inversion. Reversal of the linear order of the genes in a segment of a chromosome.
Irreversibility. The inability of an evolving group of organisms (or a structure of an organism) to return to an ancestral condition—the theory of irreversibility is that a given structure or adaptation which has been lost in evolution cannot be restored exactly to its prior condition.
Isoalleles. Alleles that produce such slight phenotypic differences that special techniques are required to reveal their presence.
Isolate. A population or group of populations that is separated from other populations.
Isolating mechanisms. Properties of individuals that prevent successful interbreeding with individuals that belong to different populations.
Isophenes. Lines on a map that connect points of equal expression of a clinally varying character.
Karyotype. The chromosome complement.
Kinship selection. The selection of characters (and the underlying genes) because they increase the probabilities of survival of close relatives.
Lamarckism. The theory, advocated by Lamarck, that evolution is brought about by volition or by environmental induction.

Levels of integration. Levels of complexity in structures, patterns, or associations at which new properties emerge that could not have been predicted from the properties of the component parts.

Linkage. The occurrence of genes on the same chromosome. The closer they are on the chromosome, the more tightly they are linked, that is, the less likely they will be separated by crossing over.

Locus. The location of a given gene on a chromosome.

Ludwig theorem. The theory that new genotypes can be added to a population if they can utilize new components of the environment (occupy a new subniche) even if they are inferior in the ancestral niche.

Luxuriance. Somatic vigor of hybrids (heterozygotes) that does not add to their fitness.

Macroevolution. See Transpecific evolution.

Macrogenesis. The sudden origin of new types by saltation.

Mechanical isolation. Reproductive isolation owing to mechanical incompatibility of male and female genitalic structures.

Meiosis. Two consecutive special cell divisions in the developing germ cells characterized by the pairing and segregation of homologous chromosomes. The resulting gametes will have reduced, that is, haploid, chromosome sets.

Meiotic drive. A force able to alter the mechanics of meiotic cell division in such a manner that the two kinds of gametes produced by a heterozygote do not occur with equal frequency.

Melanism. An unusual darkening of color owing to increased amounts of black pigment—sometimes a racial character; sometimes restricted to a certain percentage of individuals in a population, giving rise to polymorphism.

Mendelism. Particulate inheritance. See also De Vriesianism.

Meristic variation. Variation in characters that can be counted, like number of vertebrae, scales, fin rays, and so forth.

Metacentric. Having the centromere somewhere along the chromosome, but not at or near the tip; characteristic of chromosomes that are J- or V-shaped in metaphase.

Metamorphosis. A drastic change of form during development, as when a tadpole changes into a frog, or an insect larva into an imago.

Metaphase. The middle stage of cell division when the chromosomes line up (paired in meiosis) in the equatorial plane of the spindle.

Mimetic polymorphism. Polymorphism (best known in Lepidoptera) in which the various morphs resemble other species distasteful or dangerous to a predator—often restricted to females.

Mitosis. The division and separation of chromosomes during cell division.

Modifiers. Genes that affect the phenotypic expression of genes at other loci.

Monoecious. Having male and female sex organs in the same flower; corresponding to hermaphroditism in animals.

Monophagous. Adapted to subsist on one kind of food; specialized on a single host species.

Monotypic species. A species containing only a single (= the nominate) subspecies.

Morph. Any of the genetic forms (individual variants) that account for polymorphism.

Morphism. Polymorphism.

Morphology. The description and study of structural characteristics, particularly those on the surface of the body.

Mullerian mimicry. Similarity among several species that are distasteful or poisonous to a predator.
Multifactorial. Controlled by several gene loci.
Multivalent. An attaching to each other of several chromosomes (bivalent = two chromosomes).
Mutation. A change in the genetic material; most often a change in a single gene (gene mutation), consisting of a replacement, duplication, or deletion of one or several base pairs in the DNA.
Mutationism. De Vriesianiam.
Myrmecophily. The utilization by other insects, mostly beetles, of ant colonies as domicile and source of food.
Neo-Darwinism. Weismann's theory of evolution; sometimes, any modern evolutionary theory featuring natural selection.
Neontology. The study of Recent organisms—antonym of paleontology.
Neoteny. The elimination of metamorphosis into the adult stage, breeding taking place in the larval or juvenile stage.
Niche (ecological). The constellation of environmental factors into which a species (or other taxon) fits; the outward projection of the needs of an organism, its specific way of utilizing its environment.
Nondimensional species. The species concept characterized by the noninterbreeding of two coexisting demes, uncomplicated by the dimensions of space and time.
Nondisjunction. The failure of separation of paired chromosomes at meiosis and their passage to the same spindle pole, resulting in unequal chromosome numbers in the daughter cells.
Normalizing selection. The removal by selection of all genes that produce deviations from the normal (= average) phenotype of a population.
Nucleotide. A subunit of the DNA and RNA molecules, consisting of phosphoric acid, a base (purine or pyrimidine), and a sugar.
Oligogenic. Of a character, determined by few genes.
Oligolectic. Of bees, collecting the pollen of only a few kinds of flowers.
Oligophagous. Feeding on few species of food plants.
Ontogeny. The development of the individual, particularly the embryogenesis.
Open population. A population freely exposed to gene flow and subject to much input of alien genes owing to immigration.
Organization effect. An interaction among adjacent loci owing to some features of organization of the chromosomes.
Orthogenesis. Evolution of phyletic lines following a predetermined rectilinear pathway, the direction not being determined by natural selection.
Outbreeding. Crossing with genetically different, not closely related individuals, particularly with members of different populations.
Overdominance. Superiority of the heterozygote over both kinds of homozygotes.
Pair formation. The process (displays, etc.) by which the pair bond is established in species (particularly of birds) in which both parents participate in the raising of the young.
Panmictic. Of populations, randomly interbreeding, the whole population or species forming a single deme.
Paracentric inversion. An inversion that does not include the centromere.
Parapatric. Of populations or species, geographically in contact but not overlapping and rarely or never interbreeding.

Parthenogenesis. The development of eggs without fertilization.

Particulate inheritance. Mendel's theory that the genetic factors received from mother and father do not blend or fuse, but retain their integrity from generation to generation.

Penetrance. The frequency with which a (dominant or homozygous) gene manifests itself in the phenotype; most genes have 100-percent penetrance.

Pericentric inversion. An inversion that includes the centromere.

Peripheral isolate. A population isolated at or beyond the periphery of the continuous species range.

Phage. A bacterial virus.

Phenocopy. A modification of the phenotype (owing to special environmental conditions) that resembles a change of the phenotype caused by a mutation.

Phenodeviants. Phenotypes that deviate from the population (or species) mean (or norm), owing to special gene combinations, for instance extreme homozygosity.

Phenon (*pl. phena*). A sample of phenotypically similar individuals; a phenotypically reasonably uniform sample.

Phenotype. The totality of characteristics of an individual (its appearance) as a result of the interaction between genotype and environment.

Philopatry. The drive (tendency) of an individual to return to (or stay in) its home area (birthplace or other adopted locality).

Phylogeny. The history of the lines of evolution in a group of organisms; the origin and evolution of higher taxa.

Phylum (*pl. phyla*). One of the higher taxonomic categories, like Chordata or Arthropoda.

Plasmagenes. Genetic factors located in the cytoplasm (outside the nucleus).

Pleiotropy. The capacity of a gene to affect several characters, that is, several aspects of the phenotype.

Pleistocene refuges. Favorable areas south of the borders of the ice, where species and populations survived periods of glaciation.

Ploidy. A term referring to the number of chromosome sets.

Poikilothermal. Ectothermal.

Polygenes. Genes that jointly with several or many other genes control a character.

Polymorphism. The simultaneous occurrence of several discontinuous phenotypes or genes in a population, with the frequency even of the rarest type higher than can be maintained by recurrent mutation.

Polyphagous. Feeding on many different kinds of food, such as species of host plants.

Polyphenism. The occurrence in a population of several phenotypes, the differences between which are not the result of genetic differences.

Polypheny. Polyphenism.

Polyploidy. A condition in which the number of chromosome sets in the nucleus is a multiple (greater than 2) of the haploid numbers.

Polytopic subspecies. A subspecies composed of widely separated but phenotypically identical populations.

Polytypic. Of a category, containing two or more immediately subordinate categories, for instance a species with several subspecies.

Population, local. The community of potentially interbreeding individuals at a given locality.

Populationist. One who regards variable phenomena as populations of individuals (or individual events) with calculable statistics; antonym of Typologist.

Position effect. The difference in the phenotypic expression of a gene caused by a change in its spatial relation to other genes on the chromosome.

Preadaptation. The possession of the necessary properties to permit a shift into a new niche or habitat. A structure is preadapted if it can assume a new function without interference with the original function.

Preformism. The belief that the egg (or sperm or zygote) contains a preformed adult in miniature, to be "unfolded" during development.

Procaryota. Those microorganisms (viruses, bacteria, blue-green algae) that lack well-defined nuclei and meiosis.

Pseudoalleles. Genes at closely adjacent loci that react physiologically in many ways as if they were alleles and between which crossing over is rare.

Pseudogamy. Parthenogenetic development of the egg cell after the egg membrane has been penetrated by a male gamete.

Pseudopolyploidy. A numerical relation of chromosome sets in groups of related species that leads to their erroneous interpretation as polyploids.

Random fixation. The loss of an allele (and fixation of the other allele) in a population owing to accidents of sampling.

Rassenkreis. The German equivalent of *polytypic species*—not "a circle of races."

Recessiveness. The failure of a gene to express its presence in the phenotype of the heterozygote.

Recombination. The reshuffling of the parental genes during meiosis through crossing over.

Reductionism. The belief that complex phenomena can be entirely explained by reducing them to the smallest possible component parts and by explaining these.

RNA. Ribonucleic acid.

Salivary chromosomes. Giant chromosomes (with highly specific patterns of dark and light bands) in the salivary glands of the larvae of certain kinds of dipterans (flies, mosquitoes, midges, and the like).

Saltation. A change by a leap across a discontinuity.

Secondary intergradation. The intergradation or hybridization of two distinct and previously isolated populations (or groups of populations) along a zone of secondary contact.

Selfing. Self-fertilizing in hermaphrodites or monoecious plants.

Semigeographic speciation. The splitting apart of species along lines of secondary intergradation or along lines of strong ecological contrast.

Semispecies. The component species of superspecies (Mayr); also, populations that have acquired some, but not yet all, attributes of species rank; borderline cases between species and subspecies.

Sibling species. Morphologically similar or identical populations that are reproductively isolated.

Sickle-cell anemia. An anemia due to a hemoglobin mutation found mostly in tropical areas and lethal to homozygotes.

Somatic. Relating to the body; of cells that are not germ cells.

Speciation. The splitting of a phyletic line; the process of the multiplication of species; the origin of discontinuities between populations caused by the development of reproductive isolating mechanisms.

Species. A reproductively isolated aggregate of interbreeding populations.

Species group. A group of closely related species, usually with partially overlapping ranges.

Species recognition. The exchange of appropriate (species-specific) stimuli and responses between individuals (particularly during courtship).

Specific modifier. A gene that has the specific and perhaps exclusive function of modifying the expression of a gene at another locus.

Spontaneous generation. The spontaneous origin of organisms from inert matter.

Stabilizing selection. The elimination by selection of all phenotypes deviating too far from the population mean, and hence also of genes producing such deviating phenotypes.

Sterile. Incapable of producing viable gametes.

Stochastic process. A process the various outcomes of which can be predicted with a specified probability (versus deterministic process).

Subspecies. An aggregate of local populations of a species inhabiting a geographic subdivision of the range of the species and differing taxonomically from other populations of the species.

Substitutional load. The cost to a population of replacing an allele by another in the course of evolutionary change.

Substrate race. A local race selected to agree in its coloration with that of the substrate, for example, a black race on a lava flow.

Supergene. A chromosome segment that is protected against crossing over and is transmitted as if it were a single gene.

Supernumeraries. Small chromosomes, often variable in number and partially inert genetically, that are not part of the regular chromosome set.

Superoptimal stimuli. Sensory stimuli to which an animal responds more strongly than to the natural stimuli for which the response has been selected.

Superspecies. A monophyletic group of entirely or essentially allopatric species that are either morphologically too different to be included in a single species or demonstrate their reproductive isolation in a zone of contact.

Suppressor gene. A gene that suppresses the phenotypic expression of a gene at another locus.

Switch gene. A gene that causes the epigenotype to switch to a different developmental pathway.

Symbiont. An organism that exists in a relationship of mutual benefit with another organism.

Sympatric speciation. Speciation without geographic isolation; the acquisition of isolating mechanisms within a deme.

Sympatry. The occurrence of two or more populations in the same area; more precisely, the existence of a population in breeding condition within the cruising range of individuals of another population.

Synthetic lethals. Lethal chromosomes derived from normally viable chromosomes by recombination (as a result of crossing over).

Synthetic theory. The current evolutionary theory, which is a synthesis of the best components of many previously proposed theories, with mutation and selection as the basic elements.

Systemic mutation. A mutation, postulated by R. Goldschmidt, that would fundamentally reorganize the germ plasm, and permit the origin of wholly new types of organisms.

Taxon (*pl. taxa*). A population or group of populations (taxonomic group) that is sufficiently distinct to be worthy of being distinguished by name and to be ranked in a definite category.

Taxonomy. The theory and practice of classifying organisms.

Telomere. The postulated material at the tip of a chromosome, having certain properties not found in the rest of the chromosome.

Teratology. The study of structural abnormalities, especially monstrosities and malformations.

Territory. An area defended by an animal against other members of its species (and occasionally members of other species).

Tetraploid. A polyploid with four haploid chromosome sets, normally the result of the doubling of the diploid chromosome number.

Thelytoky. Parthenogenesis of the type in which females give rise to female progeny without fertilization.

Transduction. Transfer of genetic information from one bacterium to another by means of bacteriophage.

Transformation. Transfer of genetic information in bacteria by means of extracellular DNA.

Transient polymorphism. Polymorphism existing during the period when an allele is being replaced by a superior one.

Translocation. The shift of a segment of a chromosome to another chromosome.

Transpecific evolution. Evolutionary phenomena and processes beyond the species level, such as the origin of new higher taxa, new organs, evolutionary trends, the extinction of faunas, and the like.

Typologist. One who disregards variation and regards the members of a population as replicas of the "type," the Platonic *eidos*.

Variance. A sample statistic relating to deviations from the mean.

Variety. An ambiguous term of classical (Linnaean) taxonomy for a heterogeneous group of phenomena including nongenetic variations of the phenotype, morphs, domestic breeds, and geographic races.

Vector. Carrier, particularly an animal that transmits a pathogen (protozoan, bacterium, virus) from one host to another.

Zygote. A fertilized egg; the cell (individual) that results from the fertilization of an egg cell.

Bibliography

Alexander, R. D. 1967. Acoustical communication in arthropods. *Ann. Rev. Entomol.* 12:495–526.

——— 1968. Life cycle origins, speciation, and related phenomena in crickets. *Quart. Rev. Biol.* 43:1–41.

——— 1969. Comparative animal behavior and systematics. In *Systematic Biology*, Nat. Acad. Sci. Publ. no. 1692:494–520.

——— and R. S. Bigelow. 1960. Allochronic speciation in field crickets, and a new species, *Acheta veletis*. *Evolution* 14:334–346.

——— and T. E. Moore. 1962. The evolutionary relationships of 17-year and 13-year cicadas, and three new species (Homoptera, Cicadidae, *Magicicada*). *Misc. Publ. Mus. Zool. Univ. Mich.* no. 121:1–59.

——— and E. S. Thomas. 1959. Systematic and behavioral studies on the crickets of the *Nemobius fasciatus* group (Orthoptera: Gryllidae: Nemobiinae). *Ann. Entomol. Soc. Amer.* 52:591–605.

Allee, W. C., A. E. Emerson, O. Park, T. Park, and K. P. Schmidt. 1949. *Principles of Animal Ecology*. Saunders, Philadelphia.

Allen, J. A. 1877. The influence of physical conditions in the genesis of species. *Radical Rev.* 1:108–140.

Amadon, D. 1947. Ecology and the evolution of some Hawaiian birds. *Evolution* 1:63–68.

——— 1950. The Hawaiian honeycreepers (Aves, Drepaniidae). *Bull. Amer. Mus. Nat. Hist.* 95:151–262.

——— 1964. The evolution of low reproductive rates in birds. *Evolution* 18:105–110.

Andrewartha, H. G., and L. C. Birch. 1954. *The Distribution and Abundance of Animals*. Univ. of Chicago Press, Chicago.

Ashmole, N. P. 1963. The regulation of numbers of tropical oceanic birds. *Ibis* 103b:458–473.

Avery, O. T., C. M. MacLeod, and M. McCarty. 1944. Studies on the chemical nature of the substance inducing transformation of pneumococcal types. Induction of transformation by a desoxyribonucleic acid fraction isolated from pneumococcus type III. *J. Exp. Med.* 79:137–158.

Ayala, F. J. 1968. Genotype, environment and population numbers. *Science* 162:1453–1459.

Bacci, J. 1950. Osservazioni sulla sessualità dei Nereimorfi (Anellidi Policheti). *Boll. Zool.* 17:55–61.

Baker, H. G., and G. L. Stebbins, eds. 1965. *The Genetics of Colonizing Species*. Academic Press, New York.

Baldwin, J. M. 1896. A new factor in evolution. *Amer. Nat.* 30:441–451, 536–553.

Barber, H. S. 1951. North American fireflies of the genus *Photuris. Smithsonian Misc. Coll.* 117:1–58.

Barnicot, N. A. 1959. Climatic factors in the evolution of human populations. *Cold Spring Harbor Symp. Quant. Biol.* 24:115–129.

Barr, T. C. 1967. Observations on the ecology of caves. *Amer. Nat.* 101:475–492.

Barth, E. K. 1968. The circumpolar systematics of *Larus argentatus* and *Larus fuscus,* etc. *Medd. Zool. Mus. Oslo* (*Nytt Mag. Zool.*) 15, suppl. 1:1–50.

Basrur, V. R., and K. H. Rothfels. 1959. Triploidy in natural populations of the black fly *Cnephia mutata* (Malloch). *Can. J. Zool.* 37:571–589.

Bastock, M. 1967. *Courtship: An Ethological Study.* Aldine, Chicago.

Bateman, A. J. 1950. Is gene dispersion normal? *Heredity* 4:353–363.

Batten, C. A., and R. J. Berry. 1967. Prenatal mortality in wild-caught house mice. *J. Anim. Ecol.* 36:453–463.

Beardmore, J. A., Th. Dobzhansky, and O. A. Pavlovsky. 1960. An attempt to compare the fitness of polymorphic and monomorphic experimental populations of *Drosophila pseudoobscura. Heredity* 14:19–33.

Bergmann, C. 1847. Über die Verhältnisse der Wärmeökonomie der Thiere zu ihrer Grösse. *Göttinger Stud.* pt. 1:595–708.

Bessey, C. E. 1908. The taxonomic aspect of the species. *Amer. Nat.* 42:218–224.

Blair, A. P. 1941. Variation, isolating mechanisms, and hybridization in certain toads. *Genetics* 26:398–417.

Blair, W. F. 1964. Isolating mechanisms and interspecies interactions in anuran amphibians. *Quart. Rev. Biol.* 39:333–344.

——— 1965. Amphibian speciation. In H. E. Wright and D. G. Frey, eds., *The Quaternary of the United States,* Princeton Univ. Press, Princeton.

——— and M. J. Littlejohn. 1960. Stage of speciation of two allopatric populations of chorus frogs (*Pseudacris*). *Evolution* 14:82–87.

Blanchard, B. D. 1941. The white-crowned sparrows (*Zonotrichia leucophrys*) of the Pacific seaboard: environment and annual cycle. *Univ. Calif. Publ. Zool.* 46:1–178.

Bock, W. 1959. Preadaptation and multiple evolutionary pathways. *Evolution* 13:194–211.

Bowen, W. W. 1968. Variation and evolution of Gulf Coast populations of beach mice, *Peromyscus polionotus. Bull. Florida State Mus.,* Biol. Ser. 12:1–91.

Brinkmann, R. 1929. Statistisch-biostratigraphische Untersuchungen an mitteljurassischen Ammoniten über Artbegriff und Stammesentwicklung. *Abhandl. Ges. Wiss. Göttingen, Math. Nat. Kl.* (N.F.) 13:1–249.

Britten, R. J., and E. H. Davidson. 1969. Gene regulation for higher cells: a theory. *Science* 165:349–357.

——— and D. E. Kohne. 1968. Repeated sequences in DNA. *Science* 161:529–540.

Brncic, D. 1954. Heterosis and the integration of the genotype in geographical populations of *Drosophila pseudoobscura. Genetics* 39:77–88.

Brooks, J. L. 1950. Speciation in ancient lakes. *Quart. Rev. Biol.* 25:30–176.

——— 1957a. The systematics of North American *Daphnia. Mem. Connecticut Acad. Arts Sci.* 13:1–180.

——— 1957b. The species problem in freshwater animals. In E. Mayr, ed., *The Species Problem,* American Assoc. Adv. Sci. Publ. no. 50:81–123.

Brown, W. J. 1945. Food plants and distribution of the species of *Calligrapha* in Canada, with descriptions of new species (Coleoptera, Chrysomelidae). *Can. Entomol.* 77:117–133.

——— 1956. The new world species of *Chrysomela* L. (Coleoptera: Chrysomelidae). *Can. Entomol.* 88, suppl. 3:1–54.

——— 1959. Taxonomic problems with closely related species. *Ann. Rev. Entomol.* 4:77–98.

Brown, W. L., and E. O. Wilson. 1956. Character displacement. *Syst. Zool.* 5:49–64.

Brues, A. M. 1954. Selection and polymorphism in the A-B-O blood groups. *Amer. J. Phys. Anthropol.* 12:559–597.

——— 1963. Stochastic tests of selection in the A-B-O blood groups. *Amer. J. Phys. Anthropol.* 21:287–299.

——— 1964. The cost of evolution vs. the cost of not evolving. *Evolution* 18:379–383.

——— 1969. Genetic load and its varieties. *Science* 164:1130–1136.

Bush, G. L. 1969. Sympatric host race formation and speciation in frugivorous flies of the genus *Rhagoletis* (Diptera, Tephritidae). *Evolution* 23:237–251.

Busnel, R. G. 1963. *Acoustic Behavior of Animals.* Elsevier, Amsterdam and New York.

Cain, A. J. 1958. Logic and memory in Linnaeus' system of taxonomy. *Proc. Linn. Soc. London* 169:144–163.

——— and J. D. Currey. 1963. Area effects in *Cepaea. Phil. Trans.* (B) 246:1–81.

——— and P. M. Sheppard. 1950. Selection in the polymorphic land snail *Cepaea nemoralis. Heredity* 4:275–294.

Cameron, A. W. 1958. Mammals of the islands in the Gulf of St. Lawrence. *Nat. Mus. Can. Bull.* no. 154:1–165.

Cantrall, I. J. 1943. The ecology of the Orthoptera and Dermaptera of the George Reserve, Michigan. *Misc. Publ. Mus. Zool. Univ. Mich.* 54:1–182.

Carpenter, G. D. H. 1949. *Pseudacraea eurytus* (L.) (Lep. Nymphalidae): a study of a polymorphic mimic in various degrees of speciation. *Trans. Roy. Entomol. Soc. London* 100:71–133.

Carson, H. L. 1958a. Increase in fitness in experimental populations resulting from heterosis. *Proc. Nat. Acad. Sci.* 44:1136–1141.

——— 1958b. The population genetics of *Drosophila robusta. Adv. Genet.* 9:1–40.

——— 1959. Genetic conditions which promote or retard the formation of species. *Cold Spring Harbor Symp. Quant. Biol.* 24:87–105.

——— 1965. Chromosomal morphism in geographically widespread species of *Drosophila.* In H. G. Baker and G. L. Stebbins, eds., *The Genetics of Colonizing Species,* Academic Press, New York.

——— F. E. Clayton, and H. D. Stalker. 1967. Karyotypic stability and speciation in Hawaiian *Drosophila. Proc. Nat. Acad. Sci.* 57:1280–1285.

——— D. E. Hardy, H. T. Spieth, and W. S. Stone. 1970. The evolutionary biology of the Hawaiian Drosophilidae. *Evol. Biol.* suppl. 437–543.

Caspari, E. 1949. Physiological action of eye color mutants in moths *Ephestia kühniella* and *Ptychopoda seriata. Quart. Rev. Biol.* 24:185–199.

Chetverikov, S. S. 1926. On certain features of the evolutionary process from the viewpoint of modern genetics. *J. Exp. Biol.* (Russian) 2:3–54. Also 1961, *Proc. Amer. Phil. Soc.* (English) 105:167–195.

Clarke, B. 1962. Balanced polymorphism and the diversity of sympatric species. In *Taxonomy and Geography,* Syst. Assoc. Publ. no. 4:47–70.

——— and J. Murray. 1969. Ecological genetics and speciation in land snails of the genus *Partula. Biol. J. Linn. Soc.* 1:31–42.

Clay, T. 1949. Some problems in the evolution of a group of ectoparasites. *Evolution* 3:279–299.

Cody, M. L. 1965. A general theory of clutch size. *Evolution* 20:174–184.

Connell, J. H. 1959. An experimental analysis of interspecific competition in natural populations of intertidal barnacles. *Proc. XV Int. Congr. Zool., London,* 290–293.

Cooke, F., and F. G. Cooch. 1968. The genetics of polymorphism in the goose *Anser coerulescens. Evolution* 22:289–300.

Coon, C. S. 1962. *The Origin of Races.* Alfred A. Knopf, New York.

——— 1965. *The Living Races of Man.* Alfred A. Knopf, New York.

Cooper, K. W. 1953. The ecology, predation and competition of *Ancistrocerus antilope* (Panzer). *Trans. Amer. Entomol. Soc.* 79:13–35.

Cott, H. B. 1940. *Adaptive Coloration in Animals.* Methuen, London.

Cousin, G. 1967. Quelques points de vue sur l'hybridation chez les animaux. *Bull. Soc. Zool. France* 92:441–485.

Creed, E. R. 1966. Geographic variation in the two-spot ladybird in England and Wales. *Heredity* 21:57–72.

Crow, J. F. 1961. Mechanisms and trends in human evolution. *Daedalus* summer:416–431.

Da Cunha, A. B., and Th. Dobzhansky. 1954. A further study of chromosomal polymorphism in *Drosophila willistoni* in its relation to the environment. *Evolution* 13:389–404.

Darlington, C. D., and K. Mather. 1949. *The Elements of Genetics.* Allen and Unwin, London.

Darwin, C. 1859. *On the Origin of Species by Means of Natural Selection, or the Preservation of Favored Races in the Struggle for Life.* John Murray, London. [1964. Facsimile edition. E. Mayr, ed. Harvard Univ. Press, Belknap Press, Cambridge, Mass.]

Dawson, R. W. 1931. The problem of voltinism and dormancy in the polyphemus moth (*Telea polyphemus* Cramer). *J. Exp. Zool.* 59:87–131.

DeBach, P., and R. A. Sunby. 1963. Competitive displacement between ecological homologues. *Hilgardia* 34:105–166.

Dethier, V. G. 1954. Evolution of feeding preferences in phytophagous insects. *Evolution* 8:33–54.

Dilger, W. C. 1956. Hostile behavior and reproductive isolating mechanisms in the avian genera *Catharus* and *Hylocichla. Auk* 73:313–353.

Dobzhansky, Th. 1937. *Genetics and the Origin of Species,* 1st ed. Columbia Univ. Press, New York.

——— 1947. Adaptive changes induced by natural selection in wild populations of *Drosophila. Evolution* 1:1–16.

——— 1951. *Genetics and the Origin of Species,* 3rd. ed. Columbia Univ. Press, New York.

——— 1954. Evolution as a creative process. *Caryologia* suppl:435–449.

——— 1955. A review of some fundamental concepts and problems of population genetics. *Cold Spring Harbor Symp. Quant. Biol.* 20:1–15.

——— 1956. What is an adaptive trait? *Amer. Nat.* 90:337–347.

——— 1959. Variation and evolution. *Proc. Amer. Phil. Soc.* 103:252–263.

——— 1960. Evolution and environment. In S. Tax, ed., *The Evolution of Life,* Univ. of Chicago Press, Chicago. Copyright 1960 by the Univ. of Chicago Press.

——— 1962. *Mankind Evolving.* Yale Univ. Press, New Haven and London.

——— 1968. On some fundamental concepts of Darwinian biology. *Evol. Biol.* 2:1–34.

——— and P. Ch. Koller. 1938. An experimental study of sexual isolation in *Drosophila. Biol. Zentralbl.* 58:589–607.

——— and O. A. Pavlovsky. 1957. Indeterminate outcome of certain experiments on *Drosophila* populations. *Evolution* 7:198–210.

——— 1961. A further study of fitness of chromosomally polymorphic and monomorphic populations of *Drosophila pseudoobscura. Heredity* 16: 169–179.

Dougherty, E. C. 1955. The origin of sexuality. *Syst. Zool.* 4:145–169.

Dowdeswell, W. H. 1956. Isolation and adaptation in populations of the Lepidoptera. *Proc. Roy. Soc., London* (B) 145:322–329.

Dufour, L. 1844. Anatomie générale des Diptères. *Ann. Sci. Nat.* 1:244–264.

Dunn, L. C. 1964. Abnormalities associated with a chromosome region in the mouse. *Science* 144:260–263.

Ehrlich, P. R. 1968. *The Population Bomb.* Ballantine Books, New York.

Ehrman, L. 1967. Further studies on genotype frequency and mating success in *Drosophila. Amer. Nat.* 101:415–424.

——— and C. Petit. 1968. Genotype frequency and mating success in the *willistoni* species group of *Drosophila. Evolution* 22:649–658.

Eisentraut, M. 1949. Die Eidechsen der spanischen Mittelmeerinseln und ihre Rassenaufspaltung im Lichte der Evolution. *Mitt. Zool. Mus. Berlin* 26:1–228.

Ellerman, J. B., and T. C. S. Morrison-Scott. 1951. *Checklist of Palaearctic and Indian Mammals 1758 to 1946.* British Museum (Nat. Hist.), London.

Elton, C. S. 1958. *The Ecology of Invasions by Animals and Plants.* Methuen, London.

Emlen, J. T., Jr., and G. B. Schaller. 1960. Distribution and status of the mountain gorilla. *Zoologica* 45:41–52.

Falconer, D. S. 1960. *Introduction to Quantitative Genetics.* Oliver and Boyd, Edinburgh and London.

Fisher, R. A. 1930. *The Genetical Theory of Natural Selection.* Clarendon Press, Oxford.

Fitch, W. M., and E. Margoliash. 1970. The usefulness of amino acid and nucleotide sequences in evolutionary studies. *Evol. Biol.* 4:67–109.

Ford, E. B. 1945. Polymorphism. *Biol. Rev.* 20:73–88.

——— 1964. *Ecological Genetics.* Methuen, London.

——— 1965. *Genetic Polymorphism.* M.I.T. Press, Cambridge, Mass.

Frank, F. 1956. Beiträge zur Biologie der Feldmaus, *Microtus arvalis* (Pallas). II. Laboratoriumsergebnisse. *Zool. Jahrb.* (*Syst.*) 84:32–74.

Freeman, T. N., M. R. MacKay, I. M. Campbell, C. E. Cox, S. G. Smith, and G. S. Walley. 1953. Studies on the spruce and jack-pine budworms. *Can. Entomol.* 85:121–152.

——— 1967. On coniferophagous species of *Choristoneura* (Lepidoptera, Tortricidae) in North America. I–VI. *Can. Entomol.* 99:449–506.

Fryer, G. 1960. Some controversial aspects of speciation of African cichlid fishes. *Proc. Zool. Soc. London* 135:569-578.

Fuller, J. L., and W. R. Thompson. 1960. *Behavior Genetics.* John Wiley, New York and London.

Gilmour, J. S. L., and J. W. Gregor. 1939. Demes: a suggested new terminology. *Nature* 144:333–334.

Glass, B. 1957. In pursuit of a gene. *Science* 126:683–689.

Gloger, C. L. 1833. *Das Abändern der Vögel durch Einfluss des Klimas.* Breslau.

Goethe, F. 1955. Vergleichende Beobachtungen zum Verhalten der Silbermöwe (*Larus argentatus*) und der Heringsmöwe (*Larus fuscus*). *Acta XI Int. Ornithol. Congr. Basel 1954,* 577–582.

Goldschmidt, R. 1940. *The Material Basis of Evolution.* Yale Univ. Press, New Haven.

Gompertz, T. 1968. Results of bringing individuals of two geographically isolated forms of *Parus major* into contact. *Vogelwelt,* Beih. 1:63–92.

Gontcharoff, M. 1951. Biologie de la régénération et de la reproduction chez quelques *Lineidae* de France. *Ann. Sci. Nat., Zool.* 13, ser. 11:149–235.

Gorman, G. C., and L. Atkins. 1969. The zoogeography of Lesser Antillean *Anolis* lizards—an analysis based upon chromosomes and lactic dehydrogenase. *Bull. Mus. Comp. Zool.* 138:53–80.

Gottesman, I. I. 1963. Heritability of personality: a demonstration. In G. A. Kimble, ed., Psychological Monographs: General and Applied, *Amer. Phil. Soc.* 77:1–21.

Grant, P. R. 1965. A systematic study of the terrestrial birds of the Tres Marias Islands, Mexico. *Postilla* no. 90:1–106.

Grant, V. 1963. *The Origin of Adaptations.* Columbia Univ. Press, New York and London.

——— 1964. *The Architecture of the Germplasm.* John Wiley, New York, London, Sydney.

Greenewalt, C. H. 1968. *Bird Song: Acoustics and Physiology.* Smithsonian Inst. Press, Washington.

Greenwood, P. H. 1965. The cichlid fishes of Lake Nabugabo, Uganda. *Bull. Brit. Mus.* (*Nat. Hist.*) 12:315–357.

Gulick, J. T. 1873. On diversity of evolution under one set of external conditions. *Linn. Soc. J., Zool., London* 11:496–505.

——— 1905. Evolution, racial and habitudinal. *Carnegie Inst. Wash. Publ.* no. 25:1–265.

Gustafsson, A. 1953. The cooperation of genotypes in barley. *Hereditas* 39:1–18.

Haffer, J. 1967. Speciation in Colombian forest birds west of the Andes. *Amer. Mus. Novitates* no. 2294:1–57.

——— 1969. Speciation in Amazonian forest birds. *Science* 165:131–137.

Hagen, D. W. 1967. Isolating mechanisms in threespine sticklebacks (*Gasterosteus*). *J. Fish. Res. Bd. Can.* 24:1637–1692.

Haldane, J. B. S. 1932. *The Causes of Evolution.* Longmans and Green, New York.

——— 1949a. Human evolution: past and future. In G. L. Jepsen, E. Mayr, and G. G. Simpson, eds., *Genetics, Paleontology, and Evolution,* Princeton Univ. Press, Princeton.

——— 1949b. Disease and evolution. *Ricerca Sci.* suppl. 19:68–76.

——— 1954a. The statics of evolution. In J. Huxley, A. C. Hardy, and E. B. Ford, eds., *Evolution as a Process,* Allen and Unwin, London.

——— 1954b. The measurement of natural selection. *Caryologia* suppl. 6:480–487.

——— 1955. Population genetics. *New Biol.* 18:34–51.

——— 1956. The relation between density regulation and natural selection. *Proc. Roy. Soc., London* (B) 145:306–308.

——— 1957. The cost of natural selection. *J. Genet.* 55:511–524.

Hall, B. P., and R. E. Moreau. 1970. *Atlas of Speciation of African Birds.* British Museum (Nat. Hist.), London.

Hamilton, T. H. 1961. The adaptive significances of intraspecific trends of variation in wing length and body size among bird species. *Evolution* 15:180–195.

Hamilton, W. D. 1964a. The genetical evolution of social behavior. I. *J. Theoret. Biol.* 7:1–16.

——— 1964b. The genetical evolution of social behavior. II. *J. Theoret. Biol.* 7:17–52.

——— 1967. Extraordinary sex ratios. *Science* 156:477–488.

Hardin, G. 1960. The competitive exclusion principle. *Science* 131:1291–1297.

——— 1968. The tragedy of the commons. *Science* 162:1243–1248.

Harland, S. C. 1936. The genetical conception of the species. *Biol. Rev.* 11:83–112.

Harrison, G. A. 1959. Environmental determination of the phenotype. In A. J. Cain, ed., *Function and Taxonomic Importance,* Syst. Assoc. Publ. no. 3:81–86.

Hesse, R., W. C. Allee, and K. P. Schmidt. 1951. *Ecological Animal Geography,* 2nd ed. John Wiley, New York.

Hessler, R. R., and H. L. Sanders. 1967. Faunal diversity in the deep-sea. *Deep-Sea Res.* 14:65–78.

Hinde, R. A. 1959. Behaviour and speciation in birds and lower vertebrates. *Biol. Rev.* 34:85–128.

——— 1966. *Animal Behaviour.* McGraw-Hill, New York.

Hirsch, J., ed. 1967. *Behavior-genetic Analysis.* McGraw-Hill, New York.

Hoesch, W. 1953. Uber die Rassenbildung der s.w.-afrikanischen Bodenvögel unter Berücksichtigung von Wasserabhängigkeit, Niederschlagsmenge und Bodenfärbung. *J. Ornithol.* 94:274–281.

——— 1956. Das Problem der Farbübereinstimmung von Körperfarbe und Untergrund. *Bonn. Zool. Beitr.* 7:59–83.

Howell, F. C. 1952. Pleistocene glacial ecology and evolution of "classic Neandertal" man. *Southwestern J. Anthropol.* 8:337–410.

——— 1957. The evolutionary significance of variation and varieties of "Neanderthal" man. *Quart. Rev. Biol.* 32:330–347.

Hubbs, C. L. 1922. Variations in the number of vertebrae and other meristic characters of fishes correlated with the temperature of water during development. *Amer. Nat.* 56:360–372.

——— 1955. Hybridization between fish species in nature. *Syst. Zool.* 4:1–20.

——— 1961. Isolating mechanisms in the speciation of fishes. In W. F. Blair, ed., *Vertebrate Speciation,* Univ. of Texas Press, Austin.

——— and R. R. Miller. 1943. Mass hybridization between two genera of Cyprinid fishes in the Mohave Desert, California. *Papers Mich. Acad. Sci.* 28:343–378.

Hubby, J. L., and R. C. Lewontin. 1966. The number of alleles at different loci in *Drosophila pseudoobscura. Genetics* 54:577–594.

Hutchinson, G. E. 1965. *The Ecological Theater and the Evolutionary Play.* Yale Univ. Press, New Haven and London.

Huxley, J. 1942. *Evolution, the Modern Synthesis.* Allen and Unwin, London.

——— 1953. *Evolution in Action.* Harper, New York.

Jacob, F., and J. Monod. 1961. Genetic regulatory mechanisms in the synthesis of proteins. *J. Mol. Biol.* 3:318–356.

Janzer, D. H. 1967. Why mountain passes are higher in the tropics. *Amer. Nat.* 101:233–249.

Jensen, A. R. 1969. How much can we boost IQ and scholastic achievement. *Harvard Educ. Rev.* 31:1–123.

Johansen, H. 1955. Die Jennissei-Faunenscheide. *Zool. Jahrb.* (*Syst.*) 83:185–322.

John, B., and K. R. Lewis. 1966. Chromosome variability and geographic distribution in insects. *Science* 152:711–721.

Jordan, K. 1905. Der Gegensatz zwischen geographischer und nichtgeographischer Variation. *Z. wiss. Zool.* 83:151–210.

——— 1938. Where subspecies meet. *Novit. Zool.* 41:103–111.

Kalela, O. 1944. Zur Frage der Ausbreitungstendenz der Tiere. *Ann. Zool. Soc. Vanamo* 10:1–23.

Keast, A. 1961. Bird speciation on the Australian continent. *Bull. Mus. Comp. Zool.* 123:305–495.

——— 1969. Competitive interactions and the evolution of ecological niches as illustrated by the Australian honeyeater genus *Melithreptus* (Meliphagidae). *Evolution* 22:762–784.

Kettlewell, H. B. D. 1961. The phenomenon of industrial melanism in Lepidoptera. *Ann. Rev. Entomol.* 6:245–262.

Key, K. H. L. 1968. The concept of stasipatric speciation. *Syst. Zool.* 17:14–22.

Kimura, M. 1955. Stochastic processes and distribution of gene frequencies under natural selection. *Cold Spring Harbor Symp. Quant. Biol.* 20:33–53.

——— 1960. Optimum mutation rate and degree of dominance as determined by the principle of minimum genetic load. *J. Genet.* 57:21–34.

——— 1969. The rate of molecular evolution considered from the standpoint of population genetics. *Proc. Nat. Acad. Sci.* 63.1181–1188.

King, J. L., and T. H. Jukes. 1969. Non-Darwinian evolution. *Science* 164:788–798.

Kinne, O. 1954. Die *Gammarus*-Arten der Kieler Bucht. *Zool. Jahrb. (Syst.)* 82:405–496.

Kitzmiller, J. B., G. Frizzi, and R. H. Baker. 1967. Evolution and speciation within the *maculipennis* complex of the genus *Anopheles*. In J. W. Wright and R. Pal, eds., *Genetics of Insect Vectors of Disease,* Elsevier, Amsterdam.

Klauber, L. M. 1944. The California king snake: a further discussion. *Amer. Midl. Nat.* 31:85–87.

Kojima, K., and K. M. Yarbrough. 1967. Frequency-dependent selection at the esterase G-locus in *Drosophila melanogaster. Proc. Nat. Acad. Sci.* 57:645–649.

Kontkanen, P. 1953. On the sibling species in the leafhopper fauna of Finland (Homoptera, Auchenorrhyncha). *Arch. Soc. Vanamo* 7:100–106.

Korringa, P. 1958. Water temperature and breeding throughout the geographical range of *Ostrea edulis*. In L. Fage and P. Drach, eds., *Biologie comparée des espèces marines, I.U.B.S.,* ser. B, no. 24:1–17.

Kosswig, C., and W. Villwock. 1964. Das Problem der intralakustrischen Speziation im Titicaca—und im Lanaosee. *Verh. Deut. Zool. Ges.* 7:95–102.

Kullenberg, B. 1961. Studies in *Ophrys* pollination. *Zool. Bidrag Uppsala* 34:1–340.

Kurtén, B. 1959. Rates of evolution in fossil mammals. *Cold Spring Harbor Symp. Quant. Biol.* 24:205–215.

——— 1968. *Pleistocene Mammals of Europe.* Aldine, Chicago.

Lack, D. 1944. Ecological aspects of species-formation in Passerine birds. *Ibis* 86:260–286.

——— 1945. The ecology of closely related species with special reference to cormorant (*Ph. carbo*) and shag (*P. aristotelis*). *J. Anim. Ecol.* 14:12–16.

——— 1947. *Darwin's Finches.* Cambridge Univ. Press, London.

——— 1949. The significance of ecological isolation. In G. L. Jepsen, E. Mayr, and G. G. Simpson, eds., *Genetics, Paleontology, and Evolution,* Princeton Univ. Press, Princeton.

——— 1954. *The Natural Regulation of Animal Numbers.* Clarendon Press, Oxford.

——— 1966. *Population Studies of Birds*. Oxford Univ. Press, Oxford.
——— and H. N. Southern. 1949. Birds on Tenerife. *Ibis* 91:607–626.
Lamotte, M. 1951. Recherches sur la structure génétique des populations naturelles de *Cepaea nemoralis* (L.). *Bull. Biol. France* suppl. 35:1–239.
——— 1959. Polymorphism of natural populations of *Cepaea nemoralis*. *Cold Spring Harbor Symp. Quant. Biol.* 24:65–86.
Lanyon, W. E., and W. N. Tavolga, eds. 1960. *Animal Sounds and Communication.* AIBS Publ. no. 7.
Leopold, A. S. 1944. The nature of heritable wildness in turkeys. *Condor* 46:133–197.
Lerner, I. M. 1954. *Genetic Homeostasis.* John Wiley, New York.
——— 1959. The concept of natural selection: a centennial view. *Proc. Amer. Phil. Soc.* 103:173–182.
Lewis, H. 1962. Catastrophic selection as a factor in speciation. *Evolution* 16:257–271.
——— 1966. Speciation in flowering plants. *Science* 152:167–172.
Lewontin, R. C. 1962. Interdeme selection controlling a polymorphism in the house mouse. *Amer. Nat.* 96:65–78.
——— 1967a. Population genetics. *Ann. Rev. Genet.* 1:37–70.
——— 1967b. The principle of historicity in evolution. *Wistar Symp. Monogr.* no. 5:81–94.
——— and J. L. Hubby. 1966. Amount of variation and degree of heterozygosity in natural populations of *Drosophila pseudoobscura*. *Genetics* 54:595–609.
L'Héritier, Ph., and G. Teissier. 1937. Elimination des formes mutantes dans les populations de Drosophiles. Cas des Drosophiles "bar." *Compt Rend. Soc. Biol., Paris* 124:880.
Linsley, E. G., and J. W. MacSwain. 1958. The significance of floral constancy among bees of the genus *Diadasia* (Hymenoptera, Anthophoridae). *Evolution* 12:219–223.
Littlejohn, M. J. 1969. The systematic significance of isolating mechanisms. In *Systematic Biology,* Nat. Acad. Sci. Publ. no. 1692.
——— and R. S. Oldham. 1968. Rana pipiens complex: mating call structure and taxonomy. *Science* 162:1003–1005.
Lloyd, J. E. 1966. Studies on the flash communication system in *Photinus* fireflies. *Misc. Publ. Mus. Zool. Univ. Mich.* no. 130:1–95.
Lloyd, M., and H. S. Dybas. 1966. The periodical cicada problem. I and II. *Evolution* 20:133–149, 466–505.
Lorković, Z. 1953. Spezifische, Semispezifische und Rassische Differenzierung bei *Erebia tyndarus* Esp. I und II. *Rad l'Acad. Yougoslave* 294:269–309, 315–358.
Lowe-McConnell, R. H. 1969. Speciation in tropical freshwater fishes. *Biol. J. Linn. Soc.* 1:51–75.
Ludwig, W. 1950. Zur Theorie der Konkurrenz. Die Annidation (Einnischung) als fünfter Evolutionsfaktor. *Neue Ergeb. Probleme Zool., Klatt-Festschrift* 1950:516–537.
MacArthur, R. H. 1958. Population ecology of some warblers of north-eastern coniferous forests. *Ecology* 39:599–619.
——— 1965. Patterns of species diversity. *Biol. Rev.* 40:510–533.
——— 1969. Patterns of communities in the tropics. *Biol. J. Linn. Soc.* 1:19–30.
Manning, A. 1959. Comparison of mating behavior in *Drosophila melanogaster* and *Drosophila simulans*. *Behaviour* 15:123–146.

——— 1965. *Drosophila* and the evolution of behavior. In *Viewpoints in Biology,* 4, Butterworth's, London.

Marler, P. 1957. Specific distinctiveness in the communication signals of birds. *Behaviour* 11:13–39.

——— 1960. Bird songs and mate selection. In W. E. Lanyon and W. N. Tavolga, eds., *Animal Sounds and Communication,* AIBS Publ. no. 7:348–367.

Marshall, J. T., Jr. 1948. Ecological races of song sparrows in the San Francisco Bay region. Part II. Geographic variation. *Condor* 50:233–256.

——— 1960. Interrelations of Abert and Brown towhees. *Condor* 62:49–64.

Maslin, T. P. 1968. Taxonomic problems in parthenogenetic vertebrates. *Syst. Zool.* 17:219–231.

Mather, K. 1943. Polygenic inheritance and natural selection. *Biol. Rev.* 18:32–64.

——— 1953. The genetical structure of populations. *Symp. Soc. Exp. Biol.* 7:66–95.

——— 1955. Polymorphism as an outcome of disruptive selection. *Evolution* 9:52–61.

——— and B. J. Harrison. 1949. The manifold effect of selection. *Heredity* 3:1–52, 131–162.

Matthew, W. D. 1915. Climate and evolution. *Ann. N. Y. Acad. Sci.* 24:171–318.

Matthey, R. 1964. Evolution chromosomique et spéciation chez les *Mus* du sous-genre *Leggada* Gray 1837. *Experientia* 20:1–9.

Mayfield, H. 1960. *The Kirtland's Warbler.* Cranbrook Inst. Sci. Bull. no. 40, Bloomfield Hills, Michigan.

Maynard Smith, J. 1964. Group selection and kin selection. *Nature* 201:1145–1147.

——— 1968. Evolution in sexual and asexual populations. *Amer. Nat.* 102:469–474.

Mayr, E. 1942. *Systematics and the Origin of Species.* Columbia Univ. Press, New York.

——— 1945. Symposium on age of the distribution pattern of the gene arrangements in *D. pseudoobscura.* Introduction and some evidence in favor of a recent date. *Lloydia* 8:69–83.

——— 1947. Ecological factors in speciation. *Evolution* 1:263–288.

——— 1948. The bearing of the new systematics on genetical problems. The nature of species. *Adv. Genet.* 2:205–237.

——— 1949. Speciation and systematics. In G. L. Jepsen, E. Mayr, and G. G. Simpson, eds., *Genetics, Paleontology, and Evolution,* Princeton Univ. Press, Princeton. Copyright 1949 Princeton Univ. Press.

——— 1950. Taxonomic categories in fossil hominids. *Cold Spring Harbor Symp. Quant. Biol.* 15:109–118.

——— 1954. Change of genetic environment and evolution. In J. Huxley, ed., *Evolution as a Process,* Allen and Unwin, London.

——— 1955. Integration of genotypes: synthesis. *Cold Spring Harbor Symp. Quant. Biol.* 20:327–333.

——— 1957. Species concepts and definitions. In E. Mayr, ed., *The Species Problem,* Amer. Ass. Adv. Sci. Publ. no. 50:1–22.

——— 1958. Behavior and systematics. In A. Roe and G. G. Simpson, eds., *Behavior and Evolution,* Yale Univ. Press, New Haven.

——— 1959a. Isolation as an evolutionary factor. *Proc. Amer. Phil. Soc.* 103:221–230.

——— 1959b. Darwin and the evolutionary theory in biology. In *Evolution and Anthropology: A Centennial Appraisal,* Anthropol. Soc. Washington, 3–12.

——— 1959c. Agassiz, Darwin, and evolution. *Harvard Library Bull.* 13:165–194.

——— 1960. The emergence of evolutionary novelties. In S. Tax, ed., *Evolution after Darwin,* Univ. of Chicago Press, Chicago.

——— 1963. *Animal Species and Evolution.* Belknap Press of Harvard University Press, Cambridge, Mass.

——— 1965a. The nature of colonizations in birds. In H. G. Baker and G. L. Stebbins, eds., *The Genetics of Colonizing Species,* Academic Press, New York.

——— 1965b. Selektion und die gerichtete Evolution. *Die Naturwissenschaften* 52:173–180.

——— 1968. Illiger and the biological species concept. *J. Hist. Biol.* 1:163–178.

——— 1969. *Principles of Systematic Zoology.* McGraw-Hill, New York.

——— and E. T. Gilliard. 1952. The ribbon-tailed bird of paradise (*Astrapia mayeri*) and its allies. *Amer. Mus. Novitates* no. 1551:1–13.

——— and C. Vaurie. 1948. Evolution in the family Dicruridae (birds). *Evolution* 2:238–265.

McAtee, W. L. 1937. Survival of the ordinary. *Quart. Rev. Biol.* 12:47–64.

McCabe, T., and B. D. Blanchard. 1950. *Three Species of* Peromyscus. Rood Associates, Santa Barbara.

McClintock, B. 1961. Some parallels between gene control systems in maize and bacteria. *Amer. Nat.* 95:265–277.

Mecham, J. S. 1961. Isolating mechanisms in anuran amphibians. In W. F. Blair, ed., *Vertebrate Speciation,* Univ. of Texas Press, Austin.

Medawar, P. B. 1960. *The Future of Man: The Reith Lectures.* Methuen, London.

Meise, W. 1928. Die Verbreitung der Aaskrähe (Formenkreis *Corvus corone* L.). *J. Ornith.* 76:1–203.

Mengel, R. M. 1964. The probable history of species formation in some northern wood warblers (Parulidae). *The Living Bird* 3:9–43.

Mettler, L. E. 1957. Studies on experimental populations of *Drosophila arizonensis* and *Drosophila mojavensis. Univ. Texas Publ.* no. 5721:157–181.

Michener, C. D. 1947. A revision of the American species of *Hoplitis* (Hymenoptera, Megachilidae). *Bull. Amer. Mus. Nat. Hist.* 89:257–318.

Milkman, R. 1960. Potential genetic variability of wild pairs of *Drosophila melanogaster. Science* 131:225–226.

——— 1961. The genetic basis of natural variation. III. Developmental lability and evolutionary potential. *Genetics* 46:25–38.

——— 1967. Heterosis as a major cause of heterozygosity in nature. *Genetics* 55:493–495.

Miller, A. H. 1956. Ecological factors that accelerate formation of races and species of terrestrial vertebrates. *Evolution* 10:262–277.

——— 1960. Adaptation of breeding schedule to latitude. *Proc. XII Int. Ornith. Congr., Helsinki, 1958,* 513–522.

Miller, R. R. 1950. Speciation in fishes of the genera *Cyprinodon* and *Empetrichthys,* inhabiting the Death Valley region. *Evolution* 4:155–163.

——— 1961. Speciation rates in some fresh-water fishes of western North America. In W. F. Blair, ed., *Vertebrate Speciation,* Univ. of Texas Press, Austin.

Montalenti, G. 1958. Perspectives of research on sex problems in marine animals. In A. A. Buzzati-Traverso, ed., *Perspectives in Marine Biology,* Univ. of Calif. Press, Berkeley and Los Angeles.

Moore, J. A. 1949. Patterns of evolution in the genus *Rana.* In G. L. Jepsen, G. G. Simpson, and E. Mayr, eds., *Genetics, Paleontology and Evolution.* Princeton Univ. Press, Princeton.

——— 1952a. Competition between *Drosophila melanogaster* and *Drosophila simulans.* I. Population cage experiments. *Evolution* 6:407–420.

——— 1952b. Competition between *Drosophila melanogaster* and *Drosophila simulans.* II. The improvement of competitive ability through selection. *Proc. Nat. Acad. Sci.* 38:813–817.

——— 1954. Geographic and genetic isolation in Australian amphibia. *Amer. Nat.* 88:65–74.

——— 1957. An embryologist's view of the species concept. In *The Species Problem,* Amer. Ass. Adv. Sci. Publ. no. 50:325–338.

Moreau, R. E. 1966. *The Bird Faunas of Africa and Its Islands.* Academic Press, New York.

Moynihan, M. 1968. Social mimicry; character convergence vs. character displacement. *Evolution* 22:315–331.

Muller, H. J. 1940. Bearings of the Drosophila work on systematics. In J. Huxley, ed., *The New Systematics,* Clarendon Press, Oxford.

——— 1950. Our load of mutations. *Amer. J. Human Genet.* 2:111–176.

Murray, B. G. 1967. Dispersal in vertebrates. *Ecology* 48:975–978.

Myers, G. S. 1960. The endemic fish fauna of Lake Lanao, and the evolution of higher taxonomic categories. *Evolution* 14:323–333.

Neel, J. V., F. M. Salzano, P. C. Junqueira, F. Keiter, and D. Maybury-Lewis. 1964. Studies on the Xavante Indians of the Brazilian Mato Grosso. *Human Genet.* 16:52–140.

Nelson, J. S. 1968. Hybridization and isolating mechanisms between *Catostomus commersonii* and *C. macrocheilus. J. Fish. Res. Bd. Can.* 25:101–150.

Newell, N. D. 1967. Revolutions in the history of life. *Geol. Soc. Amer.,* spec. paper 89:63–91.

Nichols, D. 1959. Changes in the chalk heart-urchin *Micraster* interpreted in relation to living forms. *Phil. Trans. Roy. Soc., London* (B) 242:347–437.

Niethammer, G. 1940. Die Schutzanpassung der Lerchen. In W. Hoesch and G. Niethammer, eds., *Die Vogelwelt Deutsch-Suedwestafrikas, J. Ornith., Sonderheft* 88:75–83.

——— and H. Kramer. 1966. Zwillingsarten. *Fortschritte der Zool.* 18:31–34.

Oliver, J. A., and C. E. Shaw. 1953. The amphibians and reptiles of the Hawaiian Islands. *Zoologica* 38:71–73.

Orians, G. H., and M. F. Willson. 1964. Interspecific territories of birds. *Ecology* 45:736–745.

Osche, G. 1952. Systematik und Phylogenie der Gattung *Rhabditis* (Nematoda). *Zool. Jahrb.* (*Syst.*) 81:190–280.

Parsons, P. A., and W. F. Bodmer. 1961. The evolution of overdominance: natural selection and heterozygote advantage. *Nature* 190:7–12.

Patterson, B., and R. Pascual. 1969. Evolution of mammals on southern continents. V. The fossil mammal fauna of South America. *Quart. Rev. Biol.* 43:409–451.

Patterson, J. T., and W. S. Stone. 1952. *Evolution in the Genus* Drosophila. Macmillan, New York.

Perdeck, A. C. 1957. The isolating value of specific song patterns in two sibling species of grasshoppers (*Chorthippus brunneus* Thunb. and *C. biguttulus* L.). *Behaviour* 12:1–75.

Perutz, M. F., and H. Lehmann. 1968. Molecular pathology of human haemoglobin. *Nature* 219:902–909.

Petersen, B. 1947. Die geographische Variation einiger Fennoskandischer Lepidopteren. *Zool. Bidrag Uppsala* 26:325–531.

Petit, C. 1958. Le déterminisme génétique et psycho-physiologique de la compétition sexuelle chez *Drosophila melanogaster*. *Bull. Biol.* 92:248–329.

——— 1968. Le rôle des valeurs sélectives variables dans le maintien du polymorphisme. *Bull. Soc. Zool. France* 93:187–208.

Phillips, J. C. 1915. Experimental studies of hybridization among ducks and pheasants. *J. Exp. Zool.* 18:69–143.

Pierce, G. W. 1948. *The Songs of Insects.* Harvard Univ. Press, Cambridge, Mass.

Platt, A. P., and L. P. Brower. 1968. Mimetic versus disruptive coloration in intergrading populations of *Limenitis arthemis* and *astyanax* butterflies. *Evolution* 22:699–718.

Prakash, S., and J. L. Hubby. 1969. Patterns of genic variation in central, marginal, and isolated populations of *Drosophila pseudoobscura*. *Genetics* 61:841–858.

Prakash, S., and R. C. Lewontin. 1968. Direct evidence of coadaptation in gene arrangements of Drosophila. *Proc. Nat. Acad. Sci.* 59:398–405.

Prosser, C. L., and F. A. Brown. 1961. *Comparative Animal Physiology*, 2nd ed. W. B. Saunders, Philadelphia.

Radovanovíc, M. 1959. Zum Problem der Speziation bei Inseleidechsen. *Zool. Jahrb.* (*Syst.*) 86:395–436.

——— 1965. Experimentelle Beiträge zum Problem der Kompetition. *Verh. Deutsch. Zool. Ges., Jena* 534–540.

Rand, A. L. 1948. Glaciation, an isolating factor in speciation. *Evolution* 2:314–321.

Rau, P. 1946. The nests and the adults of colonies of *Polistes* wasps. *Ann. Entomol. Soc. Amer.* 39:11–27.

Ray, C. 1960. The application of Bergmann's and Allen's rules to the Poikilotherms. *J. Morphol.* 106:85–108.

Reid, J. A. 1953. The *Anopheles hyrcanus* group in south-east Asia (Diptera: Culicidae). *Bull. Entomol. Res. London* 44:5–76.

Remington, C. L. 1968. Suture-zones of hybrid interaction between joined biota. *Evol. Biol.* 2:321–428.

Rendel, J. M. 1967. *Canalisation and Gene Control.* Academic Press, New York.

——— 1968. Genetic control of a developmental process. In R. C. Lewontin, ed., *Population Biology and Evolution*, Syracuse Univ. Press, Syracuse.

Rensch, B. 1933. Zoologische Systematik und Artbildungsproblem. *Verh. Deut. Zool. Ges., Köln, 1933,* 19–83.

——— 1938. Bestehen die Regeln klimatischer Parallelität bei der Merkmalsausprägung von homöothermen Tieren zu Recht? *Arch. Naturg.* (N.F.) 7:364–389.

——— 1960. *Evolution above the Species Level.* Columbia Univ. Press, New York.

Richards, O. W. 1927. Sexual selection and allied problems in the insects. *Biol. Rev.* 2:298–364.

Robertson, A. 1968. The spectrum of genetic variation. In R. C. Lewontin, ed., *Population Biology and Evolution*, Syracuse Univ. Press, Syracuse.

Robson, G. C., and O. W. Richards. 1936. *The Variations of Animals in Nature.* Longmans and Green, London.

Roe, A., and G. G. Simpson, eds. 1958. *Behavior and Evolution.* Yale Univ. Press, New Haven.

Romer, A. S. 1957. Origin of the amniote egg. *Sci. Monthly* 85:57–63.

Rosen, D. E. 1960. Middle-American poeciliid fishes of the genus *Xiphophorus*. *Bull. Florida State Mus. Biol. Sci.* 5:57–242.

Rosin, S., J. K. Moor-Jankowski, and M. Schneeberger. 1958. Die Fertilität im Bluterstamm von Tenna (Hämophilie B). *Acta Genet.* 8:1–24.

Ross, H. 1957. Principles of natural coexistence indicated by leafhopper populations. *Evolution* 11:113–129.

Salomonsen, F. 1951. The immigration and breeding of the fieldfare (*Turdus pilaris* L.) in Greenland. *Proc. Xth Int. Ornith. Congr. Uppsala, 1950,* 515–526.

Schindewolf, O. H. 1936. *Paläontologie, Entwicklungslehre und Genetik.* Borntraeger, Berlin.

——— 1950. *Grundfragen der Paläontologie.* Schweizerbart, Stuttgart.

Schmalhausen, I. I. 1949. *Factors of Evolution: The Theory of Stabilizing Selection.* Blakiston, Philadelphia.

Schmidt, J. 1918. Racial studies in fishes. I. Statistical investigations with *Zoarces viviparus* L. *J. Genet.* 7:105–118.

Schnetter, M. 1951. Veränderungen der genetischen Konstitution in natürlichen Populationen der polymorphen Bänderschnecken. *Zool. Anz.* suppl. 15:192–206.

Schull, W. J., and J. V. Neel. 1965. *The Effects of Inbreeding on Japanese Children.* Harper and Row, New York.

Seiger, M. B. 1967. A computer simulation study of the influence of imprinting on population structure. *Amer. Nat.* 101:47–57.

Seiler, J. 1961. Untersuchungen über die Entstehung der Parthenogenese bei *Solenobia triquetrella* F. R. (Lepidoptera, Psychidae). III. Z. *Vererb.-Lehre* 92:261–316.

Selander, R. K. 1965. Avian speciation in the Quaternary. In H. E. Wright, Jr., and D. G. Frey, eds., *The Quaternary of the United States,* Princeton Univ. Press, Princeton.

——— 1966. Sexual dimorphism and differential niche utilization in birds. *Condor* 68:113–151.

——— 1969. The ecological aspects of the systematics of animals. In *Systematic Biology,* Nat. Acad. Sci. Publ. no. 1692:213–247.

Sewertzoff, A. N. 1931. *Morphologische Gesetzmässigkeiten der Evolution.* Gustav Fischer, Jena.

Sheppard, P. M. 1952a. Natural selection in two colonies of the polymorphic land snail *Cepaea nemoralis. Heredity* 6:233–238.

——— 1952b. A note on non-random mating in the moth *Panaxia dominula* (L.). *Heredity* 6:239–241.

——— 1958. *Natural Selection and Heredity.* Hutchinson, London.

Shields, J. 1962. *Monozygotic Twins.* Oxford University Press, London.

Short, L. 1969. Taxonomic aspects of avian hybridization. *Auk* 86:84–105.

Sibley, C. G. 1954. Hybridization in the red-eyed towhees of Mexico. *Evolution* 8:252–290.

——— 1957. The evolutionary and taxonomic significance of sexual dimorphism and hybridization in birds. *Condor* 59:166–191.

——— 1961. Hybridization and isolating mechanisms. In W. F. Blair, ed., *Vertebrate Speciation,* Univ. of Texas Press, Austin.

Simons, E. L., and D. R. Pilbeam. 1965. Preliminary revision of the Dryopithecinae (Pongidae, Anthropoidea). *Folia Primat.* 3:81–152.

Simpson, G. G. 1944. *Tempo and Mode in Evolution.* Columbia Univ. Press, New York.

——— 1952. Probabilities of dispersal in geologic time. *Bull. Amer. Mus. Nat. Hist.* 99:163–176.

——— 1953. *The Major Features of Evolution.* Columbia Univ. Press, New York.

——— 1959a. The nature and origin of supraspecific taxa. *Cold Spring Harbor Symp. Quant. Biol.* 24:255–271.

——— 1959b. Mesozoic mammals and the polyphyletic origin of mammals. *Evolution* 13:405–414.

——— 1960. The history of life. In S. Tax, ed., *The Evolution of Life,* Univ. of Chicago Press, Chicago.

——— 1961. *Principles of Animal Taxonomy.* Columbia Univ. Press, New York.

——— 1964. *This View of Life.* Harcourt, Brace and World, New York.

Smith, J. M., see Maynard Smith

Smith, N. G. 1966. Evolution of some arctic gulls (*Larus*): an experimental study of isolating mechanisms. *A. O. U. Ornith. Monogr.* 4:1–99.

Sonneborn, T. M. 1957. Breeding systems, reproductive methods, and species problems in protozoa. In E. Mayr, ed., *The Species Problem,* Amer. Assoc. Adv. Sci. Publ. no. 50:155–324.

Spencer, W. P. 1947. Genetic drift in a population of *Drosophila immigrans. Evolution* 1:103–110.

Spickett, S. G., and J. M. Thoday. 1966. Regular responses to selection. III. *Genet. Res.* 7:96–121.

Spiess, E. B. 1968. Experimental population genetics. *Ann. Rev. Genet.* 2:165–208.

——— B. Langer, and L. D. Spiess. 1966. Mating control by gene arrangements in *Drosophila pseudoobscura. Genetics* 54:1139–1149.

Spieth, H. T. 1958. Behavior and isolating mechanisms. In A. Roe and G. G. Simpson, eds., *Behavior and Evolution,* Yale Univ. Press, New Haven.

Staiger, H. 1954. Der Chromosomendimorphismus beim Prosobranchier *Purpura lapillus* in Beziehung zur Ökologie der Art. *Chromosoma* 6:419–478.

Standfuss, M. 1896. *Handbuch der paläarktischen Gross-Schmetterlinge für Forscher und Sammler.* Gustav Fischer, Jena.

Stebbins, G. L., Jr. 1950. *Variation and Evolution in Plants.* Columbia Univ. Press, New York.

——— 1960. The comparative evolution of genetic systems. In S. Tax, ed., *Evolution after Darwin,* Univ. of Chicago Press, Chicago.

Steinberg, A. G., H. K. Bleibtren, T. W. Kurczynski, A. O. Martin, and E. M. Kurczynski. 1967. Genetic studies on an inbred human isolate. *Proc. 3rd. Int. Congr. Human Genetics* 267–289.

Stephens, S. G., M. M. Green, E. B. Lewis, J. R. Laughnan, and C. Stormont. 1955. Pseudoallelism and the theory of the gene. *Amer. Nat.* 89:65–122.

Stewart, T. D. 1960. Form of the pubic bone in Neanderthal man. *Science* 131:1437–1438.

Sturtevant, A. H. 1929. The genetics of *Drosophila simulans. Carnegie Inst. Wash. Publ.* no. 399:1–62.

Suomalainen, E. 1950. Parthenogenesis in animals. *Adv. Genet.* 3:193–253.

Svärdson, G. 1950. The Coregonid problem. II. Morphology of two Coregonid species in different environments. *Inst. Freshwater Res., Drottningholm* no. 31:151–162.

——— 1961. Young sibling fish species in northwestern Europe. In W. F. Blair, ed., *Vertebrate Speciation,* Univ. of Texas Press, Austin.

Sved, J. A. 1968. Possible rates of gene substitution in evolution. *Amer. Nat.* 102:283–293.

——— T. E. Reed, and W. F. Bodmer. 1967. The number of balanced polymorphisms that can be maintained in a natural population. *Genetics* 55:469–481.

Talbot, M. 1934. Distribution of ant species in the Chicago region, with reference to ecological factors and physiological toleration. *Ecology* 15:416–439.

Tåning, A. V. 1952. Experimental study of meristic characters in fishes. *Biol. Rev.* 27:169–193.

Tax, S., ed. 1960. *Evolution after Darwin*, vol. 2, *The Evolution of Man*. Univ. of Chicago Press, Chicago.

Teissier, G. 1954. Conditions d'équilibre d'un couple d'alléles et supériorité des hétérozygotes. *Compt. Rend.* 238:621–623.

Terman, L. M., and M. H. Oden. 1959. *The Gifted Group at Mid-Life*. Stanford Univ. Press, Stanford.

Thoday, J. M., and J. B. Gibson. 1962. Isolation by disruptive selection. *Nature* 193:1164–1166.

Thorpe, W. H. 1945. The evolutionary significance of habitat selection. *J. Anim. Ecol.* 14:67–70.

Thorson, G. 1950. Reproductive and larval ecology of marine bottom invertebrates. *Biol. Rev.* 25:1–45.

Thorsteinson, A. J. 1960. Host selection in phytophagous insects. *Ann. Rev. Entomol.* 5:193–218.

Timoféeff-Ressovsky, N. W. 1940. Zur Analyse des Polymorphismus bei *Adalia bipunctata. Biol. Zentralbl.* 60:130–137.

Tinbergen, N. 1951. *The Study of Instinct*. Clarendon Press, Oxford.

——— 1954. The origin and evolution of courtship and threat display. In J. Huxley, A. C. Hardy, and E. B. Ford, eds., *Evolution as a Process*, Allen and Unwin, London.

Utida, S. 1957. Population fluctuation, an experimental and theoretical approach. *Cold Spring Harbor Symp. Quant. Biol.* 22:139–151.

Van Valen, L. 1965. Selection in natural populations. III. Measurement and estimation. *Evolution* 19:514–528.

Vaurie, C. 1951. Adaptive differences between two sympatric species of nuthatches (*Sitta*). *Proc. Xth Int. Ornith. Congr., Uppsala, 1950,* 163–166.

Vavilov, N. I. 1926. Studies on the origin of cultivated plants. *Bull. Appl. Bot. Plant. Breed., Leningrad* 16:1–248.

——— 1951. The origin, variation, immunity and breeding of cultivated plants. *Chron. Bot.* 13:1–364.

Vecht, J. van der 1953. The carpenter bees (*Xylocopa* Latr.) of Celebes, with notes on some other Indonesian *Xylocopa* species. *Idea* 9:57–69.

Vetukhiv, M. 1956. Fecundity of hybrids between geographic populations of *Drosophila pseudoobscura. Evolution* 10:139–146.

Voous, K. H. 1955. Origin of the avifauna of Aruba, Curacao, and Bonaire. *Acta XI Int. Ornith. Congr., Basel 1954,* 410–414.

Waddington, C. H. 1957. *The Strategy of the Genes*. Allen and Unwin, London.

——— 1960. *The Ethical Animal*. Allen and Unwin, London.

Wagner, R. P. Nutritional differences in the *mulleri* group. *Univ. Texas Publ.* no. 4920:39–41.

Wahrman, J., and A. Zahavi. 1955. Cytological contributions to the phylogeny and classification of the rodent genus *Gerbillus. Nature* 175:600–602.

Walker, P. M. B. 1968. How different are the DNAs from different animals. *Nature* 219:228–232.

Walker, T. J. 1964. Cryptic species among sound-producing ensiferan Orthoptera (Gryllidae and Tettigoniidae). *Quart. Rev. Biol.* 39:345–355.

Wallace, A. R. 1889. *Darwinism: An Exposition of the Theory of Natural Selection*. Macmillan, London.

Wallace, B. 1955. Inter-population hybrids in *Drosophila melanogaster. Evolution* 9:302–316.

——— 1956. Studies on irradiated populations of *Drosophila melanogaster. J. Genet.* 54:280–293.

——— 1958. The role of heterozygosity in *Drosophila* populations. *Proc. Xth Int. Congr. Genet.* 1:408–419.

——— 1959. The influence of genetic systems on geographical distribution. *Cold Spring Harbor Symp. Quant. Biol.* 24:193–204.

——— 1963. Genetic diversity, genetic uniformity, and heterosis *Can. J. Genet. Cytol.* 5:239–253.

——— 1968a. *Topics in Population Genetics.* W. W. Norton, New York.

——— 1968b. Polymorphism, population size, and genetic load. In R. C. Lewontin, ed., *Population Biology and Evolution,* Syracuse Univ. Press, Syracuse.

Wallgren, H. 1954. Energy metabolism of two species of the genus *Emberiza* as correlated with distribution and migration. *Acta Zool. Fenn.* 84:1–110.

Warner, R. E. 1968. The role of introduced diseases in the extinction of the endemic Hawaiian avifauna. *Condor* 70:101–120.

Washburn, S. L., and C. S. Lankaster. 1968. The evolution of hunting. *Perspectives Human Evol.* 1:213–229.

Watson, J. D. 1965. *Molecular Biology of the Gene.* Benjamin, New York.

——— and F. H. C. Crick. 1953. Genetical implications of the structure of deoxyribonucleic acid. *Nature* 171:964.

Weismann, A. 1902. *Vorträge über Descendenztheorie.* Gustav Fischer, Jena.

Westoll, T. S. 1949. On the evolution of the Dipnoi. In G. L. Jepsen, E. Mayr, and G. G. Simpson, eds., *Genetics, Paleontology and Evolution,* Princeton Univ. Press, Princeton.

White, M. J. D. 1954. *Animal Cytology and Evolution,* 2nd ed. Cambridge Univ. Press, Cambridge, Eng.

——— 1957a. Some general problems of chromosomal evolution and speciation in animals. *Survey Biol. Progress* 3:109–147.

——— 1957b. Cytogenetics of the grasshopper *Moraba scurra.* II. Heterotic systems and their interaction. *Austral. J. Zool.* 5:305–337.

——— 1959. Speciation in animals. *Austral. J. Sci.* 22:32–39.

——— 1968. Models of speciation. *Science* 159:1065–1070.

——— 1969. Chromosomal rearrangements and speciation in animals. *Ann. Rev. Genet.* 3:75–98.

——— 1970. Heterozygosity and genetic polymorphism in parthenogenetic animals. *Evol. Biol.* suppl. 237–267.

Wickler, W. 1967. Vergleichende Verhaltensforschung und Phylogenetik. In G. Heberer, ed., *Evolution der Organismen,* 3 Aufl. Gustav Fischer, Stuttgart.

——— 1968. *Mimicry in Plants and Animals.* World Univ. Library, New York.

Williams, G. C. 1966. *Adaptation and Natural Selection.* Princeton Univ. Press, Princeton.

Williams, R. J. 1956. *Biochemical Individuality.* John Wiley, New York.

Wilson, E. O., and W. H. Bossert. 1963. Chemical communication among animals. *Rec. Progr. Hormone Res.* 19:673–716.

Wolda, H. 1967. The effect of temperature on reproduction in some morphs of the land snail *Cepaea nemoralis* (L.). *Evolution* 21:117–129.

Wright, S. 1931. Evolution in Mendelian populations. *Genetics* 16:97–159.

——— 1943. An analysis of local variability of flower color in *Linanthus Parryae. Genetics* 28:139–156.

——— 1956. Modes of selection. *Amer. Nat.* 90:5–24.

——— 1960. Physiological genetics, ecology of populations, and natural selection. In S. Tax, ed., *The Evolution of Life* Univ. of Chicago Press, Chicago.

Wynne-Edwards, V. C. 1962. *Animal Dispersion in Relation to Social Behaviour.* Oliver and Boyd, Edinburgh and London.

——— 1965. Self-regulating systems in populations of animals. *Science* 147:1543–1548.

Zimmerman, E. C. 1948. *Insects of Hawaii,* vol. 1. Univ. of Hawaii Press, Honolulu.

——— 1960. Possible evidence of rapid evolution in Hawaiian moths. *Evolution* 14:137–138.

Index